Studies in Surface Science and Catalysis 5

CATALYSIS BY ZEOLITES

Studies in Surface Science and Catalysis

Explanation of the cover design

The figure gives a pictorial representation of surface analysis techniques. Eight basic input probes are
considered, which give rise to one or more of four types of particles that leave the surface carrying
information about it to a suitable detector. The input probes can be particle beams of electrons, ions,
photons, or neutrals or non-particle probes such as thermal, electric fields, magnetic fields or sonic
surface waves. All of the input probes (with the exception of magnetic fields) give rise to emitted
particle beams, i.e. electrons, ions, photons, or neutrals. The various surface analysis techniques can
therefore be classified according to the type of input probe and the type of emitted particle (e.g.
electrons in, ions out; thermal in, neutrals out, etc.). In analyzing the emitted particles, one can
consider four possible types of information: identification of the particle, spatial distribution, energy
distribution and number. Any or all of these forms of information are then used to develop a better
understanding of the surface under study.

Studies in Surface Science and Catalysis 5

CATALYSIS BY ZEOLITES

Proceedings of an International Symposium
organized by the Institut de Recherches sur la Catalyse — CNRS — Villeurbanne
and sponsored by the Centre National de la Recherche Scientifique,
Ecully (Lyon), September 9—11, 1980

Editors

B. Imelik, C. Naccache, Y. Ben Taarit, J.C. Vedrine, G. Coudurier and H. Praliaud

ELSEVIER SCIENTIFIC PUBLISHING COMPANY
Amsterdam — Oxford — New York   1980

ELSEVIER SCIENTIFIC PUBLISHING COMPANY
335 Jan van Galenstraat
P.O. Box 211, 1000 AE Amsterdam, The Netherlands

Distributors for the United States and Canada:

ELSEVIER/NORTH-HOLLAND INC.
52, Vanderbilt Avenue
New York, N.Y. 10017

Library of Congress Cataloging in Publication Data
Main entry under title:

Catalysis by zeolites.

(Studies in surface science and catalysis ; 5)
Includes index.
1. Catalysis. 2. Zeolites. I. Imelik, B.
II. Series.
QD505.C39   541.3'95   80-19538
ISBN 0-444-41916-0

ISBN 0-444-41916-0 (Vol. 5)
ISBN 0-444-41801-6 (Series)

Printed in The Netherlands

CONTENTS

FOREWORD

Research work involving zeolites covers a wide area in physics and chemistry. The interdisciplinary aspect of works undertaken in various ranges - synthesis, crystallisation, structure and modification, adsorption and catalysis- has lead to rapid progress in our understanding of observed phenomena. In view of the large number of annually published data in any one area and also due to the increasingly specialized character of research, it turned out to be necessary to organize periodically an "International Conference on Zeolites" so as to promote rapid exchange of ideas and knowledge.

Our objective, upon organizing this Colloquium, was to get together French and foreign specialists interested in "catalysis by zeolites" in order to examine the research progress and the industrial prospectives in the restricted area of catalysis.

As catalysts, zeolites offer a theoretical incentive interest as to the investigation of the nature of active sites, reactions, pathways, etc... and an industrial advantage as to the development of more selective and economic new processes in the field of petroleum chemistry and in organic synthesis. It is indeed important to notice that in the field of zeolites applied and basic research are closely associated. Fundamental research concerning zeolite synthesis, crystal growth, modifications... greatly influences the applied research and conversely the development of new industrial processes "Catalyses" the interest and enthusiasm of academic research scientists.

A number of existing processes, such as catalytic cracking, methanol conversion, isomerization of xylenes, alkylation of aromatics, exemplify this interaction between academic and applied research. In particular, the synthesis of the ZSM-5 zeolite by Mobil scientists, which has originated a number of catalytic processes in hydrocarbon synthesis, has insufflated a new impetus to the industrial as well as academic research.

An important property of zeolites is their behaviour as solid electrolytes in which the exchanged cations exhibit physico-chemical properties comparable to what is known in solution. This property has originated numerous applications in catalysis : oxidation, carbonylation, isomerization, etc...

Finally, the stabilization of small metal particles within the zeolite cavities has allowed the use of these materials in typically metal catalysed reactions such as hydrogenation, Fischer-Tropsch Synthesis, etc...

We intended to devote this Colloquium to topics of zeolite catalysis : acidic catalysis, catalysis by transition metal ions and catalysis by metals with the hope of achieving better understanding of these topics in the light of the discussions that we wish to be numerous and vivid.

The Organizing Committee is very grateful to the Centre National de la Recherche Scientifique which has funded this Colloquium in the frame of "Colloques Internationaux du CNRS" which are held every year.

The Organizing Committee wishes to express their gratitude to all authors and to the participants at this meeting.

Thanks of the Organizing Committee are due to Dr. Troyanowski, Secrétaire Général de la Société de Chimie-Physique for his interest in the publication of this volume.

PREFACE

Les recherches faisant intervenir les zéolithes couvrent un très large domaine de la
Chimie et de la Physique. L'aspect pluridisciplinaire des travaux entrepris dans ce domaine
- synthèse, modification, cristallisation, structure, adsorption et catalyse - a permis
d'enregistrer des progrès rapides dans la compréhension des phénomènes observés. Compte-
tenu du nombre important des travaux publiés annuellement dans chaque discipline et de la
spécialisation de plus en plus poussée des recherches, il s'est avéré indispensable d'orga-
niser périodiquement une "Conférence Internationale sur les Zéolithes" pour favoriser des
échanges rapides d'idées et de connaissances.

En organisant ce Colloque, notre objectif était de réunir autour du thème "Catalyse par
les zéolithes" les spécialistes français et étrangers afin de dresser un bilan critique
de l'état d'avancement des recherches et des perspectives d'applications industrielles dans
le domaine bien particulier de la catalyse.

En tant que catalyseurs, les zéolithes offrent un intérêt théorique : connaissance des
mécanismes catalytiques, de la nature des sites actifs etc... et un intérêt industriel :
développement de procédés nouveaux économiques et sélectifs en pétrochimie comme en chimie
fine. Ces deux aspects se recouvrent étroitement ; il est en effet important de remarquer
qu'il n'existe pas de frontière entre la recherche fondamentale et la recherche appliquée.
Les travaux fondamentaux concernant la synthèse des zéolithes et leur modification, les
mécanismes de cristallisation... ont une grande importance pour la recherche appliquée et
inversement, le développement de nouveaux procédés industriels "catalyse" l'intérêt et
l'enthousiasme des chercheurs fondamentalistes.

De nombreux procédés existant constituent un exemple de cette interaction entre la
recherche fondamentale et la recherche appliquée.

C'est le cas des procédés de catalyse acide tels que craquage catalytique, conversion
du méthanol, isomérisation des xylènes, alkylation des aromatiques etc... et particulière-
ment le cas de la zéolithe ZSM 5. Cette zéolithe, synthétisée par les chercheurs de Mobil
est à l'origine de nouveaux procédés catalytiques de synthèse d'hydrocarbures dont l'intérêt
est tel que les recherches fondamentales universitaires et industrielles se multiplient.

Une propriété importante des zéolithes est de se comporter comme un électrolyte solide
dans lequel les ions introduits par échange ont des propriétés physicochimiques comparables
à celles qu'ils ont en solution. Cette propriété est à l'origine de nombreuses applications
en catalyse : oxydation, carbonylation, isomérisation, etc...

Enfin, la stabilisation de petites particules métalliques au sein des cavités zéolithi-
ques a permis d'appliquer tous les aspects de la catalyse par les métaux : hydrogénation,
réaction de Fischer-Tropsch etc...

Nous avons voulu développer au cours de ce colloque ces trois aspects de la catalyse par
les zéolithes : catalyse acide, catalyse par les ions, catalyse par les métaux, en espérant
mieux les comprendre à la lumière des discussions que nous souhaitons nombreuses et animées.

Le Comité d'Organisation est très reconnaissant au Centre National de la Recherche Scientifique qui a financé ce colloque dans le cadre des "Colloques Internationaux du CNRS" qu'il organise chaque année.

Le Comité d'Organisation remercie Monsieur Troyanowski, Secrétaire Général de la Société de Chimie-Physique pour son aide à la réalisation de ce volume.

Enfin, que tous les auteurs des communications présentées au cours de ce colloque trouvent ici les remerciements des organisateurs.

B. IMELIK, C. NACCACHE, Y. BEN TAARIT, J.C. VEDRINE,
G. COUDURIER, H. PRALIAUD

B. Imelik *et al.* (Editors), *Catalysis by Zeolites*
© 1980 Elsevier Scientific Publishing Company, Amsterdam — Printed in The Netherlands

SYNTHESIS, REACTIONS AND INTERACTIONS OF OLEFINS, AROMATICS AND ALCOHOLS IN MOLECULAR
SIEVE CATALYST SYSTEMS

PAUL B. VENUTO

Mobil Research and Development Corporation, Research Department, P.O. BOX 900, Dallas,
Texas 75221

It is impressive, in reviewing the extensive literature related to zeolite catalysis,
to note the numerous references to reactions involving olefins and aromatics as reactants,
products or intermediates. More recently, reaction of alcohols, notably methanol, over a
new class of shape selective zeolites has assumed major importance. There exists a
hierarchy of olefin-zeolite and aromatic zeolite interactions, that ranges from diffusion
and the mildest adsorption effects through bimolecular condensation, cracking, and very
severe reorganizations of molecular structure. Because the pores of zeolites are uniform
and of sizes characteristic of simple organic molecules, diffusion ---- and chemical
reaction ---- are greatly influenced by both variations in size, shape and polarity of the
guest molecule and in the configuration and geometry of the host crystal. Olefins and
aromatics also play a major role ---- over a wide range of conditions ---- in the formation
of coke.

OLEFIN-ZEOLITE INTERACTIONS

At low temperatures, hydrogen bonding of the π-electron systems of simple olefins with
the protons in hydrogen zeolites has been observed. At higher severities, however, proton transfer and isotopic exchange can occur. The following list gives a brief overview
of olefin reactions that have been observed over crystalline zeolite catalysts, notably
synthetic faujasites ("X" - and "Y" - type):

Olefin Reactions	Product
C=C bond migration	isomer
Hydrogen-deuterium isotope exchange	deutero-olefin
Polymerization	higher olefin
Alkyl migration	isomer
Catalytic cracking	olefin + paraffin
Paraffin alkylation	higher paraffin
Intermolecular hydrogen-transfer	aromatic + paraffin
Aromatic alkylation	alkylaromatic
Addition of H-X	addition products
Carbonylation (reaction with CO)	unsaturated aldehyde
Reaction with aldehyde	diene

Pore mouth catalysis and reverse molecular size selectivity ("the faujasite trap")
have been observed in the reactions of low molecular weight olefins at relatively low
temperatures. The critically important role of olefins in intermolecular hydrogen
transfer reactions and in formation of coke over a wide range of conditions cannot be
overemphasized.

AROMATIC-ZEOLITE INTERACTIONS

Under mild conditions, hydrogen bonding of the π-electron systems of benzenoid rings with hydroxyl groups of acidic faujasites has been observed. In mordenite at 25°C, counterdiffusion of benzene and cumene cannot occur. In faujasite, however, while these same reactants can counterdiffuse, diffusion rates of aromatics are still highly sensitive to size and molecular polarity. Of course strong adsorption (high $\Delta H_{adsorption}$) on zeolite surfaces is well known, and actual fragmentation (cracking) of benzene rings over mordenite has been reported. In electrophilic hydrogen-deuterium exchange, transfer of deuterium from zeolitic OD groups to carbon of aromatic rings occurs.

The reaction pathways of alkylaromatics over acidic zeolites is determined by a complex, temperature-dependent interplay of thermodynamics and kinetics. In faujasite-type systems at temperatures below about 250°C, ring-positional isomerization and transalkylation are coupled and mechanistically related by diarylalkane intermediates. Above 300°C, isomerization is intramolecular and involves 1, 2-shifts. In mordenite, distinct shape-selective effects have been observed in transalkylation, where bulky, symmetrical polyalkyl-benzene formation is inhibited because there is insufficient space in the narrow tubular channels to form the required transition state. An adsorption-diffusion disguise has been reported in a kinetic analysis of the isomerization of o-xylene over a zeolite catalyst at 150 - 320°C.

OLEFIN-AROMATIC INTERACTIONS

Much evidence, including organic reactivity/selectivity patterns, analogies to Friedel-Crafts catalysis, catalyst activation characteristics, poisoning/promotion effects and catalyst physicochemical studies, suggests that many reactions of aromatics such as olefin-aromatic alkylation proceed via Brönsted site-catalyzed carbonium ion mechanisms over acidic zeolites such as synthetic faujasites. To achieve effective molecular engineering of these reactions, however, one must take into account complicated perturbations in the intrinsic chemical reactivity patterns imposed by the constraints of the rigid, porous crystal lattice network. A Langmuir-Rideal mechanism rationalizes many faujasite-catalyzed aromatic olefin alkylations, and there are some significant analogies to reactions over macroreticular sulfonic acid-type ion-exchange resins. However, since mass transfer in zeolites is in the "configurational regime", temperature-, adsorption-, and other disguises may result. Product desorption from zeolite crystallite was shown to be rate limiting in one study. The complex side reactions of olefin reactants (via polymerization and intermolecular hydrogen transfer) not only seriously distort observed product distributions but also cause catalyst decay by reverse molecular size selectivity effects. At very high temperatures (400 - 600°C), an interface between ionic and radical mechanisms can be encountered in zeolite catalyzed reactions.

OLEFIN-FORMING REACTIONS

The major olefin-forming reactions commonly encountered over zeolite catalysts include:

1. Dehydrogenation
2. β-Elimination

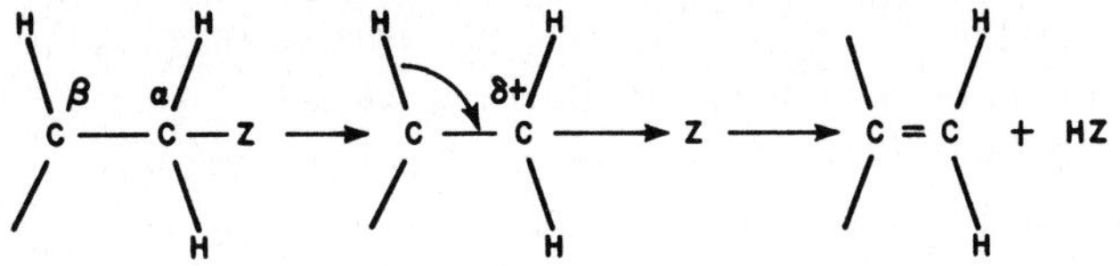

3. β-Scission

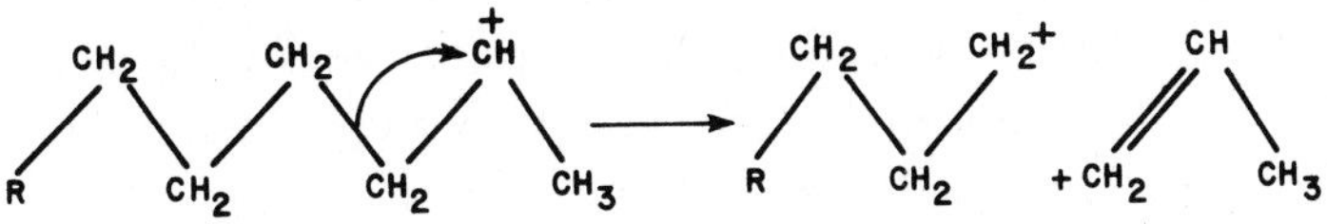

4. Aromatic dealkylation

5. Reaction of "one-carbon fragments"

Zeolites exhibit little intrinsic dehydrogenation activity, and significant catalysis of this type depends on the presence of a metallic component. β-elimination, β-scission (cracking) and aromatic dealkylation over zeolites are well known and have been studied extensively. The group of reactions that leads to formation of C=C bonds from "one-carbon fragments" is of particular interest. This includes formation of low molecular weight olefins from reaction of CH_3SH or CH_3OH over modified synthetic faujasites, formation of stilbenes from benzyl-type mercaptan systems, and the generation of light olefins from thermal decomposition of intracrystalline tetramethylammonium ions in Y-type zeolites and offretite.

AROMATIC-FORMING REACTIONS

Among aromatic-forming reactions, those that arise from inter-molecular hydrogen transfer reactions in catalytic cracking, i.e., the "gasoline stabilization reaction", coke-formation in cracking, and carbonyl condensation reactions are of great interest. In the last category, formation of heterocyclics and large aromatics of unusual shape is notable:

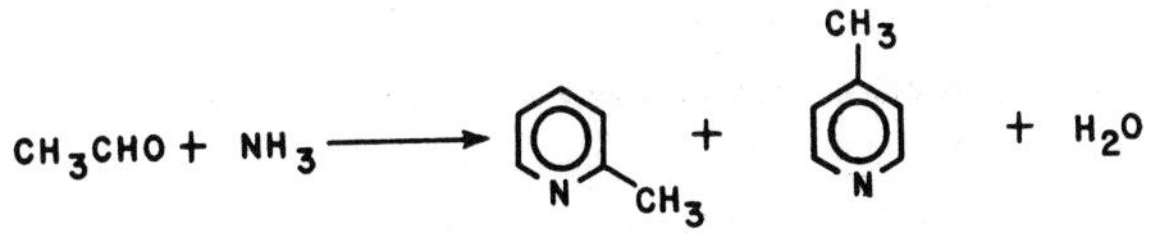

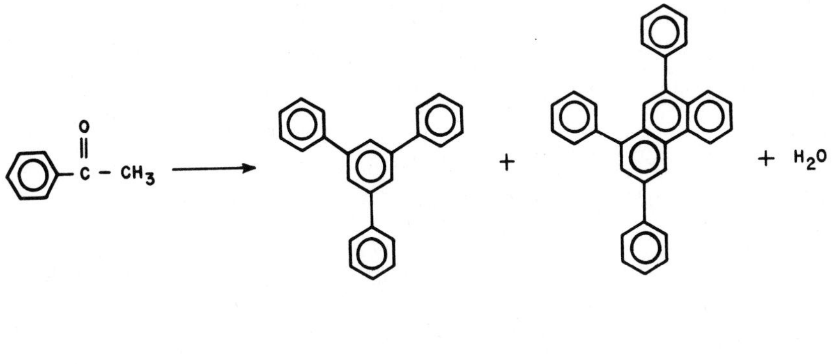

RECENT DEVELOPMENTS: FORMATION AND REACTION OF AROMATICS OVER ZSM-5-TYPE ZEOLITES

A new class of zeolites, the ZSM-5-type system, has an unusual channel structure with 10-membered ring openings (ref. 1, 2) and shows unusual shape selective properties. Prominent among the reactions catalyzed by these systems is the conversion of methanol and other oxygen compounds to hydrocarbons (ref. 3-7) mainly in the gasoline ($C_4 - C_{10}$) boiling range. Detailed reaction analyses have been conducted (ref. 8-10). Synthesis gas has been converted to aromatic hydrocarbons in the presence of polyfunctional catalyst systems containing a ZSM-5 zeolite component (ref. 11, 12). Further, even bio-mass compounds such as rubber latex, corn oil, castor oil and jojoba oil have been converted to high quality fuel with this catalyst system (ref. 13). Shape selective properties have been studied in detail (ref. 14-16). Among other transformations, para-directed aromatic conversions (ref. 17) and octane enhancement by post-reforming shape selective cracking (ref. 18) have been reported. These catalyst systems can also be used to obtain improved selectivities in process applications involving benzene-toluene-xylene (BTX)-type reactions and distillate dewaxing.

REFERENCES

1   G. T. Kokotailo, S. L. Lawton, D. H. Olson and W. M. Meier, Nature, 272 (1978) 437-438.
2   G. T. Kokotailo, P. Chu, S. L. Lawton and W. M. Meier, Nature, 275 (1978) 119-120.
3   S. L. Meisel, J. P. McCullough, C. H. Lechthaler and P. B. Weisz, Chem. Technol., 6 (1976) 86-89.
4   J. J. Wise and A. J. Silvestri, Oil & Gas J., 74 (1976) 140-142.
5   C. D. Chang and A. J. Silvestri, J. Catalysis, 47 (1977) 249-259.
6   C. D. Chang, W. H. Lang and R. L. Smith, J. Catalysis, 56 (1978) 169-173.
7   C. D. Chang, J. C. W. Kuo, W. H. Lang, S. M. Jacob, J. J. Wise and A. J. Silvestri, I & EC Proc. Design & Dev., 17 (1978) 255-260.
8   M. G. Bloch, R. B. Callan and J. H. Stockinger, J. Chromatogr. Sci., 15 (1977) 504-512.
9   P. Dejaifve, J. H. C. van Hooff, and E. G. Derouane, A.C.S. Div. Pet. Chem. Prepr., 24 (1979) 286-303.
10   E. G. Derouane, P. Dejaifve, J. B. Nagy, J. H. C. van Hooff, B. P. Spekman, C.Naccache and J. C. Védrine, C. R. Acad. Sc. Paris, t. 284 (1977) 945-948.
11   C. D. Chang, W. H. Lang and A. J. Silvestri, J. Catalysis, 56 (1979) 268-273.
12   P. D. Caesar, J. A. Brennan, W. E. Garwood and J. Ciric, J. Catalysis, 56 (1979) 274-278.
13   P. B. Weisz, W. O. Haag, and P. G. Rodewald, Science, (206) (1979) 57-58.
14   D. E. Walsh and L. D. Rollmann, J. Catalysis, 56 (1979) 195-197.
15   L. D. Rollmann and D. E. Walsh, J. Catalysis, 56 (1979) 139-140.
16   N. Y. Chen and W. E. Garwood, J. Catalysis, 52 (1978) 453-458.
17   N. Y. Chen, W. W. Kaeding and F. G. Dwyer, J. Am. Chem. Soc., 101 (1979) 6783-6784.
18   W. E. Garwood and N. Y. Chen, A.C.S. Div. Pet. Chem. Prepr., 25 (1980) 84-89.

B. Imelik *et al.* (Editors), *Catalysis by Zeolites*
© 1980 Elsevier Scientific Publishing Company, Amsterdam — Printed in The Netherlands

NEW ASPECTS OF MOLECULAR SHAPE-SELECTIVITY : CATALYSIS BY ZEOLITE ZSM-5

ERIC G. DEROUANE,
Facultés Universitaires de Namur, Laboratoire de Catalyse,
Rue de Bruxelles, 61, B-5000-Namur. Belgium.

ABSTRACT.

The molecular shape-selective catalytic properties of zeolite ZSM-5 are discussed and compared to those of more classical shape-selective catalysts. Configurational diffusion effects and transition state restrictions are found to play predominant roles in several reactions. They explain molecular sieving effects among aliphatics (of varying length and degree of branching) and alkyl aromatics; they are the basis for various para-directed aromatic reactions and shape-selective cracking, they also account for the high-resistance to coke deposition of ZSM-5 based catalysts. In the methanol-to-gasoline conversion, the zeolite seems remarkably free from major counterdiffusion limitations which may be explained by the existence of preferential diffusion paths depending on the nature of the diffusing species.

1. INTRODUCTION TO SHAPE-SELECTIVE CATALYSIS

Building-in catalytically active sites within the intracrystalline cavities and pores of zeolites is the basis for molecular shape-selective catalysts. Shape-selective catalysis which was first reported some 20 years ago by Weisz and Frilette [1] may be achieved by virtue of geometric factors, coulombic field interactions, and diffusional effects [2].

Reactant selectivity is observed when only a fraction of the reactant has access to the active sites because of molecular sieving effects, while *product selectivity* occurs when only some product species with proper dimensions (or shape) can diffuse out of the zeolite intracrystalline volume. *Restricted transition state selectivity* will take place when certain reactions will be prevented as the transition state necessary for them to proceed will not be reached due to steric and space restrictions.

Recent reviews describing and discussing shape-selectivity in catalysis have been published by Csicsery [3], Derouane [4] and Weisz [5]. Molecular shape-selective catalysis is illustrated in fig. 1 which sketches the above-mentionned possibilities.

Diffusion will of course play a role of paramount importance. Those molecules with high diffusivity will react preferentially and selectively while molecules which are excluded from the zeolite interior (their diffusivity is hence zero) will only react on the external non-selective surface of the zeolite. Products with high diffusivity will be preferentially desorbed while the bulkier molecules will be converted and equilibrated to smaller molecules which will diffuse out, or eventually to larger (partially dehydrogenated) species which will block the pores. The latter will lead to a progressive deactivation of the catalyst by carbonaceous residues laydown (i.e., coking).

Shape-selective effects in catalysis by zeolites are usually predominant when *configu-rational diffusion* is essential, i.e., a diffusion regime in which there must exist a con-tinuous matching of size, shape and configuration between the diffusing species and the available free space *in* the host catalyst, i.e. its pores, cages or channels.

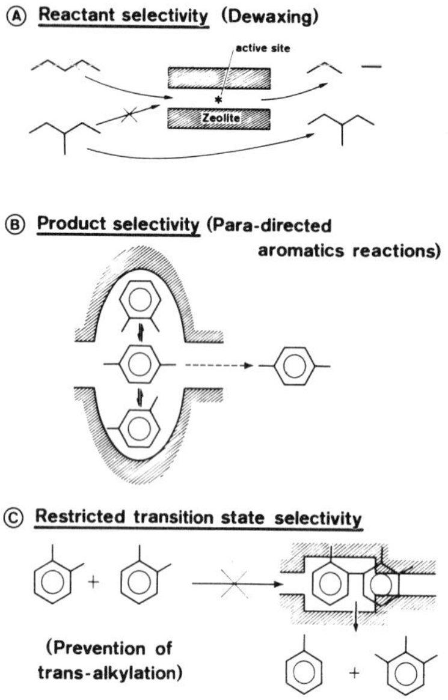

Ⓐ **Reactant selectivity** (Dewaxing)

Ⓑ **Product selectivity** (Para-directed aromatics reactions)

Ⓒ **Restricted transition state selectivity**

(Prevention of trans-alkylation)

Fig. 1. Schematic representation of molecular shape-selectivity effects.

2. CATALYTIC SITES

Active sites in shape-selective catalysts are most commonly acidic (Brönsted or Lewis) sites [3]. Bifunctional catalysts may however be obtained by the addition of a hydrogena-ting-dehydrogenating component, usually a metal such as platinum, palladium or nickel [11]. Selectoforming, or molecular shape-selective hydrocracking of n-paraffins using a (Ni,H)-erionite catalyst is an example of the latter [12]. The addition of such a metallic compo-nent usually retards also coking and aging by pore blocking due to coke deposition.

A formal distinction should also be made between catalytic sites on the outer surface of the zeolite, which will not show shape-selectivity effects, and those inside the chan-nels. Typically, for crystallites of about 1 μm, the external surface area will represent 1-2% of the total zeolite BET surface area [13].

Smaller crystallites with larger outer surface may then be expected to be less shape-selec-tive than larger crystals. This is illustrated in table 1.

Durene is indeed a too bulky molecule to be formed inside zeolite ZSM-5 and, as expected, it is observed in greater yields when the average crystallite size is decreased.

TABLE 1. Effect of crystallite size on the formation of durene in the conversion of methanol by zeolite H-ZSM-5 [14].

T(°C) = 371	WHSV (hr.$^{-1}$) = 2	Si/Al = 92
Crystallite size (μm)		Wt.% durene in hydrocarbon product
∿0.02 2-5		5.9 2.6

3. ACIDIC SITES IN ZEOLITE H-ZSM-5

Several papers have dealt with the characterization of the acidic properties of H-ZSM-5
by a variety of techniques among which IR and ESR spectroscopy [15,16] and microcalorime-
try [16,17].

H-ZSM-5 contains acidic hydroxyl groups similar to those existing on H-mordenite and has
a slightly higher acid strength [15]; the number and strength of the acidic sites were also
found to decrease with decreasing aluminium content [17].
Differences in acidity and activity, however, cannot explain the rather unique catalytic
behavior of ZSM-5 as for example a variety of other zeolitic [13,18-22] and non-zeolitic
[20,22-24] catalysts were also found to convert methanol to hydrocarbons . The unique properties of H-ZSM-5 for that particular process are the unusual product distribution [25,26]
and its high resistance to coking [27]. Both observations are explained by molecular shape-
selectivity.

4. SHAPE-SELECTIVE PROPERTIES OF ZEOLITE ZSM-5

The accessability to the catalytic sites in ZSM-5 is best viewed by considering its
channel system as schematized in fig. 2.

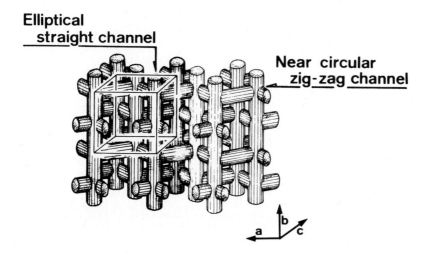

Fig. 2. Channel system in zeolite ZSM-5 [6,7]

8

The ZSM-5 framework contains two types of intersecting channels : one type is straight, has elliptical (0.51-0.58 nm) openings, and runs parallel to the b-axis of the orthorhombic unit cell, while the other has near-circular (0.54-0.56 nm) openings, is sinusoïdal (zig-zag) and directed along the a-axis.

The particular shape-selective properties of ZSM-5 result from the conjunction of four different, although structurally interrelated features :

4.1. a channel (or pore) opening consisting of 10-membered oxygen rings [6,7] which is in-termediate between that of classical shape-selective zeolites (such as erionite, ferrierite, gmelinite, chabazite, or zeolite A) and that of large pore zeolites (such as faujasite, mor-denite and fault-free offretite) as shown in fig. 3. Some shape-selective properties of ZSM-5 have been described by Chen and Garwood [8]. ZSM-5 accepts, by decreasing order of preference, normal paraffins, isoparaffins, other monomethyl-substituted paraffins, mono-cyclic aromatic hydrocarbons (eventually substituted by no more than three methyl groups), and to a much smaller extent dimethyl-substituted paraffins.

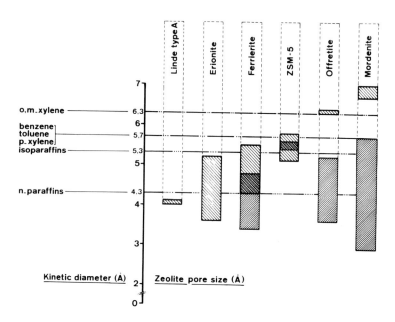

Fig. 3. Correlation between pore size(s) of various zeolites [6] and kinetic diameter of some molecules.

4.2. the presence of channel intersections (or intersecting elements) which offer a free space of larger dimensions (about 0.9 nm); the latter may play a distinct role in the or-dering of simple molecules [9] and could be the locus for the catalytic activity [10]. Namely, Valyon et al. [9] have shown that two $n-C_3$ to C_5 aliphatic molecules can be adsor-bed simultaneously at each intersecting element compared to only one for the isoparaffins and n-hexane.

4.3. *the absence of cages* along the channels; such cages which offer a larger available space may be detrimental to catalytic activity, as shown in the case of erionite [28], being the preferential locus for the formation of carbonaceous residues.

4.4. *the occurrence of slightly differentiated channel networks*. Aromatics and branched paraffins were indeed found to preferentially adsorb in the linear channels which are elliptical. This may lead to preferential diffusion paths and eventually prevent major counterdiffusion effects [29].

Reactant selectivity has been observed in a variety of reactions including cracking and hydrocracking [8,30,31], distillate dewaxing [32] and the upgrading of Fischer-Tropsch synthesis mixtures [33].

Product selectivity is of course of major importance in the methanol-to-hydrocarbons process [25,26,34], the isomerization of xylenes [35], and the prevention of internal coking [27].

Restricted transition state selectivity is essential to describe and justify the high resistance to coking of ZSM-5 type materials [27] and their extreme selectivity (no transalkylation) in the isomerization of xylenes.

Diffusional limitations affect the apparent reaction rates of related hydrocarbons [8] and explain to some extent the absence of major counterdiffusion limitations in the conversion of light molecules to aromatics by the occurrence of a molecular traffic control effect [29] because of slight differences in the channel shapes and sizes.

The intrinsic shape-selective characteristics of ZSM-5 catalysts may be improved to some extent. Indeed, reactions occur on both the outer and the inner surface of the crystallites, the inner surface being shape-selective while the outer surface is not.

Deactivation of the outer surface (or minimizing its contribution by considering larger crystallites) leads to catalysts which form less durene and less coke and also show more activity for para-directed aromatics reactions. Table 1 confirms that the formation of durene decreases for larger crystallites [14]. The formation of durene is also partially prevented by imbedding the crystallites in an alumina free binder [36], poisonning the surface with a siloxane-metacarborane polymer [37], or coating the crystallites with aluminum-free isostructural shells [38], modifications which all decrease the outer surface catalytic activity.

5. SHAPE-SELECTIVITY IN THE ADSORPTION AND DESORPTION OF SMALL MOLECULES BY H-ZSM-5

Table 2 summarizes adsorption data for various simple hydrocarbon molecules.

From the knowledge of the adsorbed amounts, using molecular dimensions and assuming an end-to-end configuration of the adsorbed molecules [9], one can evaluate the channel length occupied per unit cell by the adsorbates. A theoretical channel length can be calculated from the published structure of ZSM-5 [6,7], amounting to a total value of 8.8 nm for both types of channels and of 5.9 nm for the linear-elliptical channels only (taking into account the extra-space available at intersections). The analysis of these data clearly shows that :

i. linear aliphatics have access to both channel systems,

ii. isoaliphatic compounds experience steric hindrance effects which may restrict their adsorption and diffusion in the sinusoidal channel system,

iii. aromatic compounds and other methyl-substituted aliphatics have a strong preference
    for diffusion and/or adsorption in the linear and elliptical channels.
These observations agree with the logical assumption that flat and large molecules will
prefer to diffuse in the wider elliptical channels [29].

In addition, the adsorption of linear paraffins, isoparaffins, toluene and p-xylene occurs also more readily than that of di-methyl-paraffins or o,m-xylenes, indicating that
diffusion restrictions are strongly imposed on the latter molecules [39]. These observations are most relevant to the discussion of shape-selectivity effects in the methanol-to-
gasoline conversion.

TABLE 2. Adsorption of hydrocarbons by zeolite ZSM-5

Hydrocarbons	Total length (nm) of adsorbate in pores per unit cell (a)	Reference(b)
Propane	8.48	9
n-Butane	8.51	9
n-Pentane	8.95	9
	8.79	29
n-Hexane	8.15	9
	7.79	39
	8.07	40
	8.38	29
i-Butane	7.30	9
i-Pentane	7.50	9
	7.35	41
3-Methyl-pentane	5.42	39
	6.09	29
Toluene	5.65	39
	5.79	41
p-Xylene	5.72	39
	5.82	29

a. Total crystallographic channel length (estimated from published structures [6,7])
   = 8.8 nm; effective length (taking into account additional space at intersecting
   elements) of linear-elliptical channels = 5.9 nm.
b. Data adapted using molecular adsorbate lengths and a molecular weight of 6200 for
   H-ZSM-5.

6. SHAPE-SELECTIVE EFFECTS IN THE METHANOL-TO-HYDROCARBONS CONVERSION

The methanol-to-hydrocarbons (and eventually ligth olefins) conversion reaction has been
shown to occur sequentially : methanol is first converted to dimethylether and ligth molecular weigth olefins, the latter are oligomerized or alkylated by methanol, and ultimately
dehydrocyclized to aromatics and hydrogenated to aliphatics [25,26,42]. Counterdiffusion
effects do not appear to limit the conversion rate, which needs of course to be explained
in terms of the diffusion pathways available in zeolite ZSM-5. Compared to other acidic
zeolites as catalysts, H-ZSM-5 leads to high yields in isoparaffins (C_{7-}) and aromatics
(C_{6-10}) and shows a unique resistance to coke formation (the latter will be discussed in
another section of this paper).

The following "integrated" sequence of steps could eventually explain the detail of the methanol to hydrocarbons conversion over zeolite H-ZSM-5 [10] :

1. C_2-C_3 olefins are primary reaction products [25,26,42,43] ,
2. Ethylene and propylene can lead to a mixture of C_4-C_6 olefins by
    - *alkylation* with methanol, dimethyl ether, or paraffins, and
    - *oligomerization* with other olefins,
3. These C_4-C_6 *olefins* will contain a high percentage of *isoolefins* because of :
    - the preferential formation of secondary carbenium ion intermediates which are more stable and,
    - shape-selective constraints which prevent the formation of di-methyl-substituted aliphatics. Other mono-methyl-substituted olefins will also be present to some extent as formed directly or by isomerization.
4. By analogy to paraffins, linear olefins will diffuse faster than isoolefins in channels of the zeolite; aliphatics will also diffuse faster than olefins, being less polarisable, and therefore less influenced by the electric fields inside the zeolite,
5. *Two linear* C_2-C_5 *molecules* can be accomodated simultaneously at the *channels intersections* by contrast to only one iso-C_4, iso-C_5 or C_6 molecule [9]) where strong acid sites are presumably located. C_2-C_5 *linear olefins* will oligomerize at such sites by the mechanisms which describe the conjunct polymerization of olefins.
6. Oligomerization of the C_3-C_5 linear olefins cannot occur in the zeolite channels as it will involve secondary carbenium ions of which the electrophilic site will not be readily accessible due to the channel dimensions nor will isoolefins oligomerize jointly as two isoaliphatic molecules cannot be located simultaneously at the channel intersections [9].
7. *Bulky* C_6-C_{10} *oligomers* (*at the channel intersections*) *are aromatized* through dehydrogenation-cyclization reactions which :
    - release hydrogen (in hydrogen transfer reactions),
    - lead to aromatic hydrocarbons having smaller critical dimension than the sum of those of their precursor olefins.
8. Aromatic compounds may undergo secondary alkylation or isomerization reactions (see section on aromatic reactions in this paper).
9. Isoolefins which diffuse less readily in the zeolite channel system and show less oligomerization probability are preferentially converted to isoparaffins in hydrogen transfer reactions (from aromatization reactions).
10. Aromatic hydrocarbons with critical dimensions smaller than their olefinic precursors (see 7), and isoparaffins can then diffuse easily towards the external surface of the zeolite because :
    - of the tri-dimensional nature of the channel network [6] and
    - their preference for diffusion in the linear-elliptical channels.

If such a scheme proves to be true, the product distribution from the methanol (and eventually other light hydrocarbons conversion) would appear to be intimately controlled by the shape-selective properties of the zeolite. *Product and restricted transition state* selectivities would be predominant while the absence of major counterdiffusion limitations could be explained by a *molecular traffic control* effect in which the (light) reactants would enter the zeolite by the zig-zag channels and the products (isoparaffins and aromatics) would diffuse out by the straight channels [29].

Clues to explain the unique behavior of ZSM-5 as catalyst for this reaction would then be :
  - an end-to-end configuration of the adsorbed molecules in the channels,
  - the possibility to achieve only selected bimolecular reactions at the channel inter-
    sections, and
  - the occurrence of preferential diffusion pathways.

7. SHAPE-SELECTIVE EFFECTS IN THE CARBON FORMATION IN ZEOLITES

Coking (i.e., the laydown of a carbonaceous residue) of a zeolite is a shape-selective
reaction intimately controlled by its pore(s) size(s) and geometry. Coke, which should be
considered as a mixture of hydrogen deficient residues, originates mainly from aromatics
[44] and/or olefins [45] and aromatic-ring alkylation or hydrogen-transfer reactions are
highly important contributors to coke deposition.

A general simplified scheme for coke deposition from aromatics and/or olefins (and paraffins, if a dehydrogenating function is present) on zeolites is shown in figure 4.

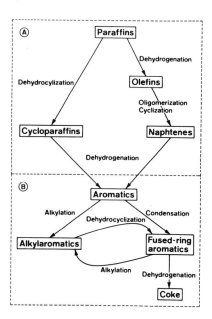

Fig. 4. Scheme for coke deposition in zeolites.

Rollmann and Walsh [27,28,46,47] have made a remarkable contribution to the understanding of coke deposition in a variety of zeolites, including the new ZSM-5 type shape-selective catalyst. The alkylation of aromatics (B in fig. 4) was found to be the decisive step
in coke formation over zeolites such as mordenite and Linde type Y [46] , i.e. large pore
zeolites. In a later stage, these alkylaromatics form fused-rings products which by dehydrogenation lead to coke. However, paraffins may also participate to coke deposition [46],
by forming precursor-cycloparaffins. An impressive correlation between coking activity and
shape-selectivity (as measured by the ratio of the cracking rates for n-hexane and 3-methyl-
pentane was recently proposed [47] and found to hold over 2 (in the amount of coke formed)
to 3 (in cracking rate ratio) orders of magnitude.

Table 3 summarizes some coke yield data for selected zeolites [28,27].

The low coking activity observed for erionite and ferrierite (as compared to mordenite) stems from high restrictions towards the formation of cycloparaffins (A in fig. 4) because of their narrow size pores [28]. As seen from table 3, these materials have openings in the 4-5 Å range which will only accept n-paraffins. The higher coke yield on erionite with respect to ferrierite is probably due to the fact that ferrierite has a neat tubular channel network while erionite also presents rather large cavities (6.3 x 13 Å) where coke formation may occur more easily. By contrast, mordenite and ZSM-5 catalysts can also accept aromatic molecules. It has been clearly demonstrated in the latter case that alkylaromatics, when formed, cannot react further in ZSM-5 (by cyclization, dehydrogenation, further alkylation, etc.) and lead to coke deposition, because of the smaller dimensions of the ZSM-5 channels. The latter explains the low coking activity of ZSM-5 type materials for which coke is probably deposited essentially on the outer-surface.

TABLE 3. Structural effects on coke deposition in zeolites [27,28]

Zeolite	Pore Size	Feed	Coke Yield (g/100 g feed conversion)
Erionite	3.6 - 5.2	Hexane isomers + [28] benzene + toluene	0.14
Ferrierite	4.3 - 5.5		0.03
Mordenite	6.7 - 7.0		0.3
		benzene + hexane [27]	∿9
ZSM-5	5.1 - 5.8		∿ 0.22

It is then concluded that the various coking activities of zeolites do not result from different mechanisms or origins for the coke formation but well from variations in structural constraints.

8. CRACKING AND HYDROCRACKING

The removal of n-paraffins from liquid reformates increases their octane number; similarly, the selective hydrocracking of n-paraffins in jet fuels, kerosene, or heating oil improves their characteristics with respect to viscosity, pour point, and freezing point. Such processes are typically referred to as selectoforming [12] or dewaxing [32].

Chen and Garwood [48-50] have reported on the use of (Ni,H)-erionite as an hydrocracking catalyst. Straight chain hydrocarbons only can enter the erionite framework and in a study of the hydrocracking of $n-C_4$ to $n-C_{16}$ hydrocarbons, a relative cracking activity pattern showing maxima at C_6 and C_{10-11} was observed [49]. It may be attributed to a "*cage effect*", analogous in essence to the "*window effect*" discussed by Gorring [51]. The erionite cage structure forces interactions between specific reactant molecules and the internal surface of the catalyst by increasing their residence time in the intracrystalline space when their length matches that of the framework cages.

Turning to ZSM-5, and concentrating on the effects of chain length and molecular shape, one observes the following trends in the cracking rates of the paraffins as seen from table 4 [8] :

a. $n-C_7 > n-C_6 > n-C_5$

b. linear > 2-methyl > 3-methyl > dimentyl- or ethyl-substituted.

Hence cracking activity increases which chain length (as expected) but decreases for bulkier molecules indicating that the rate is severely controlled by configurational diffusion. Pore aperture has more importance for ZSM-5 than molecular screening or channel tortuosity by contrast to the former observations for erionite, e.g., the occurrence of the "cage" or "window" effect.

These unique shape-selective properties of ZSM-5 are the basis for the Mobil Distillate Dewaxing (MDDW) process [32] in which the pour-point of gas-oil distillate is greatly improved by selective cracking of normal, and to a smaller extent iso-, paraffins. The new process also shows more resistance to carbon formation and lower catalyst deactivation than others.

TABLE 4. Shape selectivity effects on the relative cracking rates of C_5-C_7 paraffins
(T = 340°C; H-ZSM-5 ; 35 atm ; LHSV = 1.4)

Paraffin	Relative cracking rate
	1
	0.52
	0.38
	0.09
	0.71
	0.38
	0.22
	0.09

9. REACTIONS OF AROMATIC COMPOUNDS

A variety of processes has been described which uses ZSM-5-based catalysts for the synthesis or isomerization of C_8 aromatics. These are alkylation reactions, with particular attention to the synthesis of ethylbenzene (the MOBIL-BADGER process) [52] , the disproportionation and alkylation of toluene [53,54], and the isomerization of the xylenes [35,54] . Figure 5 shows how they are interrelated (thermodynamic equilibrium yields for the xylenes are indicated in parentheses).

The high-efficiency (activity and selectivity) of ZSM-5 catalysts for these reactions stems from a variety of features :

a. a high acidity leading to a high activity for alkylation and isomerization reactions,
b. a pore dimension that greatly favors the diffusion of para-xylene (the diffusivity of para-xylene is at least three orders of magnitude higher than that of ortho- or meta-xylene) and excludes molecules with critical dimensions higher than that of 1,3,5-trimethylbenzene [8,54] ,
c. restrictions in the nature of the transition state (which will occur in the transalkylation of aromatics) thereby preventing further conversion of ethylbenzene or the xylenes to polyalkylaromatics as well as decreasing the rate of coking (see fig. 4).

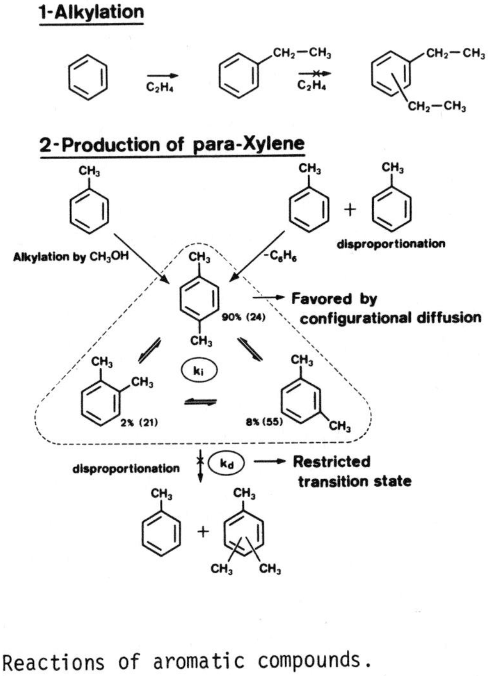

Fig. 5 Reactions of aromatic compounds.

The Mobil-Badger ethylbenzene process uses an acidic ZSM-5 catalyst [52,55] and offers performance at least equivalent to any other alternative technology (using Friedel-Crafts catalysts).

By contrast to other acidic zeolite catalysts like mordenite or faujasite which deactivate rapidly due to coke formation (in a few hours at typical reaction conditions), ZSM-5 based catalysts lead to nearly stoichiometric ethylbenzene yields and high stable activity for cycle lengths of several weeks. Ethylbenzene is barely alkylated further due to transition state restrictions and diffusional constraints.

Para-xylene can be obtained in high yields by alkylation or disproportionation of toluene [53,54] or by isomerization of the xylenes. Of major importance is of course the fact
that transalkylation cannot occur between the xylenes when ZSM-5 catalysts are used, due
to restrictions in the transition state (see fig. 5), as illustrated in table 5 by the
relative values of the isomerization (k_i) and disproportionation (k_d) rate constants measured for a variety of zeolitic catalysts. The absence of disproportionation prevents the
loss of C_8 aromatic compounds by secondary reactions, including coking.

TABLE 5. Isomerization vs. Disproportionation of the xylenes [64]

Catalyst	k_i/k_d
Faujasite, Y	10-20
Mordenite	70
ZSM-5	1000

As shown in a recent paper by Chen et al. [54] (see also table 6 which illustrate the alkylation and disproportionation of toluene), very high (~90%) yields in para-xylene are observed when the diffusional constraints are increased either geometrically (by using larger
crystals which provide longer diffusional paths) or chemically (by modification with P [56],
Sb [53,57], B [58], or Mg [59]), in which case the surface is partially deactivated and
the pore apertures and channel dimensions are reduced.

TABLE 6. Alkylation and disproportionation of toluene over ZSM-5 based catalysts [54]
(Temp. = 500-600°C; WHSV = 6-30 hr^{-1})

	Toluene alkylation by methanol (2:1)		Disproportionation of toluene	
Modification	P	LC[a]	LC[a]	Mg
Toluene conversion, wt %	21	39	13.2	10.9
Product distribution,[b] wt %				
C_6-	1.7	2.6	<0.1	0
benzene	0.1	1.9	5.5	4.9
toluene	74.1	54.0	86.8	89.2
xylene				
para	20.7	17.9	2.6	5.2
meta	0.4	14.0	3.5	0.6
ortho	0.2	7.0	1.4	0.1
others	2.2	3.3	0.1	
% xylene [c]				
para	97	46	35	88
meta	2	36	46	10
ortho	1	18	19	2

a. Large crystals = LC ; ~3 μm
b. Organic phase
c. Thermodynamic ratio : (para:meta:ortho) = (23:51:26)

These observations are rationalized by recalling that the thermodynamic equilibrium for the xylenes (easily reached in the zeolite due to its acidity or because of the presence of an hydrogenation function) is continuously displaced towards the formation of para-xylene. The latter indeed diffuses more readily towards the exterior of the cyrstallites, which reduces its relative concentration in the internal reaction volume, even more when the importance of diffusion restrictions is increased.

Two processes for the isomerization of xylenes mixtures to para-xylene have been developped using ZSM-5 based catalysts : one which takes advantage of the acidic properties of the zeolite and operates at 260-350°C and near-atmospheric pressure in the absence of hydrogen [60,61], the other which operates at high hydrogen pressure and 320-420°C using catalysts containing group VIII metals (usually NiH-ZSM-5) and also enables a partial conversion of ethylbenzene (often present to some extent as a C_8 aromatic isomer) [62,63]. Maximum yields of para-xylene are achieved in this way at varying conversion levels of ethylbenzene.

CONCLUSIONS.

Shape-selective ZSM-5 based catalysts are unique in that isoparaffins and simple monocyclic aromatics can enter their internal volume while classical shape-selective catalysts (such as erionite or Linde type A) accept only straigth chain paraffins. These properties are important for the explanation of their respective cracking and hydrocracking activity patterns.

In the methanol-to-gasoline conversion, the locus of the activity was located at the channel intersections, of which the dimensions limit the upper size of the aromatic hydrocarbons that are formed. The absence of major counterdiffusion effects is explained by "molecular traffic control", due to slight differences in the channels sizes and shapes, reactants entering by one type of channel and products diffusing out by the other.

Polyalkylaromatics, and eventually coke, are barely formed due to transition state restrictions. High-yields in para-aromatic compounds (such as p-xylene) with good stability in the catalyst activity may then be achieved when product selectivity will play an important role. Similarly, the direct synthesis of ethylbenzene from benzene and ethylene is realized with extremely high efficiency.

REFERENCES.

1 P.B. Weisz and V.J. Frilette, J. Phys. Chem., 64 (1960) 382
2 N.Y. Chen and P.B. Weisz, Chem. Eng. Progr. Symp. Ser., 63 (1967) 86
3 S.M. Csicsery, in ACS Monograph N°171, "Zeolite Chemistry and Catalysis", (J.A. Rabo,ed), ACS, Washington, D.C., 1976; p. 680
4 E.G. Derouane, in "Intercalation Chemistry", (M.S. Whittingham and A.J. Jacobson, eds.), Academic Press Inc., New York; in press
5 P.B. Weisz, in Proc. 7th Inter. Congress Catal., Tokyo, 1980, in press
6 W.M. Meier and D.H. Olson, Atlas of Zeolite Structure Types, Pub. Structure Commission of International Zeolites Association, 1978; distrib. Polycrystal Book Service, Pittsburgh, Pa., USA
7 E.M. Flanigen, J.M. Bennett, R.W. Grose, J.P. Cohen, R.L. Patton, R.M. Kirchner and J.V. Smith, Nature, 271 (1978) 512
8. N.Y. Chen and W.E. Garwood, J. Catal., 52 (1978) 453

9 J. Valyon, J. Muhalyfi, H.K. Beyer and P.A. Jacobs, Proc. Workshop on Adsorption, Berlin (D.D.R.), 1979, in press
10 E.G. Derouane and J.C. Vedrine, J. Molec. Catal., 8 (1980) 479.
11 A.P. Bolton, in ACS Monograph N° 171, "Zeolite Chemistry and Catalysis", (J.A. Rabo,ed.), ACS, Washington, D.C., 1976; p. 714
12 N.Y. Chen, J. Maziuk, A.B. Schwartz and P.B. Weisz, Oil Gas J., 66 (1968) 154
13 P.B. Venuto and P.S. Landis, Adv. Catalysis. Relat. Subj., 18 (1968) 259
14 B.P. Pelrine, U.S. Pat. 4100262, assigned to Mobil Oil Corp., (1978)
15 J.C. Védrine, A. Auroux, V. Bolis, P. Dejaifve, C. Naccache, P. Wierzchowski, E.G. Derouane, J. B.Nagy, J.P. Gilson, J.H.C. van Hooff, J.P. van den Berg and J. Wolthuizen, J. Catal., 59 (1979) 248
16 A. Auroux, P. Wierzchowski and P.C. Gravelle, Thermochimica Acata, 32 (1979) 165
17 A. Auroux, V. Bolis, P. Wierzchowski, P.C. Gravelle and J.C. Védrine, J.C.S. Faraday I, 75 (1979) 2544
18 W. Zatorski and S. Krzyzanowski, Acta Phys. Chem., 24 (1978) 347
19 M.S. Spencer and T.V. Whittam, Acta Phys. Chem., 24 (1978) 307
20 B.J. Ahn, J. Armando, G. Perot and M. Guisnet, C.R. Acad. Sci. Paris, 288C (1979) 245
21 K.V. Topchieva, A.A. Kukasov and T.V. Dao, Khimiya, 27 (1972) 628
22 D.E. Pearson, J. Chem. Soc. Chem. Comm., (1974) 397
23 L. Kim, M.M. Wald and S.G. Brandenberger, J. Org. Chem., 43 (1978) 3432
24 W.K. Bell and C.D. Chang, U.S. Pat. 3969427, assigned to Mobil Oil Corp. (1976)
25 C.D. Chang and A.J. Silvestri, J. Catal., 47 (1977) 249
26 E.G. Derouane, J. B.Nagy, P. Dejaifve, J.H.C. van Hooff, B.P. Spekman, J.C. Védrine and C. Naccache, J. Catal., 53 (1978) 40
27 D.E. Nalsh and L.D. Rollmann, J. Catal., 56 (1979) 195
28 L.D. Rollmann, J. Catal., 47 (1977) 113
29 E.G. Derouane and Z. Gabelica, J. Catal., submitted for publication
30 I. Wang, T.J. Chen, K.J. Chao and T.C. Tsai, J. Catal., 60 (1979) 140
31 P.A. Jacobs, J.B. Uytterhoeven, M. Steyns, G. Froment and J. Weitkamp, Proc. 5th Intern. Confer. Zeolites, Naples, June 1980
32 N.Y. Chen, R.L. Gorring, H.R. Ireland and T.R. Stein, Oil Gas J., 75 (1977) 165
33 See for example J.C. Kuo, U.S. Pat. 4046830, assigned to Mobil Oil Corp. (1977)
34 S.L. Meisel, J.P. McCullough, C.H. Cechthaler and P.B. Weisz, Chemtech., 6 (1976) 86
35 N.Y. Chen, W.W. Kaeding and F.G. Dwyer, J. Amer. Chem. Soc., 101 (1979) 6783
36 T.Y. Yan, U.S. Pat. 3843741, assigned to Mobil Oil Corp., (1979)
37 C.C. Chu, U.S. Pat. 3965210, assigned to Mobil Oil Corp., (1976)
38 L.D. Rollmann, U.S. Pat. 4148713, assigned to Mobil Oil Corp., (1979)
39 J.R. Anderson, K. Foger, T. Mole, R.A. Rajadhyaksha and J.V. Sanders, J. Catal., 58 (1979) 114
40 R.J. Argauer and G.R. Landolt, U.S. Pat. 3702886, assigned to Mobil Oil Corp., (1972)
41 E.G. Derouane and Z. Gabelica, to be published
42 P. Dejaifve, J.C. Védrine, V. Bolis and E.G. Derouane, J. Catal. in press 1980.
43 N.Y. Chen and W.J. Reagan, J. Catal., 59 (1979) 123
44 W.G. Appleby, J.W. Gorbson and G.M. Good, Ind. Eng. Chem. Process Des. Devel., 1 (1962) 102
45 P.B. Venuto, in "Catalysis in Organic Synthesis", (G.V. Smith, ed.), Academic Press, New York, 1977
46 D.E. Walsh and L.D. Rollmann, J. Catal., 49 (1977) 369
47 L.D. Rollmann and D.E. Walsh, J. Catal., 56 (1979) 139
48 N.Y. Chen and W.E. Garwood, Ind. Eng. Chem. Process Res. Devel., 17 (1978) 513
49 N.Y. Chen and W.E. Garwood, Advan. Chem. Ser., 121 (1973) 575
50 N.Y. Chen and W.E. Garwood, J. Catal., 53 (1978) 284
51 R.L. Gorring, J. Catal., 31 (1973) 13
52 L.B. Young, U.S. Pat. 3962364, assigned to Mobil Oil Corp., (1976)
53 S.A. Butter, U.S. Pat. 4007231, assigned to Mobil Oil Corp., (1977)
54 N.Y. Chen, W.W. Kaeding and F.G. Dwyer, J. Amer. Chem. Soc., 101 (1979) 6783
55 F.G. Dwyer, J.P. Lewis and F.H. Schneider, Chem. Eng., January 5, 1976
56 S.A. Butter and W.W. Kaeding, U.S. Pat. 3965208, assigned to Mobil Oil Corp., (1976)
57 S.A. Butter, U.S. Pat. 3979472 (1976) and Brit. Pat. 1528674 (1978), assigned to Mobil Oil Corp.
58 W.W. Kaeding, U.S. Pat. 4029716, assigned to Mobil Oil Corp.,(1977)
59 W.W. Kaeding and L.B. Young, U.S. Pat. 4034053, assigned to Mobil Oil Corp., (1977)
60 W.O. Haag and D.H. Olson, U.S. Pat. 3856871, assigned to Mobil Oil Corp., (1974)
61 D.H. Olson and W.O. Haag, U.S. Pat. 4159282, assigned to Mobil Oil Corp., (1979)
62 R.A. Morrison, U.S. Pat. 3856872, assigned to Mobil Oil Corp., (1974)
63 M.P. Nicoletti and J.F. Van Kirk, U.S. Pat. 4159283, assigned to Mobil Oil Corp.,(1979)
64 S.L. Meisel, J.P. McCullough, C.H. Lechthaler and P.B. Weisz, Leo Friend Symposium, A.C.S., Chicago, August 30, 1977

B. Imelik *et al.* (Editors), *Catalysis by Zeolites*
© 1980 Elsevier Scientific Publishing Company, Amsterdam — Printed in The Netherlands

ROLE OF BASIC SITES IN CATALYSIS BY ZEOLITES

Yoshio Ono
Department of Chemical Engineering, Tokyo Institute of Technology, Ookayama, Meguro-ku, Tokyo, 152, Japan

1 INTRODUCTION

Zeolites usually act as acidic catalysts, having very high catalytic activity for major carbonium ion hydrocarbon transformations such as cracking, alkylation and isomerization. The large scale industrial application of these reactions prompted very intensive studies on the acidic nature of zeolites, and an enormous number of data have been accumulated. On the contrary, much less attention have been directed to the zeolites as basic catalysts. There are, however, some reactions for which basic sites play a primary role in the catalysis. In this short review, the roles of basic sites in adsorption and catalysis by zeolites will be summarized. In addition, a comment on the mechanism of methanol conversion into hydrocarbons will be made.

2 CATALYSIS BY ALKALI METAL CATION EXCHANGED ZEOLITES

2.1 Zeolites as base

For acid-catalyzed reactions, hydrogen, rare-earth-, and alkaline earth-exchanged zeolites are the most active catalysts, and the parent Na and K-exchanged zeolites are inactive. However, in some reactions, NaY has a catalytic activity much higher than, or comparable with HY, e.g. the reaction of γ-butyrolactone with hydrogen sulfide(1) or primary amines(2). The closer studies of some of such reactions revealed the primary contribution of basic sites to catalyses. The first evidence for the basic nature was given by Yashima et al. in the alkylation of toluene with methanol(3).

2.2 Alkylation of toluene with methanol

The acid-catalyzed reaction of aromatic hydrocarbons with olefins result almost exclusively in addition of alkyl group to the aromatic ring, while the base-catalyzed reaction of alkyl aromatics with olefins are unique in that they allow one to enlarge the alkyl groups of arylalkane(4). Similar phenomena have been observed in the alkylation of alkylaromatics over zeolites. The alkylation of toluene with methanol over acidic zeolites produces xylenes exclusively. Sidorenko et al.(5) found, however, that the alkylation of toluene with alkali metal cation exchanged zeolites produced a mixture of xylenes, styrene and ethylbenzene at 425 and 475 °C. Especially, KX and RbX gave ethylbenzene and stylene predominantly. Yashima et al.(3) studied the reaction in more detail. Over Li-exchanged zeolites, xylenes were only products, while over Na, K, Rb, and Cs-exchanged zeolites,

styrene and ethylbenzene were produced selectively.   The trend was observed
also in the alkylation of toluene with formaldehyde.   The activity for
side-chain alkylation has a tendency to be greater for the X-type zeolites
than for the corresponding Y-type zeolites, and also depends on the basicity
of alkali metal element, that is, Na < K < Rb < Cs, except for CsX, of which
crystallinity was lost partially.   Addition of hydrogen chloride to the
reaction system promoted the aromatic ring alkylation, and inhibited the
side-chain alkylation.   On the other hand, addition of aniline poisoned the
formation of xylenes over LiY, but promoted that of styrene and ethylbenzene
over KX and RbX.   From these facts, Yashima et al. suggested that the methylation in aromatic ring and in the side-chain were caused by the catalyst
acidity and basicity, respectively.   The presence of basicity was confirmed
with an indicator method.   KX and RbX showed the color of the basic form
for cresol red and thymolphthalein while LiX did not.

The reaction of xylenes and ethylbenzene with methanol over RbX also gave
the side-chain alkylation products(3).   Similarly, alkylation of toluene
with ethylene gave cumene and α-methylstyrene(3).   Alkylation of α- and β-
naphthalene with methanol gave the corresponding ethyl naphthalenes(10).

2.3   Dehydrogenation of 2-propanol

Dehydration and dehydrogenation of alcohols are catalyzed by the acidic
and the basic sites, respectively(6), and the reactions give a diagnostic
means of knowing acid-base character of solid surfaces.   Yashima et al.(7)
carried out the reaction of 2-propanol over a series of alkali metal cation
exchanged zeolites.   Both dehydration and dehydrogenation were observed,
and the order of the catalytic activity for Y-zeolites was as follows.

    For dehydration   LiY > NaY > KY > RbY > CsY
    For dehydrogenation LiY < NaY < KY < RbY < CsY

Addition of pyridine to the reaction system poisoned the dehydration, but
did not affect the dehydrogenation.   On the other hand, addition of phenol,
an acidic compound, almost perfectly depressed the dehydrogenation activity,
and enhanced the dehydration activity.   The results again indicated the
presence of basic sites in alkali metal cation exchanged zeolites.   The
basic sites were suggested to be $(AlO_4)^-$ units in the zeolite lattice.

2.4   Ring transformation of γ-butyrolactone into γ-thiobutyrolactone

The reaction of γ-butyrolactone and hydrogen sulfide to give γ-thiobutyro-
lactone(tetrahydro-2-thiophnone) was reported by Venuto and Landis(11).

A notable feature of the reaction is that NaX is the most active catalyst
among the zeolites studied, HY and REX having only a meager activity.
We have studied the kinetics of the ring transformation over a series of
alkali metal cation exchanged Y-zeolites in order to find the mechanistic
grounds of high activity of alkaline zeolites(1).

The catalytic activity of various cation form of zeolites were compared

at 330 °C. The activities and selectivities to γ-thiobutyrolactone on
various zeolites are given in Table 1, which shows the following features of
the reaction.
  (1) Alkali metal cation exchanged zeolites are much more active than
       acidic zeolites(HY, MgY).
  (2) The catalytic activity of alkali metal cation exchanged Y-zeolites
       increases in the following order.
            LiY < NaY < KY < RbY < CsY
  (3) NaX is a more active catalyst than NaY.
These features are in sharp contrast with those found in ordinary carboniogenic reactions, for which HY is much more active than NaY and Y-zeolites
are more active than X-zeolites, and LiY being more active than CsY.

  Fig. 1 shows the effect of contact time on the yield of γ-thiobutyrolactone over CsY at 330 °C. At W/F=18.8 g h mol^{-1} (W=weight of catalyst, g,
and F=total feed, mol h^{-1}), a 99 % yield with 100 % selectivity was attained. The decay of the catalytic activity was not observed.

  The kinetics of the reaction over LiY, NaY and CsY were studied. The
rate of the reaction increases with the partial pressure of hydrogen
sulfide, and decreases with the partial pressure of γ-butyrolactone. The
rate of the reaction over LiY, NaY and CsY can be expressed by

$$r = \frac{k \, K_L K_H P_L P_H}{(1 + K_L P_L + K_H P_H)^2}$$

where r is the rate of the reaction, P_L and P_H are the partial pressures of
γ-butyrolactone and hydrogen sulfide, respectively. The rate constant, k,
changes with the increasing order LiY < NaY < CsY. The constant K_H also
changes with the increasing order LiY < NaY < CsY, while the constant K_L
does not depend much on the catalyst used. From the temperature dependence
of the rate constant, the activation energy for k is determined to be 39,
31, and 26 kcal mol^{-1} for LiY, NaY, and CsY, respectively.

  In order to gain information on the nature of the active centers, the
effects of addition of hydrogen chloride and of pyridine on the catalytic
activity were examined. Matsumoto et al.(8) reported that the cracking of
cumene over NaY is greatly enhanced by addition of hydrogen chloride to the
system. The enhancement of the catalytic activity was ascribed to the
formation of Bronsted acid sites by the interaction of hydrogen chloride
with NaY. Addition of hydrogen chloride, however, almost completely
inhibited the ring conversion. The deactivation was not caused by the
destruction of the zeolite framework, since the activity was recovered
by consecutive treatment of the catalyst with air and hydrogen at 500 °C.
Thus, it can be assumed that the deactivation is associated with the
poisoning of the active centers, presumably the basic sites, by hydrogen
chloride. The dependence of the reaction rate on the pyridine partial
pressure is given in Fig. 2, which shows that pyridine does not poison, but
enhances the catalysis. The activity increases 35 and 250 % for CsY and

LiY, respectively, with pyridine partial pressure >0.01 atm. The effect of hydrogen chloride and pyridine indicates that the acidic sites are not responsible for the catalysis, but the active centers are associated with basic sites.

A possible candidate for the basic sites in zeolites is oxygen anions bound to aluminum cations $(AlO_4)^-$. A negative charge at the site is neutralized and shielded by an alkali metal cation. However, the results suggest that the reactant molecules are accessible to and are adsorbed on the basic oxide anions at reaction conditions. An infrared study by Karge and Rasko has shown that hydrogen sulfide is dissociatively adsorbed on Na form of faujasite-type zeolites(9). They suggest that the adsorption sites are cationic sites of low coordination, where the Na^+ cations are weakly bound. The higher the Si/Al ratio, the smaller the population of such sites. Our results showing that NaX has much higher catalytic activity than NaY appears to support the idea. The order of the activity among Y-zeolites (LiY<NaY<K<RbY<CsY) may be associated with the ionic radius of the cations. The larger cations are bound more weakly to the basic sites, which, in turn, are more readily attacked by foreign molecules. The observed activation energy values can be understood on this basis. Thus, the activation energy for CsY is much smaller than for NaY and LiY, which implies that the reactant molecules are more easily activated by the former catalyst.

γ-Butyrolactone molecules have an easily polarizable carbonyl group and can be activated by the electrostatic field exerted by an alkali metal cation and/or $(AlO_4)^-$ unit. The reaction may proceed through the reaction between adsorbed hydrogen sulfide and adsorbed γ-butyrolactone, which may compete for the adsorption sites made up of the combination of M^+-(AlO_4) unit. This explains the observed Langmuir-Hinshelwood type kinetics where the adsorption of γ-butyrolactone is much easier than that of hydrogen sulfide $(K_L>K_H)$.

TABLE 1. CATALYTIC ACTIVITY OF VARIOUS ZEOLITES

Catalyst	Exchanged (%)	Conversion (%)	Yield (%)
LiY	58	27	26
NaY	—	52	51
KY	97	45	45
RbY	64	51	51
CsY	64	79	78
NaX	—	99	86
KL	—	23	22
HY	66	4	1
MgY	56	2	2

Reaction conditions; 330 °C, H_2S/lactone=6, W/F=6.26 g h mol⁻¹.

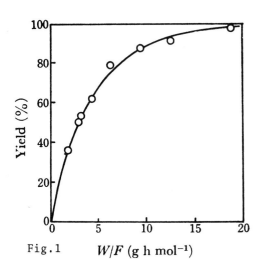

Fig.1. Effect of contact time on γ-thiobutyrolactone yield. CsY, 330°C, H_2S/lactone = 6.

Fig.1 W/F (g h mol⁻¹)

2.5 Ring transformation of tetrahydrofuran into tetrahydrothiophene(12)

For this ring conversion, too, alkaline cation exchanged zeolites have high catalytic activities, and acidic zeolites(HY, MgY) have very low activity, alkaline cation exchanged X-zeolites being more active than the corresponding cation form of Y-zeolites. These features again indicate that basic sites play an important role in the catalysis, and, in fact, hydrogen chloride inhibited the reaction completely. However, the reaction is greatly suppressed, the extent of the activity depression depending on the partial pressure of pyridine, in sharp contrast with that found in the reaction of γ-butyrolactone and hydrogen sulfide for which the pyridine does not poison but enhances the catalysis. This indicates the participation of acid in the catalysis. For the ring conversion of tetrahydrofuran into pyrolidine, only acidic zeolites are effective, indicating that the presence of the Bronsted acid sites are essential for the ring opening of tetra-hydrofuran(13, 14, 15). Since there are not intrinsic Bronsted acid sites in NaX, acidic sites should have been created by the reaction itself.

Infrared study revealed that hydrogen sulfide dissociates on NaX, produc-ing an acidic OH group (3650 cm^{-1}) and a SH group(2560 cm^{-1})(9). Thus, the interaction of basic sites produces acidic OH group, which in turn reacts with tetrahydrofuran. The reaction scheme can be summarized as

$$Na^+ + OZ^- \longrightarrow Na^+SH^- + H^+OZ^-$$

$$H^+OZ^- + \langle_O\rangle \longrightarrow \left(\langle^+_O\rangle \rightleftharpoons \langle_O^+\rangle\right) + OZ^-$$
$$\qquad\qquad\qquad\quad H \;\; (I) \;\; H$$

$$(I) + NaSH \longrightarrow \langle_S\rangle + H_2O + Na^+$$

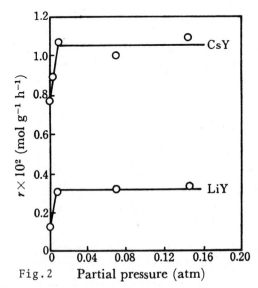

Fig.2   Partial pressure (atm)

Effect of pyridine partial pressure on the reaction rate. CsY, 300°C.

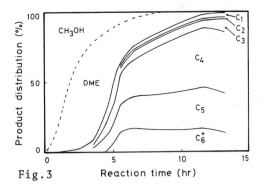

Fig.3   Reaction time (hr)

Change in product distribution with time in CH_3OH conversion over H-ZSM-5 at 239°C in a closed circulation system.

H-ZSM-5: 0.15 g, CH_3OH: 57.8 Torr.

2.6   Other reactions possibly catalyzed by basic sites

Since, as described above, the dissociation of H_2S is greatly enhanced by basic catalysts, the reactions involving H_2S as a reactant could be catalyzed by basic sites.   Reduction of nitro compounds with H_2S and the conversion of acetic anhydride to a mixture of thioacetic acid and acetic acid are probably examples of such reactions(11).   Both reactions are catalyzed by alkaline metal cation exchanged zeolites.

Ring transformations of γ-butyrolactone into 1-alkyl pyrrolidinones could

proceed in a similar mechanism as the reaction of γ-butyrolactone with hydrogen sulfide(2).   For this reaction too, alkaline cation-exchanged Y-zeolites have higher activity than HY or MgY, though CaY is more active than NaY.   The reaction proceeds over 90 % selectivity, except in the case that the NH_2R is ammonia.   The kinetic study suggested the dissociation of ammonia is caused by the basic sites.   The dissociative adsorption of ammonia was observed by infrared spectroscopy on the oxide ion of alumina surface (16).

Isakov et al. studied the aldol condensation of n-butylaldehyde to 2-ethyl-2-hexanal at 200°C(17).   The activity order was KY>NaY>HY.   Since it is known that aldol condensation is catalyzed by bases, it is very plausible that the same reaction is effected by the solid bases.

Shipperijin and Lukas found that dimerization of cyclopropenes proceeded over KA or NaA at -10°C and proposed a mechanism involving a carbanion intermediate(18).

3   BASIC CHARACTER OF DECATIONIZED ZEOLITES

The thermal transformation of NH_4Y proceeds according to the following scheme.

(I)   (II)   (III)

The intermediate form (I) is HY, which has Bronsted acidity.   The elimination of water from HY leads to the formation of Lewis sites (II), whose presence were confirmed by IR spectra of adsorbed pyridine, and basic sites (III).   Though the scheme seems to be well-established, the role of basic sites (III) in dehydrodehydroxylated zeolites in adsorption and catalysis is still scanty.   Such sites have been paid attention to as reducing sites rather than basic sites.   Turkevich and Ono reported that trinitrobenzene anion radical was formed on NH_4Y treated at 550°C(19).   Flockhart et al. found that tetracyanoethylene and dinitrobenzene as well as trinitrobenzene formed their anion radicals, and assigned the reducing sites to the structure (III)(20).   The formation of SO_2^- anion radicals are also reported for dehydrodehydroxylated HY, H-mordenite, HL(21, 22, 23).   When oxygen is introduced into SO_2^- bearing mordenite, O_2^- ions are formed.   The ESR spectrum of O_2^- ion shows the hyperfine structure due to Al nucleus, implying that the adsorption sites involves aluminium ions. While sites (III) works as reducing sites, Lewis acid sites (II) functions as oxidizing sites(21, 24, 25). The formation of cations radicals of triphenylamine, aromatic compounds are known. Flockhart et al. found that there are strong interaction between the two types of sites since the reducing power is greatly enhanced by the adsorption of certain electron-donor molecules(26). The number of anion radicals formed shows that only a small portion of the possible sites (III) are sufficiently powerful to reduce the molecules employed. The enhancement of the electron donating capacity by electron-donor molecules indicates the reinforcement of the sites (III) by the adsorption of those molecules on neighboring oxidation sites (II).

Bielanski and Dotka reported that HY (I) has an amphiprotic properties (27). Thus, the OH groups react with acid molecules such as acetic acid to form carboxylic anions and water molecules, implying that the OH groups behave as Bronsted base centers towards acid molecules.

$$zeol-OH + HOOCR \longrightarrow zeol^+ + RCOO^- + H_2O$$

4 ON THE MECHANISM OF METHANOL CONVERSION OVER ZEOLITES

Zeolites, especially ZSM-5, are effective for the conversion of methanol into hydrocarbons(28, 29). The reaction is autocatalytic one(30, 31). Fig. 3 demonstrates the time-course of the methanol conversion in a close-circulation system at 239°C(32). For the first 3h, the reaction is very slow, the hydrocarbon yield increases sharply 4-5 h after starting the reaction. The induction time is about 10 h at 219°C. The time until the sharp increase in hydrocarbon yield greatly shorten in the presence of olefins like ethylene or cis-2-butene. Thus, the mechanism is written as

$$CH_3OH \longrightarrow \text{light olefins } (C_2', C_3')$$
$$C_2' (C_3') + CH_3OH \xrightarrow{fast} \text{olefins + paraffins}$$

Thus, the most important problem in understanding the mechanism is how the first C-C bond is formed from methanol. Chang and Silvestri proposed a mechanism involving carbene-like mechanism(31).

$$CH_3OR + CH_3OR' \longrightarrow \begin{array}{c} H \diagdown \quad \diagup OR \\ CH_2 \\ H \cdots CH_2OR' \end{array} \longrightarrow CH_3CH_2OR' + ROH$$

In the mechanism, the abstraction of H and OR from CH_3OR molecules is caused by the concerted interaction of the molecules with acid and base centers.

$$\text{Basic sites} \quad \overset{|}{\underset{|}{O}}{}^- ---H^{\delta+}---CH_2---\overset{H}{\underset{|}{O}}---H^{\delta+}---\overset{|}{\underset{|}{O}} \quad \text{Bronsted acid}$$

Though the abstraction of OH (or OR) with Bronsted sites is rather easy reaction, the cleavage of C-H bond by basic sites is quite questionable, since it is unlikely that oxygen anions in zeolites are so strongly basic to activate C-H bonds. As we have seen already, hydrogen form of zeolites are inactive for base-catalyzed reactions, implying that oxygen anions in hydrogen form are too weakly basic to catalyze these reactions. The present author and coworkers rather propose a mechanism involving only Bronsted sites as active centers.

$$CH_3OH \xrightarrow[-H_2O]{} \underset{\substack{| \\ O^{\delta-} \\ | \\ zeol}}{CH_3^{\delta+}} \xrightarrow{+CH_3OH} \left[\begin{array}{c} H \\ \vdots \cdots CH_2-OH \\ CH_3 \end{array} \right]^+ \longrightarrow CH_3CH_2OH \longrightarrow C_2H_4$$

The similar mechanism are proposed in the superacid systems for polymerization or alkylation of methane(36, 37). When methanol (CD$_3$OH) was contacted with H-ZSM-5 at 150°C and evacuated, the part of methanol was converted into methoxy groups (CD$_3$O-). By heating the system for 1 h at 239°C, the methyl groups were decomposed to reproduce Bronsted sites (D$^+$). The decomposition products in gas phase was almost identical to those in the methanol conversion at 239°C. Chang and Silvestri pleaded against the superacid-like mechanism on the basis that the yield of methane is small(28). However, since the reaction is autocatalytic, any significant information on the primary process cannot be given from the distribution of products, which are obtained after the substantial fraction of methanol are consumed by the reaction with olefins. Actually, the initial hydrocarbon distribution under the conditions of the experiment in Fig. 2, methane is the most predominant hydrocarbon. Methane account for 80 % of the produced hydrocarbons (in carbon basis) at the conversion of 0.3 % (1 h reaction) and 15 % at the conversion of 2.5 % (2.5 h reaction).

The fact that strong-acid resine Nafion H and heteropolyacids are active for methane conversion also supports the mechanism proposed above(35). In conclusion, methanol conversion does not involve basic sites, but it is catalyzed solely by Bronsted acid centers.

REFERENCES

1 K. Hatada, Y. Takeyama and Y. Ono, Bull. Chem. Soc. Jpn., 50 (1977) 2517.
2 K. Hatada and Y. Ono, Bull. Chem. Soc. Jpn., 51 (1978) 448.
3 T. Yashima, K. Sata and N. Hara, J. Catal., 26 (1972) 303.
4 H. Pines and W.M. Stalick, Base-Catalyzed Reactions of Hydrocarbons and Related Compounds, Academic Press, New York, 1977, p. 240.
5 Y.N. Sidorenko, P.N. Galich, V.S. Gutyrya, V.G. Ilin and I.E. Neimark, Dokl. Akad. Nauk SSSR, 173 (1967) 132.
6 G.V. Krylov, Catalysis by Nonmetals, Academic Press, New York, 1970, p.116.
7 T. Yashima, H. Suzuki and N. Hara, J. Catal., 33 (1974) 486.
8 H. Matsumoto, K. Yasui and Y. Morita, J. Catal., 12 (1968) 84.
9 H.G. Karge and J. Rasko, J. Colloid. Interface Sci., 64 (1978) 522.
10 O.D. Konoval'chikov, P.N. Galich, V.S. Gutyrka, G.P. Lugovskaya, Kin. Katal., 9 (1968) 1387.
11 P.B. Venuto and P.S. Landis, in D.D. Eley, H. Pines and P.B. Weisz (Eds), Advances in Catalysis, Vol.18, Academic Press, New York, 1968, p. 331.
12 Y. Ono, T. Mor and K. Hatada, Acta Phys. Chem., 24 (1978) 233.
13 K. Fujita, K. Hatada, Y. Ono and T. Keii, J. Catal., 35 (1974) 325.
14 Y. Ono, K. Hatada, K. Fujita, A. Halgeri and T. Keii, J. Catal., 41 (1976) 322.
15 Y. Ono, A. Halgeri, M. Kaneko and K. Hatada, Amer. Chem. Soc. Sym. Ser., No.40, 1977, p. 596.

16 J.B. Peri, J. Phys. Chem., 69 (1969) 231.
17 Ya.I. Isakov, Kh.M. Minachev and N.Ya. Usachev, Bull. Acad. Sci. USSR, Div. Chem-Sci. (1972) 1124.
18 A.J. Schipperijin and J. Lukas, Rec. Trav. Chim. Pays-Bas., 92 (1973) 572.
19 J. Turkevich and Y. Ono, in D.D. Eley, H. Pines and P.B. Weisz (Eds), Advances in Catalysis, Vol.20, Academic Press, New York, 1969, p. 135.
20 B.D. Flockhart, L. Mcloughlin and R.C. Pink, J. Catal., 25 (1972) 305.
21 H. Tokunaga, Y. Ono and T. Keii, Bull. Chem. Soc. Jpn., 45 (1973) 3362.
22 Y. Ono, H. Tokunaga and T. Keii, J. Phys. Chem., 79 (1975) 752.
23 Y. Ono, M. Kaneko, K. Kogo, H. Takayanagi and T. Keii, JCS Faraday Trans. I, 72 (1976) 2150.
24 D.N. Stamires and J. Turkevich, J. Am. Chem. Soc., 86 (1964) 749.
25 F.R. Dollish and W.K. Hall, J. Phys. Chem., 71 (1967) 1005.
26 B.D. Flockhart, M.C. Meggarry and R.C. Pink, Adv. Chem. Ser., 121 (1973) 509.
27 A. Bielanski and J. Datka, J. Catal., 32 (1974) 183.
28 C.D. Chang and A.J. Silvestri, J. Catal., 47 (1977) 249.
29 E.G. Derouane, J.B. Nagy, P. Dejaifve, J.H.C. van Hooff, B.P. Spekman, J.C. Vedrine and C. Naccache, J. Catal., 53 (1978) 40.
30 N.Y. Chen and W.J. Reagan, J. Catal., 59 (1979) 123.
31 Y. Ono, E. Imai and T. Mori, Z. Phys. Chem. Neue Folge, 115 (1979) 99.
32 Y. Ono and T. Mori, in preparation.
33 G.A. Olah, G. Klopman, R.H. Schlosberg, J. Am. Chem. Soc., 91 (1969) 3261.
34 G.A. Olah, J.R. DeMember and J. Shen, J. Am. Chem. Soc., 95 (1973) 4952.
35 Y. Ono, T. Mori and T. Keii, 7 th International Congress on Catalysis, communication paper, 1980, Tokyo.

B. Imelik *et al.* (Editors), *Catalysis by Zeolites*

AROMATICS FORMATION FROM METHANOL AND LIGHT OLEFINS CONVERSIONS ON H-ZSM-5 ZEOLITE : MECHANISM AND INTERMEDIATE SPECIES.

VEDRINE J.C., DEJAIFVE P., GARBOWSKI[*] E.D.

Institut de Recherches sur la Catalyse, C.N.R.S.,

2, avenue Albert Einstein - F. 69626 VILLEURBANNE CEDEX - FRANCE

and (*) Laboratoire de Chimie Industrielle - Université Claude Bernard - LYON I - FRANCE

and

DEROUANE E.G.

Facultés Universitaires de Namur, Laboratoire de Catalyse

61, rue de Bruxelles - B. 5000 NAMUR - BELGIQUE

ABSTRACT

Methanol, dimethylether, ethylene, propene and butene conversions have been studied on H-ZSM-5 zeolite by gas chromatography. UV-spectroscopy study of the adsorbed carbenium ions formed by contacting the previous compounds with the zeolite and progressive heating has been carried out in details. Intermediate cyclopentenyl and cyclohexenyl charge delocalized carbocations have been detected first, followed by protonated aromatics cations and further by protonated polyalkyl aromatic and polyaromatic cations, precursors of coke. The mechanism of aromatic formation by "conjunct polymerization" of light olefins ($C_3^=$ to $C_5^=$) is discussed and assessed by the UV spectroscopy data. Shape selectivity of the zeolite is considered for explaining the high resistance of the material to coking within its narrow size channels ($\phi \approx 5.5$ Å).

1   INTRODUCTION

Potential uncertainty of supply in oil has largely stimulated interest in the production of synthesis gas ($CO + H_2$) from coal and from the enormous quantities of hydrocarbon gas in oil fields which are remote from consumer markets. Low pressure process from synthesis gas is now running in industrial plants for production of methanol which is a liquid raw material, therefore easy to be transported in conventional tankers and reliable as an energy source. A process of conversion of methanol into a mixture of hydrocarbons corresponding to gasoline with a high RON ($\approx$ 95) [1] has been commercialised recently on a new generation of crystalline alumino-silicate zeolites with an unique structure and called H-ZSM-5 [2]. In such a process the conversion reaction results in a high yield of isoparaffins and aromatics with a number of C atoms less than 10 resulting for the ZSM-5 zeolite in the important property to have a very high resistance to coke formation and therefore to aging.

Previous investigations of the methanol conversion on the H-ZSM-5 zeolite led us to propose [3] a carbenium ion mechanism to account for the formation of higher aliphatics and aromatics and to explain the formation of aromatics by conjunct polymerization of olefins [4] as suggested before by Ipatieff and Pines [5].

In the present paper gas chromatography and UV-Visible spectroscopy have been conjointly used in the conversion of methanol, dimethylether, ethylene, propene and butene. The purpose of this paper is to bring some new insight to the characterization of the mechanism of aromatics formation by determining intermediate species stable enough to be detectable by UV spectroscopy. Shape selectivity of the material is also considered for explaining catalytic results.

2   EXPERIMENTAL METHODS

Materials. High-purity grade (99 + %) methanol, dimethylether, ethylene, propene and but-1-ene were used without further purification.

The catalysts consist of the acidic form of the ZSM-5 zeolite prepared in our laboratory as previously described [3]. The acidic form is obtained by exchanging the Na cations with HCl 0.5 N at 80°C and further by calcining in air at 550°C. Experiments have been performed on samples with SiO_2/Al_2O_3 ratios equal to 28, 38, 48 and 54.5. No fundamental differences were observed for the different samples except for a kinetic point of view since the number of active acid sites decreases with the number of Al atoms [25].

UV reflectance spectra were recorded at room temperature on an Optica CF_4DR spectrometer with MgO as a reference. The samples have been activated at 400°C with a programmed temperature enhancement of 1°C min^{-1} under dry oxygen flow.

Conversion of methanol, dimethylether, ethylene, propene and but-1-ene has been studied using a fixed-bed continuous flow microreactor and nitrogen as vector gas. After 20 min., the reaction products were analyzed on-line by gas chromatography using flame-ionisation detection and two separation columns in series : a polar diglycerol on chromosorb followed by a squalane column. Calibration by comparison with pure compounds as standards, has led to good carbon balances in the products with respect to the amount of C in the feed. This indicates that coking of the catalyst is not an important feature with the ZSM-5 zeolite.

3   CATALYTIC RESULTS

By dehydration, methanol is converted into dimethylether and further into higher hydrocarbons including saturated aliphatics, olefins and aromatics.

In the investigation of the aromatization mechanism, significant information can be obtained from the analysis of the variations in the distribution of hydrocarbons : olefins are the major products at low conversion while for conversions higher than 50 % the amount of olefins decreases abruptly and paraffins (mainly $i-C_4$ and $i-C_5$) and aromatics appear simultaneously. Such an effect is presented in Fig. 1 which give the yield in olefins and saturated aliphatics versus the aromatic yield.

The conversion reactions of light olefins (ethylene, propene and but-1-ene) have been also studied and the main experimental results in the aromatics formation are summarized in Table 1. It is worthnoting that for all these compounds the major olefinic products are propene and butenes while aromatics are essentially toluene and xylenes. However, ethylene was observed to give low conversion level and C balance smaller than 100 % which is due to the formation of carbonaceous residues within the zeolite channels as we have suggested in another paper [6] at variance with other results [7].

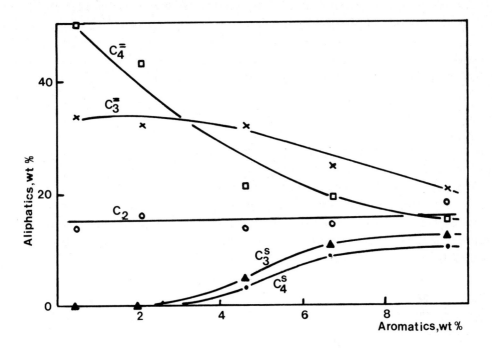

Fig. 1. Catalytic yield in aliphatics (olefins and saturated C_2-C_5) versus aromatics yield in the reaction of methanol conversion at 372°C on H-ZSM-5 zeolite.

TABLE 1

Distribution of aromatics in methanol, dimethylether and olefins reaction on H-ZSM-5 (% in C atoms).

Reagent	CH_3OH	C_2H_4	C_3H_6	C_4H_8	DME
Catalyst (0.1g) SiO_2/Al_2O_3: Temp. (°C) WHSV (hr^{-1})	38 372 10.27	38 372 4.24	38 372 6.96	54.5 379 6.63	38 372 7.0
Conversion degree (in HC) %	81.3	9.2	58.0	44.8	85.8
Aromatics (%) in H.C.	9.5	5.2	4.9	2.2	6.7
Aromatics Distribution (%)					
Benzene	0.3	0.8	2.0	5.5	0.7
Toluene	7.5	22.9	27.1	27.5	8.8
Ethyl benzene	3.7	16.7	12.8	10.3	4.4
(m+p) xylenes	65.7	41.3	43.1	45.0	64.2
o-xylene	3.7	2.5	2.5	0.7	2.9
(m+p) Et toluene	8.2	14.6	9.7	6.9	8.8
o-Et toluene	-	1.2	0.1	-	-
1-2-4-Trimethyl benzene	10.8	-	0.8	1.4	10.2
Isopropyl benzene	-	-	1.0	1.2	-
n-propyl benzene	-	-	-	-	-
C_{10} aromatics	-	-	0.8	1.5	-

The olefins reaction over acidic zeolites is known to exhibit a very complex situation [8] : many various pathways may compete such as isomerization, condensation, polymerization, cracking of polymeric intermediates, aromatization, alkylation, and finally coking of the catalyst. However, on the H-ZSM-5 zeolite conversion of light olefins is not so complex which is due to its interesting shape selectivity [9] related to its narrow size channel structure.

The conversion reactions involve acid sites of Brönsted type which have been characterized on H-ZSM-5 zeolite [10] which behaves as an ionizing solvent [11]. The catalytic results can be rationalized by the "conjunct polymerization" of olefins as observed in strongly acidic media [5] and which involves a condensation of an olefin and a carbenium ion followed by a dehydrocyclization forming cyclic C_6^+ hydrocarbons. Hydrogen transfer reactions between these cyclic hydrocarbons and carbocations from light olefins lead to aromatics and saturated C_2-C_5 aliphatics. Following such a mechanism it can easily be shown that the main aromatics should be toluene and xylenes and to a lesser extent benzene by considering olefins and carbenium ions C_2-C_5 :

$$C_2^= + C_4^+ \text{ or } C_3^= + C_3^+ \text{ or } C_4^= + C_2^+ \longrightarrow \text{benzene}$$

$$C_4^+ + C_3^= \text{ or } C_3^+ + C_4^= \text{ or } C_5^= + C_2^+ \text{ or } C_2^= + C_5^+ \longrightarrow \text{toluene}$$

$$C_4^= + C_4^+ \text{ or } C_3^= + C_5^+ \longrightarrow \text{xylenes}$$

Note at this stage that this scheme is a very simple situation ; actually, in the experimental conditions, alkylations and isomerizations of the primary aromatics occur, as experimentally observed [4, 12]. However, these reactions are relatively "clean", due to shape selectivity of the ZSM-5 resulting mainly in methyl -and to a lesser extent ethyl-substituted C_8-C_{10} aromatics.

4   UV SPECTROSCOPY RESULTS

UV spectroscopy turns out to be a very attractive technique for identifying carbocations [13, 14] in solution and has been extended for studies of adsorbed carbocations in heterogeneous catalysts [15].

Adsorption of ethylene, propene, dimethylether and methanol has been followed by UV spectroscopy at room temperature and at increasing temperatures up to get too dark and thus too UV photons absorptive materials. It clearly appears that :

(i) olefins immediately react at room temperature giving rise to an intense absorption band ca 280 nm. By heating the sample there is first a slight band shift at 100°C towards higher wavelengths ($\approx$ 295 nm) and small shoulders appearing ca 370 and 440 nm. For ethylene at 150°C the sample turns brown with broad bands at 350 and 300 nm as shown in Fig. 2, and could not be studied after heating at higher temperatures. For propene spectra may be recorded after heating at 200°C (bands at 365 and 455 nm appear in addition to the 300 nm band) and even at 300°C with broad and intense bands at 420 and 300 nm.

(ii) methanol does not react at room temperature. Only a small shift of a band assigned to Al-O band is observed, which is presumably due to the chemisorption of methanol near Al sites. Appearance of strong bands at 290 and 365 nm occurs (Fig. 3) by heating at 200°C overnight. Further heating at 250°C results in an increase of the 365 nm band and appearance of a shoulder at 260 nm. At last by heating at 300°C the sample turns brown while a broad spectrum is recorded with maxima at ca 600, ca 500, 420, 330, 270 and 240 nm.

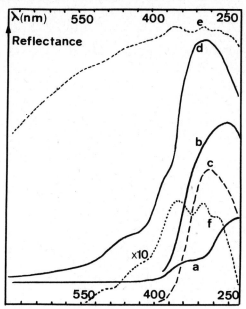

Fig. 2. UV-visible spectra recorded at room temperature after the activation of H-ZSM-5 zeolite sample at 400°C. a : initial ; b : introduction of 10 Torr of C_2H_4 at room temperature ; c : b-a ; d : after heating 1 hr at 100°C and e : 1 hr at 150°C. (SiO_2:Al_2O_3 = 28).

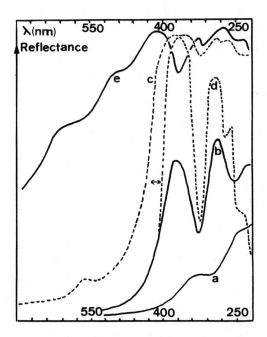

Fig. 3. UV-visible spectra recorded at room temperature after activation of H-ZSM-5 zeolite sample (SiO_2:Al_2O_3 = 28) at 400°C : a : initial ; b : introduction of 100 Torr of CH_3OH at room temperature and heating at 200°C for 1 hr ; c : heating overnight at 200°C ; d : expansion of c and zero suppressed ; e : after heating at 250°C for 1 hr.

(iii) dimethylether does not react at room temperature but reacts at 100°C i.e. at a lower temperature than methanol. It gives at 100°C a band at 290 nm and at 150°C a small band at 365 nm and a large band at 295 nm. Heating at 200°C for 1 1/2 hrs gives rise to bands at 365 and 295 nm of about equal intensity while further heating (overnight) at 200°C gives a shoulder around 420 nm and intense bands at 370 + 355 and 295 + 265 nm and a light brown color. For heating at 250°C the sample turns dark brown while intense and broad bands at 420, 330 and 280 nm are observed.

The interpretation of such electronic spectra is rather difficult because of the effect of a solid adsorbent which may affect peak positions with respect to those observed in given solvents [13, 14] and because of overlapping of bands since several carbocations are very probably formed simultaneously. However, some values of electronic spectra bands for some protonated compounds which may be formed are given in Table 2 and Fig. 4.

TABLE 2

Comparative electronic absorption bands in nm of some conjugate acid carbocations of the following compounds from refs 13, 14. Underlined values correspond to maxima in absorption intensities.

Compounds	λ values (nm)			
benzyl cumene	318	370	480	
benzene	325			
toluene	330			
1,3,5 trimethylbenzene	256	355		
1,2,3,5 tetramethylbenzene	261	365		
naphtalene	254	280	390	
anthracene	408			
naphtacene	448	592		
phenanthrene	410	510		
acenaphtene	255	264	354	420
perylene	400	462	601	
pyrene	230	380	465	

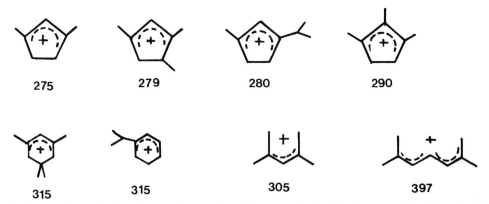

Fig. 4. Comparative electronic spectra values (in nm) of allylic-type cyclic (C_5 and C_6) and linear carbocations from ref. 13.

Analysis of UV spectra in view of such carbocations and of what may occur for H-ZSM-5 zeolite in olefins and methanol conversion led us to draw attention to the following aspects :

(i) olefins readily react with the zeolite resulting at room temperature in a cyclopentenyl carbocation with a band at 280 nm whereas for Y-type zeolite a linear π-allyl carbocation was observed ($\approx$ 310 nm) [15].

(ii) the above carbocation undergoes a structural rearrangement into a cyclohexenyl carbocation by heating at 100°C as observed from the band position shift towards higher wavenumbers (295-300 nm) and broadening of the peaks.

(iii) further heating of the samples results in the formation of alkyl aromatics as evidenced by their absorption bands in the 250-265 and 350-370 nm regions, the cyclohexenyl carbocation band at ca 300 nm being still present.

(iv) ultimate heating above 200°C results in the formation of polyalkyl aromatics and polyaromatics as evidenced by the absorption peaks at ca 600, ca 500, 420, 330, 280 and ca 260 nm.

(v) such an evolution is well observable for olefins although for C_2H_4 carbonaceous and linear residues are formed within the channels [16] and preclude UV studies above 150°C.

Higher temperatures are needed for methanol and to a lesser extent for dimethylether to react, presumably because the first limiting step is dehydration followed by the formation of the first C-C bond which will be studied in details elsewhere [17]. It is then difficult to detect the cyclopentenyl intermediate since at the temperature required with those oxygenated compounds, this intermediate is easily rearranged into the cyclohexenyl carbocations.

5   DISCUSSION

In previous works by us [3, 4] and others [8] the aromatization mechanism was postulated to correspond to conjunct polymerization [5] of olefins but the complete sequence of the reaction pathways, particularly the cyclization step, remained speculative. A possible mechanism to convert propene to propane and benzene involves an electrocyclic conversion of pentadienyl to cyclopentenyl cation [18] and rearrangements of the stable cyclopentenyl cations to the stable cyclohexenyl cation [19]. The key step of this mechanism is an hydride transfer between a carbenium ion and an hydrocarbon [20].

The mechanism we then propose takes into account previous postulations [18-20] which are supported by our UV spectroscopy data and is summarized in the following page.

After aromatization has occured the formation of carbocations may then continue resulting in alkylation and dehydrocyclization which give rise to polycondensed aromatics [21] (polyalkylnaphtalene and other higher polycyclic aromatics) which are hydrocarbons precursors of coke [22]. For large pore zeolite coking largely occurs and precludes any further reaction [19-21]. ZSM-5 zeolite is known not to deactivate easily by coking [23, 24] due to the narrow size of the channels and to the bidimensionnal structure circulation [9]. The UV spectroscopy bands observed after heating at 200°C and above present maxima ca 600, 500, 420, 330, 300 and ca 260 nm corresponding to polyalkylaromatics and polyaromatics which cannot be formed within the zeolite channels. These compounds indeed are formed at the surface of the zeolite particles and are detected by UV spectroscopy since in reflectance and for high absorbance of the surface residues only the surface layers are analyzed.

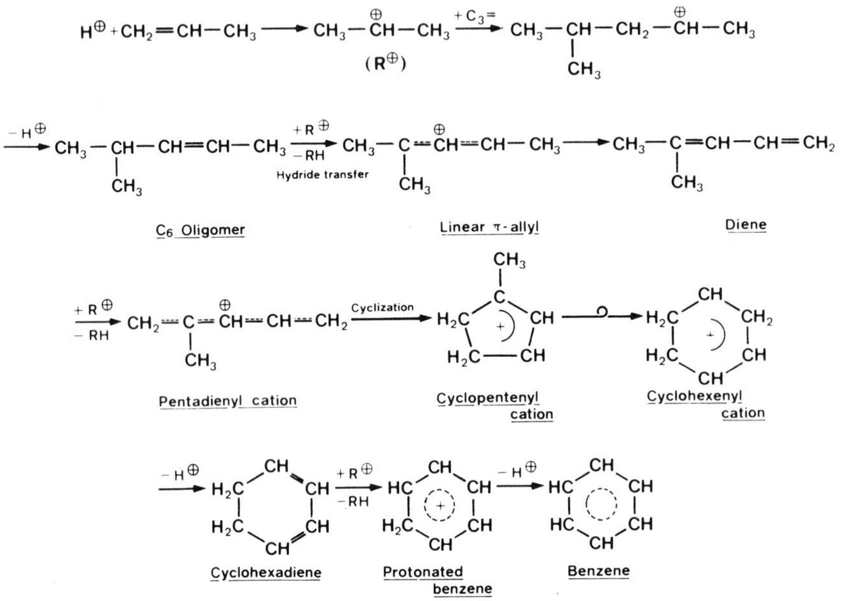

$$H^{\oplus} + CH_2{=}CH{-}CH_3 \longrightarrow CH_3{-}\overset{\oplus}{C}H{-}CH_3 \xrightarrow{+C_3^{=}} CH_3{-}CH{-}CH_2{-}\overset{\oplus}{C}H{-}CH_3$$

$(R^{\oplus})$ ... CH_3

$$\xrightarrow{-H^{\oplus}} CH_3{-}CH{-}CH{=}CH{-}CH_3 \xrightarrow[-RH]{+R^{\oplus}} CH_3{-}C{\equiv}\overset{\oplus}{C}H{=}CH{-}CH_3 \longrightarrow CH_3{-}C{=}CH{-}CH{=}CH_2$$

CH_3 (Hydride transfer) ... CH_3 ... CH_3

C6 Oligomer ... Linear π-allyl ... Diene

$$\xrightarrow[-RH]{+R^{\oplus}} CH_2{\equiv}\overset{\oplus}{C}{\equiv}CH{\equiv}CH{\equiv}CH_2 \xrightarrow{Cyclization}$$

CH_3

Pentadienyl cation ... Cyclopentenyl cation ... Cyclohexenyl cation

Cyclohexadiene ... Protonated benzene ... Benzene

6   CONCLUSIONS

UV spectroscopy allows to follow the carbocations which are formed in the aromatization pathways and shows that cyclopentenyl carbocations are formed first, followed by the formation of cyclohexenyl carbocations and further of alkylaromatics and then polyalkylaromatics and polyaromatics by increasing contact temperatures. Such carbocations are formed during the intermediate steps and an aromatization mechanism by conjunct polymerization of olefins and carbocations is proposed in good agreement with catalytic data which show that aromatics and isoparaffins yield increases when olefin yield decreases. Linear π-allyl cations are also formed in this mechanism but are not unambiguously detected in the present study presumably because of their low stability with respect to that of cyclic carbocations.

No coking occurs within the zeolite channels because of their narrow size ($\approx$ 5.5 Å) but formation of polyalkylaromatics and polyaromatics, precursors of coke, very probably occurs at the external surface of the zeolite particles, which does not hinder catalytic reaction to occur.

REFERENCES

1. S.L. Meisel, J.P. Mc Cullough, C.H. Lechthaler and P.B. Weisz, Chem. Tech., 6 (1976) 86.
   C.C. Chang and A.J. Silvestri, J. Catal., 47 (1977) 249.
2 R.J. Argauer and G.R. Landolt, US Patent 3.702.886 (1972).

3 E.G. Derouane, J.B. Nagy, P. Dejaifve, J.H.C. Van Hooff, B.P. Spekman, C. Naccache and J.C. Vedrine, C.R. Acad. Sci., Paris, 284 ser C, (1977), 945 and J. Catal. 53 (1978) 40.
4 P. Dejaifve, J.C. Vedrine, V. Bolis and E.G. Derouane, Proceed. of Amer. Chem. Soc. meeting on advances in petrochem. processes p. 286 and J. Catal., 1980, in press.
5 V.N. Ipatieff and H. Pines, Ind. Eng. Chem., 28 (1936) 684.
6 J.C. Vedrine, P. Dejaifve, C. Naccache and E.G. Derouane, Proceed of the VIIth Internat. Congress on Catal., Tokyo (1980).
7 J.R. Anderson, K. Foger, T. Mole, R.A. Rajadhyaksha and J.V. Sanders, J. Catal., 58 (1979) 114.
8 M.L. Poutsma, in J.A. Rabo (Eds), Zeolite chemistry and catalysis, ACS Monograph 171, Washington 1976, pp. 487-492.
9 E.G. Derouane and J.C. Vedrine, J. Molec. Catal., 8 (1980) 479.
10 J.C. Vedrine, A. Auroux, V. Bolis, P. Dejaifve, C. Naccache, P. Wierzchowski, E.G. Derouane, J.B. Nagy, J.P. Gilson, J.H.C. Van Hooff, J.P. Van den Berg and J.P. Wolthuizen, J. Catal., 59 (1979) 248.
11 D. Barthomeuf, J. Phys. Chem., 83 (1979) 249.
12 C.D. Chang, W.H. Lang and R.L. Smith, J. Catal., 56 (1979) 169.
13 G.A. Olah, C.U. Pittman and M.C.R. Symons, in Carboniums Ions, edit. by G.A. Olah and P. Von R. Schleyer, John Wiley and Sons, New York 1968, Vol. I, Chap. 5
14 G. Dallinga, E.L. Mackor and A.A. Verrijn Stuart, Mol. Phys., 1 (1958) 123.
15 E.D. Garbowski and H. Praliaud, J. Chim. Phys., 76 (1979) 687.
16 V. Bolis, J.C. Vedrine, J.P. Van den Berg, J.P. Wolthuizen and E.G. Derouane, J.C.S. Faraday Trans I (1980) in press.
17 P. Dejaifve, V. Ducarme, J.C. Vedrine and E.G. Derouane, J.Molec.Catal., submitted 1980.
18 T.S. Sorensen, J. Am. Chem. Soc., 89 (1967) 3782 and 3794.
19 N.C. Deno, H.G. Richey, N. Friedman, J.D. Hodge, J.J. Houser and C.U. Pittman, J. Am. Chem. Soc., 85 (1963) 2991, 2995 and 2998.
20 C.D. Nenitzescu, in G.A. Olah and P. von R. Schleyer (Eds), Carbonium Ions, Wiley and Sons, New York, 1968, Vol. II, Chap. 13.
21 P.B. Venuto, P. and L.A. Hamilton, Ind. Eng. Chem., 6 (1967) 190.
22 D.E. Walsh and L.D. Rollmann, J. Catal., 49 (1977) 369.
23 P. Wierzchowski, E.D. Garbowski and J.C. Vedrine, J. Chim. Phys., 1980, Submitted.
24 L.D. Rollmann and D.E. Walsh, J. Catal., 56 (1979), 139 and 195.
25 A. Auroux, M. Rekas, P.C. Gravelle and J.C. Vedrine, Proceed of Vth Internat. Confer. on zeolites, Napoli (1980) in press.

B. Imelik *et al.* (Editors), *Catalysis by Zeolites*
© 1980 Elsevier Scientific Publishing Company, Amsterdam — Printed in The Netherlands

PORE CONSTRAINTS AND ACID CATALYSIS IN THE NOVEL HIGH-SILICA ZEOLITES HNu-1 AND HFu-1.

J. DEWING, F. PIERCE AND A. STEWART

I.C.I. Ltd., Corporate Laboratory, Runcorn, Cheshire, England.

1.   INTRODUCTION

High silica zeolites have recently aroused considerable interest largely as a result of the description by Mobil Oil of the reaction of methanol over the zeolite ZSM-5 to give substantial yields of aromatic hydrocarbons [1]. Synthesis of this zeolite involved the use of tetra-n-propyl ammonium ions, which are left as trapped charge balancing cations in the crystalline product [2]. Use of other tetra alkyl ammonium ions has led to the synthesis of other novel high silica zeolites and in this paper we wish to describe some features of the zeolites Nu-1 [3] and Fu-1 [4] prepared in ICI Ltd. Laboratories.

Evidence of shape selectivity in methanol conversion over ZSM-5 [5] has become understandable with the publication of the detailed crystal structure of ZSM-5 and its variants [6]. Sorption data for hydrocarbons and bases indicate the very strong acidity of the zeolite hydroxyls and show a discrimination between p-xylene and o-xylene [7] which is in good agreement with the size constraint to be expected from the crystal structure. A recent report by Vedrine et al. [8] describes the interaction of pyridine with ZSM-5 as revealed by infrared spectroscopy. Two absorption bands at 3720 cm^{-1} and 3605 cm^{-1} are attributed to silanol groups terminating the lattice and the internal acidic groups respectively.

Information has already been presented on the conversion of methanol over the ICI high silica zeolite Nu-1 [9]. In this paper we describe the size constraints and number of available acid sites in both Nu-1 and Fu-1. We have used adsorption of series of bases and ethers followed both by i.r. spectroscopy and gravimetric means. Reaction of zeolitic -OD groups with propene has been used to demonstrate the selective interaction of the olefin with the acidic hydroxyl group. Various examples of acid-catalysed reactions can be explained in terms of the characterisation work and poisons were selected to indicate the situation of active sites in such catalysis.

2.   EXPERIMENTAL

2.1   Materials Used

Ammonia, methylamine, dimethylamine and methanol (BDH Ltd.) were used as supplied. Trimethylamine (BDH Ltd.), containing small quantities of the other methylamines as impurities, purified by the Hinsberg method. Pyridine

(BDH Ltd.) and 2,6-di-tbutylpyridine (K and K Fine Chemicals) were distilled and dried over 3A molecular sieve before use. Propene (polymerisation-grade quality, Air Products) was used without further treatment. Tetrahydrofuran (Hopkin and Williams Ltd.) was distilled from a mixture with sodium and benzophenone to remove water and peroxides. 1-Methylnaphthalene and p-xylene (BDH Ltd.) were dried over 3A molecular sieve before use. Deuterium oxide was supplied by Fluorochem Ltd.

The three zeolites were Fu-1, Nu-1 and ZSM-5, prepared according to the methods described in the patent literature and subsequently exchanged and calcined to yield their acid forms. The samples used here had approximate Si : Al ratios of 14, 18 and 36 for Fu-1, Nu-1 and ZSM-5 respectively.

Reactions over HFu-1 with liquid reactants were carried out by entraining the vapour in helium and passing over about 1 g. of zeolite. Propene was passed directly over the catalyst and liquid products were condensed in a vessel surrounded by an ice-bath. Products were identified by gas chromatography.

2.2   Experimental Procedures

Adsorption of various gases and vapours was followed both by infrared spectroscopy and by microbalance measurements. The i.r. observations were made with a Perkin Elmer 257 grating spectrophotometer using 15 mm. diameter discs of the pure zeolites, formed by pressing 10-20 mg. of powder at 1.75 tonnes. Discs were mounted in an internally heated vacuum cell equipped with single-crystal silicon windows, which were sealed directly to the Pyrex envelope. This cell was connected to a standard grease-free vacuum and gas handling line capable of an ultimate pressure of 10^{-6} torr.

Adsorption experiments in this cell were generally conducted in the following way. Each zeolite disc was degassed at 450°C for 16 hours and cooled to 50°C. After this treatment the pressure was normally below 10^{-5} torr. The vapour pressure of each absorbent, listed in Table 1, was controlled by means of low-temperature slush baths. After adsorption of base, the sample was evacuated at 150°C to leave only chemisorbed material. Spectra were recorded at various stages throughout these treatments.

In experiments to study exchange of propene the deuterium form of the zeolite was prepared by exchange of the protonic form with 2 torr D_2O at 150°C. Subsequent back exchange was carried out with propene at 500 torr.

Microbalance experiments were done using a Cahn RG electrobalance mounted in a conventional vacuum and gas-handling system. Each sample of weight between 25-35 mg. was degassed at 350°C prior to all adsorptions at 150°C (vapour pressures as in Table 1). After adsorption for 0.75 hours, the system was evacuated to constant sample weight, still at 150°C.

3.   RESULTS AND DISCUSSION

The hydroxyl region of the spectra of degassed HFu-1 and HNu-1 are shown in Figs. 1 and 2, together with representative spectra after base adsorption. The spectra of the clean materials both comprise a pair of hydroxyl absorptions with a common high frequency band at 3740 cm^{-1} normally attributed to surface silanol groups [10].   The lower frequency bands were at 3610-3600 and 3550 cm^{-1} for HFu-1 and HNu-1 respectively.   These bands are associated with the acid centres and observations of the perturbation of these bands by molecules of different size can provide detailed information about size constraints and perhaps the environment of the sites.   Table 1 summarises the adsorption data and indicates which bases have access to the internal groupings.   Physisorption of any of the bases perturbs the 3740 cm^{-1} band but this is reversed on simple evacuation at 50°C, indicating the uncon-

TABLE 1.

Accessibility of Acid Sites to Organic Base Probe Molecules

Probe Molecule	Vapour Pressure (torr)	Perturbation of IR Absorption Band			
		HFu-1		HNu-1	
		3740 cm^{-1}	3610 cm^{-1}	3740 cm^{-1}	3550 cm^{-1}
NH$_3$	40	+	+	+	+
MeNH$_2$	30	+	+	+	+
Me$_2$NH	5	+	+	+	-
Me$_3$N	27	+	-	+	-
Pyridine	5	+	-	+	-

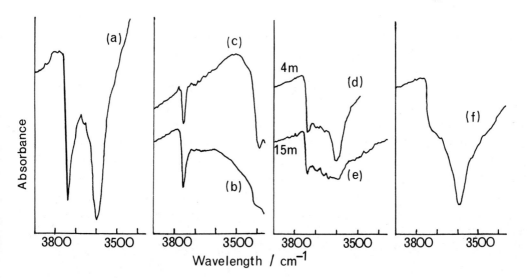

Fig. 1.   Spectra of Zeolite HFu-1.   (a) After evacuation at 450°C for 16 hrs.; and in the presence of bases (b) NH$_3$ (40 torr), (c) CH$_3$NH$_2$ (30 torr), (d) and (e) (CH3)$_2$NH (5 torr) showing slow uptake, and (f) pyridine (0.5 torr, 16 hours at 50°C).

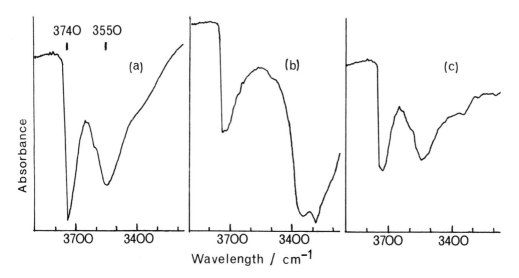

Fig. 2. Spectra of zeolite HNu-1. (a) After evacuation at 450°C for 16 hrs., (b) after NH_3 adsorption and subsequent evacuation for 1.5 hrs., and (c) in presence of 760 torr $(CH_3)_2NH$, 40 min. after addition of base.

strained nature of these silanol groups and their relatively weak acidity. This is entirely consistent with the postulate that these are lattice terminations. Studies with cyclic ethers show the same pattern [11], demonstrating a similar size constraint for access to the internal acid sites for HFu-1. Ether adsorptions were not investigated with the other zeolites, but the similarity of the size constraint for ethers and amines for HFu-1 indicates that the amine adsorption was not influenced by the formation of immobile protonated species which would have resulted in pore blocking. We conclude therefore that the amine adsorptions provide a true measure of the pore constraints appropriate to the understanding of catalytic reactions.

In the case of HFu-1 (Table 1 and Fig. 1) ammonia and monomethylamine had such ready access to the acid site that only complete reaction could be observed in a spectrometer scan started immediately after exposure to base. The corresponding ammonium ion was formed. In the unsubstituted case the species was identified by a broad absorption band at 1450 cm^{-1} as well as by the removal of the -OH absorption at 3610 cm^{-1}. Dimethylamine reacted more slowly, allowing spectra to be recorded of partially reacted samples before a fully chemisorbed state was obtained. At 50°C this process took approximately 20 minutes to complete. Times given in Fig. 1 relate to the start of each spectrometer scan. Neither trimethylamine nor pyridine could interact with the internal acid groups. Indeed pyridine did not gain access even after 48 hours at 200°C. After adsorption, evacuation at 150°C for many hours led to no recovery of the band at 3610 cm^{-1} nor to any diminution of the CH stretching bands. It was concluded that this was a suitable criterion for complete chemisorption to be applied in the micro-balance experiments.

HNu-1 showed a similar pattern except that dimethylamine was not chemisorbed and the uptake of monomethylamine was slow. Again the chemisorbed bases appeared to be stable at 150°C under evacuation. The papers of Anderson [7] and Vedrine [8] and indeed the known aromatic hydrocarbon forming reactions [1] would indicate that the smaller bases have ready access to the acid sites of HZSM-5, leading to stable chemisorbed states.

On the basis of these findings a study was undertaken of quantitative chemisorption of the various bases by gravimetry at 150°C. The results fell into the expected pattern and are given in Table 2. There is a clear

TABLE 2.
Molecules of Different Bases Chemisorbed at 150°C (DTPB = 2,6-di-t-butyl-pyridine).

Adsorbate / Adsorbent	MeNH$_2$ molecules g^{-1}	Me$_3$N molecules g^{-1}	C$_5$H$_5$N molecules g^{-1}	DTBP molecules g^{-1}
HFu-1	5.6×10^{20}	7.3×10^{19}	5.7×10^{19}	2.4×10^{19}
HNu-1	1.9×10^{20}	2.8×10^{19}	1.2×10^{19}	4×10^{18}
HZSM-5	1.8×10^{20}	2.0×10^{20}	1.9×10^{20}	4×10^{18}

difference between the uptake of those bases which can gain access to the various zeolite pore systems and those which are excluded, e.g. the difference between monomethylamine and pyridine for HFu-1 and HNu-1 and between pyridine and 2,6-di-t-butylpyridine on HZSM-5. Where bases have internal access the majority of the potential acid sites as represented by the aluminium content are available for chemisorption in HFu-1 and HZSM-5. This indicates a reasonable degree of crystal perfection in these materials. The differences shown by the various zeolites in the uptake of those bases excluded from the pore structures may reflect either different crystallite size or different distribution of aluminium between the bulk and surface of the crystals. Our results do not distinguish between these two possibilities, but we believe comparison of the uptake of admitted and excluded bases corresponds closely to the ratio of total to external acidity.

The void volume of the pore structure of HNu-1 and HFu-1 is of the order of 7 cm^3 per 100 g. as shown by water or n-hexane [9, 12]. The volume associated with the chemisorbed bases, assuming normal liquid densities is significantly less than this figure and suggests that the chemisorption is not limited by the internal voidage.

To confirm that the internal acid sites identified by base adsorption were capable of interaction with small hydrocarbons a series of experiments was done using HFu-1 exchanged with D$_2$O to yield DFu-1. This material was exposed in the infrared cell to propene, p-xylene and 1-methylnaphthalene and the exchange of the -OD groups of the zeolite followed. The exchange

44

of zeolitic -OH groups with D_2O at 50°C showed a difference in reactivity of the silanol and acidic hydroxyls, the latter reacting rapidly while the former required higher temperatures (150°C) to undergo complete reaction (Fig. 3). The frequencies of the new OD bands were 2770 and 2670 cm^{-1}.

Exposure of the D-Fu-1 to propene at 50°C led to rapid (i.e. less than 10 minutes for complete reaction) exchange of the 2760 cm^{-1} band and the regeneration of the 3610 cm^{-1} band apparently without any broadening due to adsorbed hydrocarbon. Exchange with the non-acidic surface silanol groups was as expected much slower and reaction at 150°C was required to completely remove the OD band at 2770 cm^{-1}. This did not regenerate the 3740 cm^{-1} silanol band but led to formation of saturated CH bands between 3000 and 2800 cm^{-1}. The silanol band could only be regenerated by further heating (up to 450°C), when the CH bands were gradually eliminated with a concomitant growth in the silanol absorption.

In contrast the exchange of larger hydrocarbons with D-Fu-1 was very slow, no change being detectable on 30 min. exposure to 5 torr p-xylene at 50°C. Prolonged contact at 250°C resulted in a very slow exchange which may have been due either to slow penetration of the zeolitic pore by the aromatic hydrocarbon or by transport of deuterium from the internal OD group to the lower population of external acid sites by traces of water or other low molecular weight impurities in the system. Exchange of the much larger molecule, 1-methylnaphthalene, was comparable in rate to that of p-xylene which clearly demonstrates the importance of impurities such as

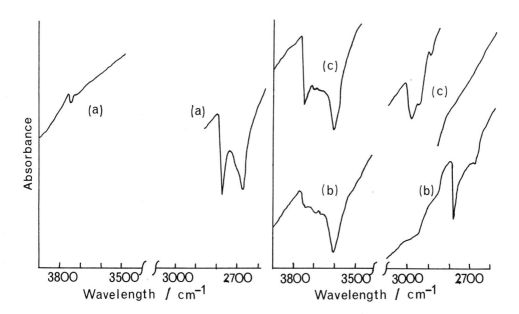

Fig. 3. The exchange of DFu-1 with propene. (a) Initial spectrum after exchange of HFu-1 with D_2O (150°C), (b) After exposure to propene (500 torr) at 50°C for 10 mins. followed by evacuation, and (c) after exposure to propene (500 torr) at 150°C for 10 mins. and evacuation.

traces of water as vehicles for the transport of internally held deuterium to the crystal surfaces. In the absence of a pool of protium in hydrocarbon molecules, exchange of the OD groups was even slower. Slow exchange did take place during continued evacuation over several days. The probable source of protium here was the glass walls of the vacuum system.

Our conclusions from these exchange experiments with Fu-1 are that propene interacts readily and completely reversibly with the internal acid hydroxyls, but more slowly and irreversibly with the external silanol groups perhaps forming alkoxy species which can be thermally decomposed. We conclude further that aromatic hydrocarbons do not gain access to the internal acid site.

It would therefore seem that the reaction of propene at these sites is defined solely by an equilibrium involving, on one side, the more favoured situation of alkene and free acid site, without the possibility of any dimerisation within the zeolite. However we observed that when propene was passed over HFu-1 in a continuous flow reactor, liquid products were formed at temperatures greater than 160°C. At 300°C condensed liquid consisted essentially of branched oligomers, these components accounting for more than 80% of the C_6 and C_9 fractions. These results are in line with early reports of olefin-polymerisations, e.g. over chabazite [13] in that it is unlikely that the observed products could have passed through the small pore systems. It must therefore be assumed that such reactions occurred on the external surface of HFu-1. The conversion of methanol over HFu-1 has been shown to yield mainly linear olefins together with dimethyl ether and water [14]. This type of distribution would be expected from a pore system with a constraint at about 5Å. Our adsorption work would indicate that an aromatic base such as pyridine or quinoline should selectively poison the external surface acid sites. Treatment of HFu-1 with either of these bases prevented the formation of liquid products from propene at temperatures of about 300°C. However methanol still underwent conversion to dimethyl ether and water at 300°C after quinoline treatment of the zeolite. When it was eventually calcined, activity of the sample for propene oligomerisation recovered.

HFu-1 was active in other acid-catalysed reactions such as dehydration of ethers. Tetrahydrofuran, a cyclic ether which was unable to perturb the acidic hydroxyl site in adsorption studies [11], gave butadiene as the major product at about 200°C. The ketone, acetone, yielded mainly isobutene and mesitylene at temperatures between 250 and 330°C together with other products of aldol-type condensation and cracking reactions. The population of sites on the surface of HFu-1 is sufficient to play an important role in its catalytic chemistry, affecting the product distributions of reactions which might take place selectively in the pore system alone. In contrast, the microbalance results for ZSM-5 show only a few sites to be on the surface of this zeolite. The nature of the pore system and the acidity within the channel system are clearly controlling factors in reactions over ZSM-5. The absence of products of size larger than durene in the methanol reaction [1,5] and resistance to deactivation through "coking" [15] bear this out.

4. CONCLUSIONS

It is clear that the chemisorption of small bases can be regarded as a means of counting acid sites since we have shown that not only is the low frequency acidic hydroxyl band totally removed but there is also a reasonable correspondence between the number of molecules chemisorbed and the number of aluminium ions in the zeolites. This chemisorption is in no way limited by the void volume of the pore structures. Larger bases such as pyridine which have ready access to HZSM-5 are excluded from both HNu-1 and HFu-1. Our studies with smaller bases demonstrate effective size constraints for HNu-1 and HFu-1 at about 4.0Å and 5.0-5.2Å respectively. These materials although high silica zeolites have properties which are significantly different to the better known ZSM-5 family.

A previous publication on HFu-1 [12] describes cracking of relatively large molecules. Our catalytic results and characterisation work show that the source of this activity is the external crystal surface which is relatively large and accommodates about 10% of the total acidity. A similar proportion is available to aromatic molecules in the case of HNu-1. The presence of catalytically significant levels of acidity on zeolite crystal surfaces is rarely considered but this work emphasises the need for complete characterisation of such properties so that selectivity factors can be properly understood.

REFERENCES

1   S.L. Meisel, J.P. McCullough, C.H. Lechthaler and P.B. Weisz, Chem. Tech., 6 (1976), 86.
2   U.S. Patent 3,702,886.
3   U.S. Patent 4,060,590.
4   German Patent Application, 1978, 2,748,276.
5   C.D. Chang and A.J. Silvestri, J. Catal. 47 (1977) 249.
6   G.T. Kokotailo, S.L. Lawton, D.H. Olson and W.M. Meier, Nature, 272 (1978) 437; G.T. Kokotailo, P. Chu, S.L. Lawton and W.M. Meier, Nature, 275 (1978) 119.
7   J.R. Anderson, K. Foger, T. Mole, R.A. Rajadhyaksha and J.V. Sanders, J. Catal., 58 (1979) 114.
8   J.C. Vedrine, A. Auroux, V. Bolis, P. Dejaifve, C. Naccache, P. Wierzchowski, E.G. Derouane, J.B. Nagy, J-P. Gilson, J.H.C. van Hooff, J.P. van den Berg, and J. Wolthuizen, J. Catal., 59 (1979) 248.
9   M.S. Spencer and T.V. Whittam, Acta Phys. et Chem., 24 (1978) 307.
10 J.W. Ward, in J.A. Rabo (Ed.), Zeolite Chemistry and Catalysis, A.C.S. Monograph 171, American Chemical Society, Washington, 1976, Ch.3, p.126.
11 J. Dewing, F. Pierce and A. Stewart, submitted for publication.
12. M.S. Spencer and T.V. Whittam, in R.P. Townsend (Ed.), Properties and Applications of Zeolites, Special Publ. No. 33, The Chemical Society, London, 1980, p.342.
13 R.M. Barrer and D.W. Brook, Trans. Faraday Soc., 49 (1953) 940.
14 Brit. Patent 1,563,345.
15 L.D. Rollmann and D.E. Walsh, J. Catal., 56 (1979) 139.

B. Imelik *et al.* (Editors), *Catalysis by Zeolites*
© 1980 Elsevier Scientific Publishing Company, Amsterdam — Printed in The Netherlands

ACTIVITY-PROMOTIONS BY $SnCl_4$ AND HCl ON DECATIONATED-Y AND Na-Y ZEOLITES

K. OTSUKA, T. IWAKURA AND A. MORIKAWA

Department of Chemical Engineering, Tokyo Institute of Technology,
Ookayama, Meguro-ku, Tokyo 152, Japan

SUMMARY

The enhancing actions of $SnCl_4$ and HCl on the catalytic activities of
H-Y and Na-Y zeolites have been studied by applying the isomerization of
cis-but-2-ene as a test reaction. The $SnCl_4$-enhanced catalytic activity
was ascribed to the OH groups responsible for the absorption band at
~3200 cm^{-1} which were transformed from the 3640 cm^{-1} OH groups through
the interaction with $SnCl_4$. The HCl-promoted activity of the dehydroxylated H-Y was attributed to the 3640 cm^{-1} OH groups newly formed through
the dissociation of HCl.

INTRODUCTION

One of the most easy and promising method for improving the catalytic
activity and selectivity of the solid catalysts is the modification of
their surfaces by the contact with various inorganic gases. These favorable actions of foreign gases to catalysis have been recognized specifically
on zeolites [1,2]. The promoting-effects of inorganic gases, such as SO_2
and NO_2, on the reactions with a carbenium-ion intermediate over X and Y
zeolites have been reported [3-6]. The favorable modification of the
surfaces by SO_2 and NO_2 was ascribed to the acidic charactor of the gases.

Preliminary experiments showed that $SnCl_4$ and HCl gases, known as a
Lewis acid and a strong Brönsted acid reagent, respectively, enhance the
catalytic activities of Y zeolites. We report here the enhancing-actions
of these gases on the carboniogenic activities of H-Y and Na-Y for the
isomerization of cis-but-2-ene. The characterization of the modified
surfaces by these gases has been demonstrated by infrared spectroscopic
studies.

EXPERIMENTAL

The H-Y was prepared by the decomposition of NH_4-Y (86%-decationized
Na-Y) at 400°C under vacuum in a quartz reactor. After the adsorption of
a small amount of $SnCl_4$ or HCl (usually 1.32 X 10^{-4} mol g^{-1}) on the zeolites
at 25°C, the isomerization of cis-but-2-ene was carried out in a gas-
circulation apparatus at the same temperature. The activity of the zeolite
for the isomerization was determined from the average quantities of the
butene-isomers produced per second, which had been calculated from the
total quantities of the products within the initial 10 min. Infrared-
spectra measurements were carried out at 25°C for the zeolites in the form
of a self-supporting wafer (3.8 ± 0.2 mg cm^{-2}).

RESULTS

   The Effects of SnCl$_4$ and HCl on the Catalytic Activity of H-Y.   For the
H-Y pretreated at 400°C, SnCl$_4$ enhanced the activity of the zeolite up to
5 times of the original activity.   On the other hand, only a weak enhancing
effect on the activity was observed for HCl (less than 10%-increase).   The
effect of HCl was improved when the zeolite had been treated at the higher
temperatures.   Figure 1 shows the activity of the zeolite, A, as a function
of the concentration of SnCl$_4$ or HCl adsorbed.   Figure 2-a shows the effects

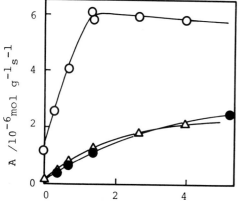

Quantity of the gas adsorbed/10^{-4}mol g^{-1}

Fig. 1.   The enhanced-activity vs.
the quantity of SnCl$_4$ or HCl adsorbed:

O, SnCl$_4$ on 400°C-treated H-Y;
△, SnCl$_4$ on 650°C-treated H-Y;
●, HCl on 650°C-treated H-Y.

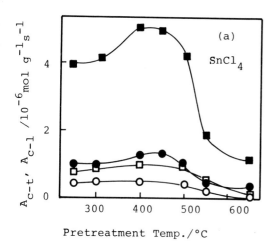

Pretreatment Temp./°C

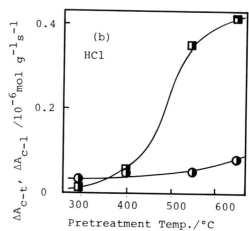

Pretreatment Temp./°C

Fig. 2.   Effect of pretreatment temperatures on the enhanced-activities:
(a) With SnCl$_4$: ■, A$_{c-t}$; ●, A$_{c-1}$.   Without SnCl$_4$: □, A$_{c-t}$; O, A$_{c-1}$.
(b) HCl-enhanced activity: ■, ΔA$_{c-t}$; ●, ΔA$_{c-1}$.   The concentrations of the
gases were 1.32 X 10^{-4}mol g^{-1}.

of pretreatment temperatures for the zeolite on the catalytic activities in the presence and absence of $SnCl_4$. A_{c-t} and A_{c-1} represent the activities based on the trans-but-2-ene and but-1-ene productions, respectively. The HCl-enhanced activities of the 650°C-pretreated H-Y, i.e., the differences in the activities obtained in the presence and absence of HCl, are plotted as functions of pretreatment temperatures in Fig. 2-b. The effect of $SnCl_4$ weakens as a rise in the pretreatment temperatures in contrast to the effect of HCl.

Ir-spectroscopic Studies on the Interactions of $SnCl_4$ and HCl with the Surface. Figure 3 shows the changes in the infrared spectra of the hydroxyl groups of the 400°C-pretreated H-Y after the addition of $SnCl_4$. The $SnCl_4$ reduced the intensity of the band at 3640 cm^{-1}, generating a new band at ~3200 cm^{-1}. A slight decrease in the intensity of the band at 3540 cm^{-1} was also observed after the addition of $SnCl_4$. The evacuation of $SnCl_4$ at 350°C restored the intensities of the two bands close to the original values, resulting in the disappearance of the 3200 cm^{-1}-band accordingly. It is noted that the restored intensities of the 3640 and 3540 cm^{-1} OH-bands never exceed the original intensities of the bands. The fact shows that the $SnCl_4$ does not increase the number of the OH groups. The addition of $SnCl_4$ onto the D_2O-treated H-Y zeolite caused the new band at ~2380 cm^{-1} OD groups which had been formed from 3640 cm^{-1} OH groups by the hydrogen-deuterium exchange. These facts suggest that the new band at 3200 cm^{-1} is transformed directly from the OH groups of 3640 cm^{-1} through their inter-action with $SnCl_4$. No evidence to prove the interaction of $SnCl_4$ with the OH groups of 3540 cm^{-1}-band has been obtained. The absorbance of the 3200 cm^{-1}-band has been plotted in Fig. 4 as a function of the pretreatment

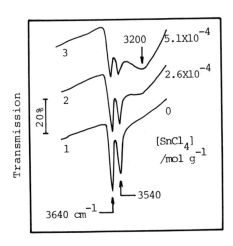

Fig. 3. The OH-band at 3200 cm^{-1} caused by $SnCl_4$.

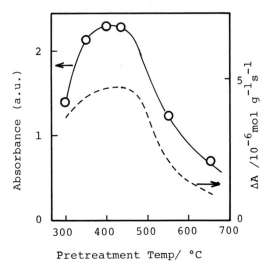

Pretreatment Temp/ °C

Fig. 4. The absorbance of 3200 cm^{-1}-band and the $SnCl_4$-enhanced activity, ΔA, as functions of pretreatment temperature: $[SnCl_4]=5.0 \times 10^{-4}$ mol g^{-1}

temperature for the zeolite. The dotted curve in the figure indicates the $SnCl_4$-enhanced activity (from Fig. 2-a). Both the absorbance and the enhanced activity depended on the treatment temperatures in a very similar way. Figure 5 shows the changes in the absorption spectra of the surface OH groups with time after the addition of cis-but-2-ene in the presence (b) and absence (a) of preadsorbed $SnCl_4$. The spectra were recorded after

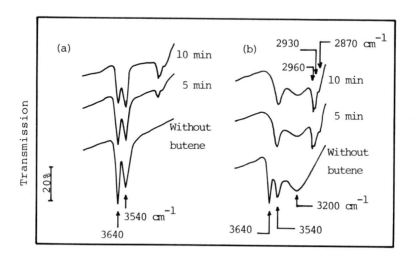

Fig. 5. The changes in the OH-bands after the addition of cis-but-2-ene: (a) Without $SnCl_4$; (b) With $SnCl_4$.

condensing the butene in the gas-phase by liquid nitrogen trap. In the absence of $SnCl_4$, the band at 3640 cm^{-1} decreased gradually with time. On the other hand, in the presence of $SnCl_4$ (b), the OH groups at 3640 cm^{-1} disappeared immediately after the introduction of cis-but-2-ene accompanying the decrease of the 3200 cm^{-1}-band; the intensity of the OH-band at 3540 cm^{-1} was not decreased by the butene. It has been confirmed that the physically adsorbed cis-but-2-ene on the zeolite without $SnCl_4$ gives the bands at 1644 and 3010 cm^{-1} due to the $\nu_{C=C}$ and $\nu_{=CH-}$, respectively. These bands were absent in the presence of $SnCl_4$, suggesting a strong interaction of the double-bond with the surface. The bands at 2960 ($\nu_{CH_3\ as.}$), 2930 ($\nu_{CH_2\ as.}$) and 2870 cm^{-1} ($\nu_{CH_3\ s.}$) observed both in the presence and absence of $SnCl_4$ are probably due to the polymers of butenes.

The addition of HCl gas onto the 400°C-pretreated H-Y left the intensities of the two OH-bands at 3640 and 3540 cm^{-1} unchanged. The band due to the physisorbed HCl (2900 cm^{-1}) was observed on the 400°C-pretreated zeolite. However, no physisorbed HCl was observed on the 650°C-pretreated H-Y. Moreover, the addition of HCl onto the 650°C-pretreated zeolite increased the intensities of the two OH-bands.

The Enhancing-actions of $SnCl_4$ and HCl Adsorbed on Na-Y and Na-X.

The HCl adsorbed on Na-Y did not induce the isomerization of butene even though the gas generated a large amount of OH groups of 3640 cm^{-1}

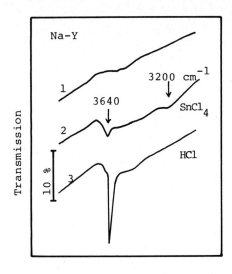

Fig. 6. The OH-bands caused by SnCl$_4$ and HCl on Na-Y:

1, background

2, after SnCl$_4$ adsorption (4.5 X 10^{-4}mol g^{-1})

3, after HCl adsorption (4.1 X 10^{-4} mol g^{-1})

(Fig. 6). In the case of SnCl$_4$, the 3640 cm^{-1} OH groups were observed, which must be generated probably through the reaction of SnCl$_4$ with chemisorbed water (Fig. 6). Though the number of the OH groups formed by SnCl$_4$ was much less than that by HCl, an activity-promotion was caused only by SnCl$_4$. The SnCl$_4$-caused band at 3200 cm^{-1} was also confirmed here (Fig. 6).

Neither HCl nor SnCl$_4$ induced the isomerization over Na-X zeolite. No production of OH groups on the zeolite was observed by the addition of the gases.

DISCUSSION

The effect of impurity water in the foreign gases used has been severely checked. The addition of water (1.2~6.3 X 10^{-5}mol g^{-1}) on the 400°C-pretreated H-Y before or after the adsorption of SnCl$_4$ (1.32 X 10^{-4}mol g^{-1}) never enhanced but rather reduced the activity of the zeolite. The addition of 1.5 X 10^{-4}mol g^{-1} of water did not change the spectra in Fig. 3. For 650°C-pretreated H-Y, water enhanced the activity of the zeolite appreciably. However, the enhancing ability of the water was less than a half of that of SnCl$_4$ or HCl. The SnCl$_4$ and HCl were purified before use by the contact with the dehydrated 13X or Na-A molecular sieves at -78°C. No impurity water in the reagents has been confirmed by mass spectroscopic analysis. Hence, the contribution of impurity water can be neglected for the enhanced-activities in Figs. 1 and 2.

Promoting-action of SnCl$_4$. It is generally accepted that the catalytic activity of H-Y zeolite is caused by the OH groups playing the role of Brönsted acid sites. A sec-butyl carbenium ion has been postulated as a common reaction intermediate in the isomerization between butenes over acidic catalysts [2,7,8]. The hydroxyl groups responsible for the 3640 cm^{-1} band assigned to O$_1$-H groups are the active sites for carboniogenic reactions especially at low temperatures [8,9]. The OH concentration decreased considerably when the zeolite had been pretreated at the temperatures

higher than 500°C, in agreement with earlier observations [10]. The
similar dependences of the activities in the presence and absence of $SnCl_4$
upon the treatment temperatures (Fig. 2-a) implies that the favorable
action of $SnCl_4$ is related to the OH groups of 3640 cm^{-1}. The band at
3200 cm^{-1} must be caused from the interaction of the 3640 cm^{-1} OH groups
with the Lewis acid $SnCl_4$ as follows:

$$
(3640 \text{ cm}^{-1}) \qquad\qquad (3200 \text{ cm}^{-1}) \ \delta+
$$

$$
\tag{1}
$$

We can expect that the OH groups thus formed have a strong acid charactor
favorable for the isomerization of butenes via a sec-butyl carbenium ion.
The absence of the $\nu_{C=C}$ and $\nu_{=CH-}$ vibration bands on the $SnCl_4$-adsorbing
surface implies a strong interaction of the double-bond of butenes with
the OH groups. A good correlation between the $SnCl_4$-enhanced activity
and the absorbance of the band at 3200 cm^{-1} suggests that the active sites
for the isomerization are the OH groups transformed from the 3640 cm^{-1} OH
groups. The kinetic curves of the isomerization of cis-but-2-ene in the
presence of $SnCl_4$ showed a rapid initial reaction during first few minutes
followed by a steady progress of the isomerization. The enhanced-activity
by $SnCl_4$ was observed not only for the initial activity but also for the
continued activity after 60 min. The 3640 cm^{-1} OH groups disappeared
during initial few minutes (Fig. 5-b), which must be resulted from the
irreversible consumption of a part of the 3200 cm^{-1} OH groups. The hydrogen
of the groups may be involved in the polymerization, cracking or hydrogen-
transfer reactions of butene which are believed to accompany the isomerization [8,11,12]. The spectra in Fig. 5-b indicate that a considerable
concentration of the 3200 cm^{-1} OH groups still remains after the disappearance of the OH groups of 3640 cm^{-1}. This fact can be explained by the
heterogeneity of the supercage OH groups of 3640 cm^{-1}-band [8], i.e., there
are two kinds of 3640 cm^{-1} OH groups, one provides the very active 3200 cm^{-1}
OH group responsible for the initial high activity but is consumed quickly
and the other is transformed to the less active 3200 cm^{-1} OH group which
gives the continued activity of the zeolite in the presence of $SnCl_4$.

The Promoting-action of HCl. The considerable enhancing-effect of HCl
emerged when the H-Y had been adsorbed at the temperatures higher than
500°C (Fig. 2-b). The spectroscopic study showed the regeneration of the
OH groups of 3640 and 3540 cm^{-1}-bands by the addition of HCl onto the
650°C-dehydroxylated zeolite. The OH groups must be supplied by the
dissociation of HCl added. The HCl-enhanced activity of the zeolite
pretreated at high temperatures can be attributed to the newly formed
3640 cm^{-1} OH groups. For 400°C-pretreated H-Y, the promoting effect of HCl

was not appreciable probably because the zeolite had enough OH groups already and no seat available for accommodating the protons of HCl. In fact, there was no OH-band increased after the addition of HCl on the zeolite.

The roughly estimated concentration of the 3640 cm^{-1} OH groups formed in the presence of 5.6×10^{-4} mol g^{-1} of HCl on the 650°C-pretreated H-Y was 1/30 of the concentration of the 400°C-pretreated zeolite. However, the activity of the former catalyst was twice as greater than that of the latter. It is believed that the treatment of H-Y at the temperatures higher than ~500°C causes the dehydroxylation of the zeolite, generating electron donor and acceptor sites [13-15]. The HCl-enhanced activity was pronounced when the zeolite had been treated at the temperatures greater than ~500°C (Fig. 2-b). We believe that the higher activity of the OH groups formed by HCl can be ascribed to the increase in the acidity of the groups by the inductive effect of the electron acceptor sites formed by the dehydroxylation of H-Y [16].

The Effects of $SnCl_4$ and HCl on Na-Y. The results of Fig. 6 shows that the 3640 cm^{-1} OH groups formed by the dissociation of HCl over Na-Y have no catalytic activity for the butene isomerization. However, the OH groups become active when a partial charge of the groups are withdrawn by the Lewis acid reagent such as $SnCl_4$ as shown by equation (1). The OH groups responsible for the 3200 cm^{-1}-band must be the active sites also in this case.

In conclusion, the formation or the presence of the supercage OH groups at 3640 cm^{-1} is a necessary but not sufficient condition for the foreign gases to promote the carboniogenic activities of the Y zeolites. The activation of the groups by the electron withdrawing-effect of the foreign gases, such as $SnCl_4$ (this work), SO_2 or NO_2 [4-6], through a direct or indirect interaction with the groups, is essential to improve the catalytic activity.

REFERENCES

1   Kh.M. Minachev and Ya.I. Isakov, Advan. Chem. Ser., 121 (1973) 451.
2   P.A. Jacobs, Carboniogenic Activity of Zeolites,Elsevier, North-Holland, Amsterdam, 1977, 158 pp.
3   Y. Ishinaga, K. Otsuka and A. Morikawa, Bull.Chem.Soc. Japan,52 (1979)933.
4   K. Otsuka, Y. Wada, K. Tanaka and A. Morikawa, Bull.Chem. Soc. Japan, 52 (1979) 3443.
5   Y. Wada, K. Otsuka and A. Morikawa, J. Catal., 62 (1980), in press.
6   K. Otsuka, K. Nakata and A. Morikawa,J.C.S. Faraday I, 76 (1980), in press.
7   J.W. Hightower and W.K. Hall, J. Phys. Chem., 71 (1967) 1014.
8   P.A. Jacobs, L.J. Declerck, L.J. Vandamme and J.B. Uytterhoeven, J.C.S. Faraday I, 71 (1975) 1545.
9   J.W. Ward, J. Catal., 11 (1968) 259.
10 J.W. Ward, Advan. Chem. Ser., 101 (1971) 380.
11 T.J. Weeks,Jr.,I.R. Ladd, C.L. Angell and A.P. Bolton, J.Catal.,33(1974)256.
12 T.J. Weeks,Jr., and A.P. Bolton, J.C.S. Faraday I, 70 (1974) 1676.
13 J.B. Uytterhoeven, L.G. Christner and W.K. Hall, J.Phys.Chem., 69 (1965) 2117.
14 C.L. Angell and P.C. Schaffer, J.Phys.Chem., 69 (1965) 3463.
15 J.W. Ward, J. Catal., 9 (1967) 225, 396.
16 J.H. Lunsford, J. Phys. Chem., 72 (1968) 4163.

B. Imelik *et al.* (Editors), *Catalysis by Zeolites*
© 1980 Elsevier Scientific Publishing Company, Amsterdam — Printed in The Netherlands

STRUCTURAL AND PHYSICOCHEMICAL ASPECTS OF ACIDIC CATALYSIS IN ZEOLITES

Denise BARTHOMEUF, Laboratoire de Catalyse Organique, L.A. 231 CNRS, Université Claude Bernard, 43 Boulevard du 11 novembre 1918, 69622 Villeurbanne.

1. INTRODUCTION

For a long time now various reviews have reported correlations between the acidic and catalytic properties of zeolites (1,2). The more recent ones (3-6) have pointed out that if general relationships are observed between proton number or acid strength and carbonium ion reactions, little progress has been made in the activity of the sites in catalysis. This also implies little progress made in the ability for predicting catalytic activity. In fact, as might be suspected, a large number of parameters may change the environment of protons and hence their reactivity. Independently of the concept of sites, it should not be forgotten that some zeolite properties (one of them being acidity), are relevant to collective framework interactions (7-13) and then in that sense depending on atoms or ions in the structure (Si, Al, OH, cations).

The various aspects of acidity related to catalytic properties will therefore be illustrated by looking at the influence and importance of the framework constituents in view of localized sites and of overall properties. Mainly protonic sites, the more important in carbonium ions formation, will be considered.

2. CORRELATIONS BETWEEN OH GROUPS AND CATALYSIS

2.1. Hydroxyl content

It is known that infrared band intensities in Y zeolites increase simultaneously with the rate of various acidic reactions when cations are exchanged for protons (o-xylene isomerization (14), butene isomerization, alcohol dehydration, cumene cracking and paraffins cracking (5). In fact, this parallel behavior does not reflect a simple quantitative correlation between the two properties (15). The integrated coefficient depends on the samples (16) and as a result the hydroxyl groups concentration cannot be easily deduced from infrared peak areas. The only important point for catalysis is to find the sites with the right strength ; only some of the hydroxyls fulfill this requirement (5,15). Jacobs (5) presented a sequence of reactions requiring increasing acid strength and decreasing number of sites : i) dehydration of alcohols, ii) isomerization of olefins, iii) alkylation of aromatics, iv) isomerization of aromatics, v) transalkylation of alkylaromatics, vi) cracking of alkylaromatics, vii) cracking of paraffins.

It is difficult to evaluate which fraction of hydroxyls is implied in each of these reactions. An indirect evaluation allowed Jacobs to postulate that for isooctane cracking not more than 0.01 % of the available supercage hydroxyls (3640 cm^{-1} OH groups) are implied (or 3 for every 10.000 cages) (5).

Another puzzling feature is that the maximum of OH band intensity is at a pretreatment temperature (~ 773K) equal to or lower than that of activity for various reactions. This temperature difference, up to 150°C, is the highest for the reactions iv) to vii) of Jacobs sequence where a strong acidity is required. Up to now no satisfactory explanation has been given (induction of a high protonic strength through Lewis sites or formation of super-active proton at high activation temperature) (5). Another hypothesis might be suggested. It was observed that in some strongly acidic Y zeolites, protons may exist which are not cha-racterized by infrared OH bands (17). It was recently observed (18,19) that a steamed morde-nite catalyst with no detectable acidic hydroxyls was active in toluene disproportionation, a reaction which requires very strong acidic sites. It can then be inferred that very strong protons may not give infrared bands because they are highly delocalized. They would only be easily detectable by their high catalytic activity.

2.2 Hydroxyl wavenumbers

It is recognized that in Y zeolites the 3560 and 3640 cm^{-1} bands are not characteristic of specific acid strengths but correspond to hydroxyls vibrating in different cages, the 3640 cm^{-1} OH groups being in supercages and 3560 cm^{-1} in sodalite cavities. Both of the hydroxyls show a range of acid strengths. The lower wavelength hydroxyls are typically ob-served in Y zeolites (8) while the higher wavelength OH groups are observed in various structures, the Si/Al ratio being the most important parameter in determining the wavenumber (5,8,11,20). The figure 1A shows that each zeolite is characterized throughout its Si/Al ratio by wavenumbers varying in a short range from weak to strong sites. For similar acid strengths the various zeolites do not show the same wavenumber (left part of figure 1A (20) and for different acid strengths zeolites have nearly the same wavenumber (right part of

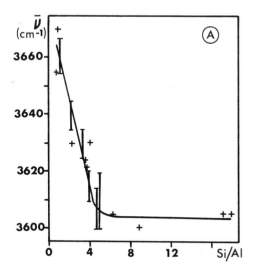
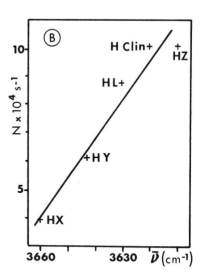

Fig.1 A - Change in wavenumber of the hydroxyl vibrating in the large cavities as a function of Si/Al ratio. From left ot right : A, GeX, X, Y, chabazite, L, Ω, dealuminated Y, dealu-minated Y, offretite, mordenite, clinoptilolite, dealuminated Y, dealuminated mordenite, ZSM-5, (see references, 11, 20 for details).

   B - Turnover number (N) at 100°C for isopropanol dehydration as a function of hydro-xyl wavenumber (after ref. 8).

figure 1A). Hence the wavenumber does not characterize primarily the acid strength. It has been suggested in the study of the left part of figure 1A (20) that the wavenumber changes, observed from low to high Si/Al ratios, are related to a "decrease in the interactions between the (AlO_4) tetrahedra, which decreases the force constant k of the hydroxyl groups considered as a harmonic oscillator". These previous results completed here in the right part of figure 1A with recent results (18,21) show that beyond a Si/Al ratio close to 6 there is no significant change in the hydroxyl wavenumber. It may then be inferred that at these low alumina content the $(AlO_4)^-$ tetrahedra are far enough to each other in the framework and no interactions occur. The maximum of proton "freedom" is then reached. The figure 1B deduced from reference (8) correlates quite significantly the turnover numbers in isopropanol dehydration to the wavenumbers.

In conclusion, the most important results of the hydroxyl studies dealing with catalysis, are concerning acid strength and proton "freedom" characterized by high wavenumbers band. The band intensities give valuable results for acidity studies but they will only be meaningful for catalysis when it is possible to evaluate which part of these hydroxyls is involved in each type of acidic catalysis. Moreover, the OH group concentration cannot be identified for many zeolites as long as the absorption coefficient is not determined.

3. THE IMPORTANCE OF ZEOLITE CATIONS IN ACIDIC CATALYSIS
3.1. Cation field

Three important main points for catalysis with zeolites are related to the cation field : modification of molecule in cation fields, increase in the acid strength and formation of new acid sites.

(i) In the modified "electrostatic theory" relating the carboniogenic activity of zeolites to the electrostatic field of cations, it is assumed that adsorbed molecules are polarized or even ionized in cage fields in order to decrease the free energy of the whole zeolite-adsorbate system (2,22). The acid sites will then, operate on molecules already modified with weakened chemical bonds. The main problem which is still open is how to quantify this influence on reaction rates. General trends are observed between selectivity in catalysis (for instance alkylation of toluene with methanol) and the fields in various alkali-metal-exchanged X and Y zeolites studied by laser Raman spectroscopy of benzene adsorption. A general correlation would imply to take into account all the interactions in the cages (23).

(ii) For a long time (15,24), the increase in acid strength in polyvalent cationic zeolites has been explained on the basis of a polarization of the hydroxyls making the protons more acidic.

(iii) A typical zeolite property, not observed in other acidic catalysts, such as amorphous silica-aluminas, is the well-known generation of protons upon water hydrolysis (3,25). It is usually explained in terms of charge interactions in the framework. A more géneral explanation may be relevant. Several schemes for the reaction have been proposed, one of them being written as :

$$\text{Zeol.Me}^{n+} + xH_2O \rightleftharpoons \text{Zeol-Me}^{(n-y)+} OH_y + yH^+ + (x-y)H_2O \qquad (1)$$

It has been observed that the equilibrium moves to the right, i.e. gives more acidity, upon zeolite hydration, cation radius decrease or the silica to alumina ratio increase (Y instead of X zeolites) (3). Also a Ca-Be-X zeolite type, a weakly acidic catalyst, gives no water dissociation (26). Studies of the hydroxyls of $\text{Zeol-Me}^{(n-y)+} OH_y$ applied to CaHY or

MgHY zeolites have shown that the OH groups of these species are basic (27). The reaction then looks like the known salt hydrolysis in aqueous solutions reaction (2), keeping in mind that the zeolite acts as the anion of a silicoaluminic acid. Similarly to reaction (1) salt hydrolysis increases at high dilution level, at high cation basicity (i.e. small radius) and is greater for strongly acidic anions. This comparison applies the rules of salt

$$Mg\ Cl_2 + 2\ H_2O \rightleftharpoons Mg\ (OH)_2 + 2\ Cl^- + 2\ H^+ \qquad (2)$$

hydrolysis to zeolites and it may be inferred that strongly acidic zeolites (dealuminated or ultrastable Y, mordenites, ZSM-5...) exchanged with basic cations would be able to easily generate protons upon reaction (1).

3.2. The role of cationic adsorption centers

There are at least two adsorption site types in zeolites : cationic and acidic sites. Catalysis on acid sites is then acting while the reactant is simultaneously adsorbed on the cations. In pure adsorption experiments, it has been shown with ethane (28b), propane (28c), pentane (27a) or cyclohexane adsorbed (28d) on X or Y zeolites that the energy of adsorption and the equilibrium constant of adsorption K_1 increase in the order :

Li < Na < K < Rb < Cs

This scale of adsorptive properties explain the range of activity in radical mechanism reactions on cationic zeolites which show a higher activity for KY than NaY zeolites in n-hexane or n-butane cracking at 773-823 K. The cations increase the reactant concentration on the surface (2,29) in the temperature range usually needed for acid catalysis.Considering the reactions occurring on acid sites, it is well-known that a reversed order of activities is observed ; HK-zeolites for example being less active than HNa ones. One wonders whether a correlation exists between the two types of sites. As an example, figure 2 reports as a function of ion-exchange, on the one side the changes in the equilibrium constant of adsorption K_1 of cyclohexane, and on the other side the activity in isooctane cracking for a series of NaHY, KHY and CaHY zeolites (28d). It is obvious that the smaller the adsorption the higher the catalytic activity whatever the cation or the ion-exchange degree. It has then been proposed (28d) that on zeolithes containing both cations and protons there is a competition for the reactant adsorption between cations and protons, the highly polarizing cations giving a small hydrocarbon adsorption, therefore a high carbonium ion number and a high catalytic activity.

The conclusion are firstly that the polarizing power of the cations is playing a role in acidic catalysis by acting on the number of potential acidic sites through an adsorption equilibrium (if the surface is not fully covered, i.e. for non-zero order reactions).

Secondly, when the reactant can be transformed on cations and on acid sites, the selectivity of the reaction will depend, among other parameters, upon the importance of the adsorption not only on cation and proton sites acting independently but, for a major part, on the competitive weight of each of these sites in the overall rate of reactant consumption.

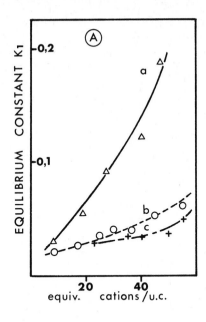
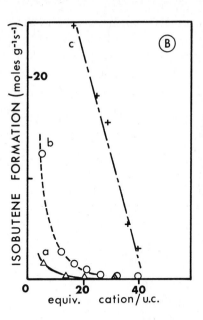

Fig. 2 Study of Y zeolites with various cations as a function of cation exchange
a : KHY, b : NaHY, c : CaHY
A- equilibrium constant of cyclohexane adsorption at 445K.
B- isobutene rate in isooctane cracking at 738K (after (28d))

4. CORRELATIONS BETWEEN ALUMINUM CONTENT AND CATALYSIS

4.1 Site heterogeneity

The origin of carboniogenic activity of zeolites is of course linked to the presence of aluminium atoms which generate protonic acidity. A recent theoritical calculation confirmed that the more acidic protons are those attached to AlO_4 tetrahedra (30). In fact all the protonic sites are not similar. This is illustrated by the modifications in catalytic properties upon aluminum removal by dealumination or steaming. The stability of such treated zeolites is increased and their catalytic activity is differently modified according to which type of aluminum atoms is concerned. In Y zeolites the location of all the framework aluminum atoms is crystallographically identical. However it has been observed that 35 % of these atoms are chemically different since they may be easily removed without changing the catalytic activity or eliminating the strong acidity. (4,31). Hence only weakly acidic sites are involved in this extraction. They differ from the strongly acidic centers only by the nature of their closest neighbors. Dempsey relates the weak acid sites to protons associated through O_1H hydroxyls with the aluminum atoms of the square faces carrying two aluminum atoms in the sodalite cage (32). No major difference is observed between chemically dealuminated or steamed zeolites. The two treatments are concerned with the same framework aluminum sittings. Quite different is the situation for mordenite zeolite. Its structure shows four different crystallographic aluminum atoms locations. It turns out that chemical dealumination extracts preferentially one aluminum type (33) while steaming removes another type of aluminum atoms (18,19). As a result figure 3 shows that the hydrochloric

60

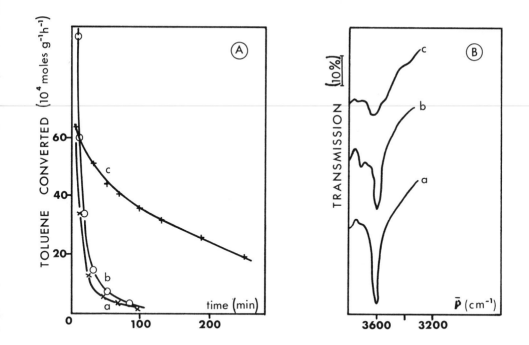

Fig. 3 Study a large port Norton Mordenite
a : dry air pretreated H mordenite, b : dealuminated H mordenite, c : steamed H mordenite
A- Toluene disproportionation at 723K as a fonction of run time (after ref. 18 and 19).
B- Hydroxyl groups after a 2 hrs evacuation.

acid dealumination at room temperature does not disturb the sites responsible for the toluene disproportionation or the acidic 3600 cm^{-1} OH band while steaming removes the major
part of detectable acidic hydroxyls and improves the catalytic activity in reducing deactivation (the sites eliminated gave a large coking but were not very important for toluene
disproportionation). Hence it is clearly evidenced from these two examples of zeolites that
the location of aluminum atoms induce specific properties to the protons attached to them.
Secondly it turns out that each zeolite structure generates typical site environment, each
site being characterized by its acidic properties, by its catalytic activity and probably
selectivity. In conclusion, except for stability and acid strength which appear to be overall properties related to the Si/Al ratio, one cannot easily predict which kind of acidity
or catalytic activity modifications will induce the pretreatment, due to the high variety of
potential aluminum sittings.

4.2. Acid strength increase

It has been known for some time that the more siliceous the zeolites, the stronger are
their acid sites. This property has been explained in terms of a comparison with usual inorganic oxyacids and the acid strength has been related to the silico-aluminic anion composition (11). Such a trend is also valid for dealuminated or steamed zeolites which have lost
a part of their framework aluminum since the simultaneous increase in acid strength and

Si/Al ratio has been shown on Y (17) or recently on ZSM-5 (34) zeolites.

5. OVERALL ACIDIC AND CATALYTIC PROPERTIES

Besides properties mainly related to the topochemistry of sites some other features relevant of general zeolite properties have been mentioned : deformation of the adsorbed molecules in the cavities fields, hydrolysis of water by polyvalent cations, changes in OH band wavenumbers, increase in acid strength and stability as the zeolites become more sili-ceous. In addition to these properties, framework i.r. wavenumbers (12) and SiK_β emission energies vary also smoothly with the overall aluminum content whatever the crystalline structure (13). The idea that zeolites have to be considered as a whole system and not as a juxtaposition of sites is growing (7-11,22,35). As far as acidic catalysis is concerned two approachs have been suggested. In the first one (7,8), a fine calculation of charge on the oxygen and hydrogen atoms and on cations using the Sanderson electronegativity model has been performed and the values of charges obtained are very valuable in establishing rela-tions with several general zeolite properties. Regarding catalysis a good correlation has been obtained between the isopropanol dehydration activity (which implies most of the hy-droxyls (5) and the proton charge. Due to the calculation method used, the evaluated charges of protons (values between 0.09 and 0.15) are connected by the authors to acid strength, the higher the charge, the higher the acid strength. This demonstrates that the turn-over number for isopropanol dehydration is the highest for strongly acidic zeolites such as H-clinop-tilolite and H-mordenite (see figure 1_B). The method also indicates a linear increase in hydrogen charge (acid strength) as the sodium cations are exchanged for protons. The advan-tage of the method is to give an average value of acid strength which allows an easy compa-rison of various zeolites to be made but it relates the larger part of acidic properties only to the proton charge, i.e. the acid strength.

A second approach is based on the existence of the collective properties observed in zeolites and by trying to find a model that could account for the larger part of them. Zeo-lites and solutions were compared and some significant correlations and conclusions could be drawn (11). The mentioned comparison with salt hydrolysis (reactions 1 and 2) and the acid strength dependence on anion, like in inorganic oxyacids, are examples of correlations. An important conclusion for catalysis is the likely existence of activity coefficients for protons which would be implied in reaction rate :

$$r = k \left[H^+ \right] \left[S \right] \frac{f_{H^+} \cdot f_S}{f_*} \qquad (3)$$

where $\left[H^+ \right]$ and $\left[S \right]$ are the concentrations of proton and substrate, f_S, f_{H^+} and f_* are the activity coefficients of the substrate, the protons and the transition state. The high proton concentration (5 to 9 H^+ per liter of ionic crystal) reinforces this hypothesis (9, 10). The activity coefficient would increase simultaneously with the Si/Al ratio (decrease in the interactions in the framework) and would then be higher at low alumina content when little or no interactions between the charges occurs. Two questions then arise : how is it possible to evaluate the activity coefficeents and what is the Si/Al limit ratio beyond which the activity coefficient will be one, i.e. no interaction will decrease the reaction rate. There is no direct answer to the first question. Efficiency coefficients of sites ob-tained from acidity (9,36a) and adsorption (36b) measurements are probably very close to

usual activity coefficient properties. Another answer to the questions is given in figure
1A. This figure expresses the importance of the interactions in the framework for the hydroxyl wavenumber. The lower the charge density in the framework (high Si/Al ratio), the lower
these interactions and then the hydroxyls wavenumber are. An activity coefficient for protons in solutions reflects the hindrance in reactivity due to charge interactions in the
medium. It is therefore proposed that the changes in hydroxyl wavenumber proceeds from a
concept very close to that of activity coefficients of protons. The OH wavenumber changes
in figure 1A would then show the trend of the variation of the activity coefficients of
zeolite protons. This means that the activity coefficient would increase for Si/Al ratio
varying from 1 to 6 and then beyond this value it would be one. Table 1 reports the aluminum atoms content for which this will occur for usual zeolites.

TABLE 1

Evaluation of Al content in various zeolites for a value of one of the proton activity
coefficient.

Zeolite	Number of tetrahedra per unit cell	Number of Al/unit cell for Si/Al~6	Number of Al/unit cell in usual form
Offretite	18	2 to 3	5
L	36	5	9
Clinopti-lolite	36	5	6
Mordenite	48	7	8
ZSM-5	96	14	1 to 2
X and Y	192	27	86(X) and 56(Y)

In faujasite zeolites the value of 27 aluminum atoms per unit cell is quite valuable
since it was already said from acidity studies (36a) that for this aluminum content "the
influence in the supercage of each $(AlO_4)^-$ tetrahedron is not perturbed by its neighbors
and each of them acts as a whole towards the acid-base reactions in the big cavity". Similarly, theoretical calculations showed that for the same Si/Al ratio there was a maximum of
aluminum sittings with no close aluminum neighbors (37). It is obvious from the two last
columns of table 1 that the gain in reaction rate by decreasing the aluminum content would
be the highest for faujasite type zeolites. In conclusion, the existence of activity coefficients is evidenced by the framework interactions. The changes of coefficients upon the
Si/Al ratio would vary like these interactions reflected in the hydroxyl wavenumbers of
figure 1A. Fro Si/Al ratios higher than 6 the activity coefficient would be one and then
have no action on catalysis.

6. CONCLUSIONS

The paper attempted to clarify the influence on catalysis of the various modifications
of zeolites (cations, Si/Al ratio...) in order to view some predictions for better catalytic
properties.

A first conclusion takes into account the results presented in order to point out the importance of acidity parameters on the various factors of the reaction rate

$$r = k \left[H^+\right]\left[S\right] f_{H}+. \ f_S \ / \ f_*$$

Influence on k. The acid strength governed by the Si/Al ratio and the polarizing action of polyvalent cations is important for activation energy. The deformation of molecules in the cation field also has to be considered in the preexponential factor.

Influence on H^+. Besides the influence of pretreatment, out of the scope of this paper (heating, evacuation, hydration, reduction of ions to metal generating protons...), the proton concentration depends on : i) first by the number of charges in the zeolites i.e. the Si/Al ratio, ii) the number of protons generated upon salt hydrolysis (action of polyvalent cations)iii) the number of sites strong enough for catalysis (Si/Al ratio, pllyvalent cations, extent of cation exchange for protons), and iv) the number of H^+ sites available for adsorption of the reactant (competitive adsorption on cations and protons).

Influence on f_H+. The higher the Si/Al ratio, the lower is f_H+. At Si/Al > 6 f_H+ = 1.

A second conclusion relates from the acidity point of view the selectivity to the acid strength, to the cage fields and to the competition for adsorption on cations and protons. This invokes the Si/Al ratio, the polarizing action of cations and the cage shape.

A third conclusion considers the trend of research in catalysis. On metal or semi-conductors collective properties were first considered to determine the overall behavior of the catalysts. It progressively appeared that localized sites are of great importance in adsorption or catalysis (defects, vacancies, crystalline faces, number of metal atoms or cations close neighbors, cation valency and coordination...) and a large part of present research is devoted to these fields. The approach for zeolites appears to be quite opposite. Starting from a fine attempt to characterize the localized active sites it is now evident that the fundamental general behavior of zeolites is governed by overall properties independent of crystallographic structure and relevant to a model close to the model of solutions. This suggests that zeolites are at the border-line between solid and liquid state : crystalline liquids ?

REFERENCES

1 J.W. Ward, Adv.Chem.Ser., 101 (1971), 380-404.
2 J.A. Rabo and M.L. Poutsma, Adv.Chem.Ser. 102 (1971) 284-314.
3 J.W. Ward, in Zeolite Chemistry and Catalysis (edited by J.A. Rabo) ACS monograph 171, ACS Washington, 1976, p. 118.
4 D. Barthomeuf, ACS Symposium Series 40 (1977) 453.
5 P.A. Jacobs, Carboniogenic activity of zeolites, Elsevier, Amsterdam, 1977, 253 pp.
6 H.W. Haynes Jr, Catal.Rev.Sci.Eng. 17(1978) 273-336.
7 W.J. Mortier, J. Catal. 55(1978) 138-145.
8 P.A. Jacobs, W.J. Mortier, J.B. Uytterhoeven, J. Inorg.Chem., 40 (1978) 1919-1923.
9 D. Barthomeuf, C.R. Acad. Sci., Paris, Ser. C, 286 (1978) 181.
10 D. Barthomeuf, Acta Phys. Chem., 24 (1978) 71.
11 D. Barthomeuf, J. Phys. Chem., 83 (1979) 249.
12 E.M. Flanigen in "Zeolite Chemistry and Catalysis" (edited by J.A. Rabo), ACS monograph 171, ACS Washington, 1976, p 80.
13 R.L. Patton, E.M. Flanigen, L.G. Dowell and D.E. Passoja, ACS, Symp. Ser., 40 (1977) 64.
14 J.W. Ward and R.C. Hansford, J. Catal., 13 (1969) 364-372.
15 J.T. Richardson, J. Catal. 9 (1967) 172 and 182 ; 11 (1968) 275.
16 A. Bielanski and J. Dakta., Bull.Acad.Pol.Sci., XXII (1974) 341-350 and J.Catal. 37 (1975), 383-386.
17 R.Beaumont, P. Pichat, D. Barthomeuf and Y. Trambouze, Catalysis (ed. Hightower), North Holland, Amsterdam, 1 (1973) 343.

18 C. Mirodatos, B.H. Ha, K. Otsuka and D. Barthomeuf, Fifth Int.Conf. on zeolites, Napo-
    li, 1980, and references therein.
19 M.L. Armando-Martin, Thesis Poitiers 1978
20 D. Barthomeuf, JCS Chem. Com. , 21 (1977) 743.
21 A. Auroux, V. Bolis, P. Wierzchowski, P.C. Gravelle and J.C. Vedrine, JCS Farad. Trans
    II 75 (1979) 2544.
22 J.A. Rabo in "Zeolite Chemistry and Catalysts" (edited by J.A. Rabo) ACS Monograph,
    ACS Washington 1976, p. 332.
23 J.J. Freeman and M.L. Unland, J. Catal., 54 (1978) 183.
24 A.E. Hirschler, J. Catal. 2 (1963) 428 and 11 (1978) 274.
25 C.J. Planck, Proc. 3rd Inter. Cong. Catal., Amsterdam, 1 (1964) 727.
26 G. Poncelet and M.L. Dubru, J. Catal., 52 (1978) 321.
27 C. Mirodatos, P. Pichat and D. Barthomeuf, J. Phys. Chem., 80 (1976) 1335.
28 a) A.V. Kiselev, Disc. Farad. Soc. 40 (1965) 205. b) A.G. Bezus, A.V. Kiselev,
    Z. Sedlacek and Pham Quang Du, Trans. Farad. Soc. 67 (1971) 468. c) O.M. Dzhigit, K.
    Karpinskii, A.V. Kiselev, K.N. Mikos and T.A. Radhmanova, Rus. J. Phys. Chem. 45 (1971)
    848. d) R. Beaumont, B.H. Ha and D. Barthomeuf, J. Catal. 40 (1975) 160.
29 M.L. Poutsma and S.R. Schaeffer, J. Phys. Chem., 77 (1973) 158.
30 W.J. Mortier, P. Geerlings, C. Van Alsenoy and H.P. Figeys, J. Phys. Chem., 83 (1979)
    855.
31 R. Beaumont and D. Barthomeuf, C.R. Acad. Sci. Paris, 272C (1971) 363,
    D. Barthomeuf and R. Beaumont, J. catal., 30 (1973) 288.
32 E. Dempsey, J. Catal., 33 (1974) 497.
33 R.W. Olsson and L.D. Rollman, Inorg. Chem., 16 (1977) 651.
34 A. Auroux, M. Rekas, P.C. Gravelle and J.C. Vedrine, in "Proceed of the Vth Int. Conf.
    on Zeolites" (edited by L.V. Rees), Heyden, London, 1980 p. 433.
35 P.H. Kasai and R.J. Bishop, K. Phys. Chem., 77 (1973) 2308.
    J.A. Rabo and P.H. Kasai, Prog. Solid State Chem., 9 (1975) 1.
36 a) R. Beaumont and D. Barthomeuf, J. Cata. 26 (1972) 218.
    b) D. Barthomeuf and B.H. Ha, JCS, Farad. Trans. I, 69 (1973) 2158.
37 R.J. Mikovski and J.F. Marshall, J. Catal., 44 (1976) 170.

B. Imelik *et al.* (Editors), *Catalysis by Zeolites*
© 1980 Elsevier Scientific Publishing Company, Amsterdam — Printed in The Netherlands

ISOBUTANE/BUTENE ALKYLATION ON CERIUM EXCHANGED X AND Y ZEOLITES

JENS WEITKAMP

Engler-Bunte-Institute, University of Karlsruhe, Richard-Willstätter-Allee 5, D-7500 Karlsruhe 1, Federal Republic of Germany

1   INTRODUCTION

While zeolite catalysts have gained tremendous importance in some petroleum refining processes, viz. catalytic cracking, hydrocracking, and isomerization, they failed to receive acceptance in the alkylation of isobutane with light olefins. In this process which yields highly branched paraffins required as gasoline components conventional liquid phase catalysts, i.e. concentrated sulfuric acid or anhydrous hydrogen fluoride are still applied.

Chemically, alkylation, isomerization, and cracking have many features in common. In particular they are catalyzed by acids and proceed via carbenium ions. Both the mechanisms and process engineering aspects of hydrocarbon conversion on acid zeolites have been treated comprehensively in recent textbooks (ref. 1-3).

In a strongly simplified manner the mechanism of isobutane/butene alkylation catalyzed by, e.g., H_2SO_4 or HF has often been interpreted in terms of the following chain reaction initiated by protonation of the alkene and hydride transfer between the resulting sec. butyl cation and isobutane:

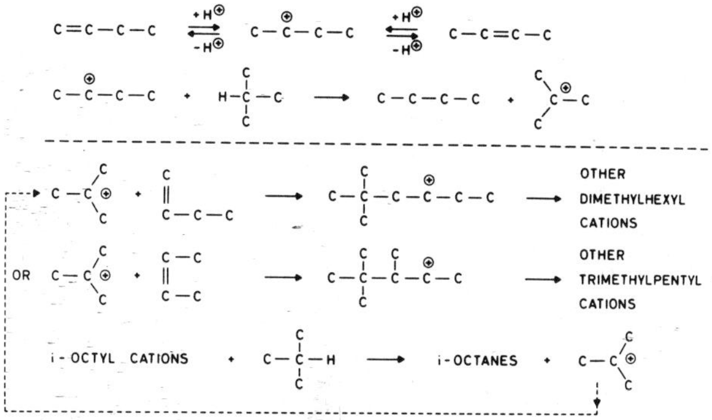

Double bond shift in the alkene is considered to be rapid and hence the tertiary butyl cation can be alkylated either by 1-butene or 2-butene regardless of which butene is used in the feed. The primary products of the alkylation steps can rearrange to some extent whereby different dimethylhexyl and trimethylpentyl cations are formed. Hydride transfer between i-octyl cations and isobutane finally yields a mixture of isooctanes and tert. butyl cation which propagates the chain. A 1:1 overall isobutane : butene stoichiometry is predicted according to this scheme.

The true mechanism of reaction is considerably more complicated and many side reactions of alkyl carbeniumions have been specified (ref. 4-7) to account for the observed products.

The principal feasibility of isobutane/alkene alkylation on zeolites was demonstrated as early as 1968 (ref. 8-10). Since then **surprisingly few** groups have published results of research work on this reaction, namely Kirsch, Potts et al. (ref. 8,9,11-13), Minachev et al. (ref. 14-16), and Schöllner et al. (ref. 17-19). From their work it is evident that zeolites although highly active initially undergo rapid deactivation due to coke forming side reactions. It was mainly this deactivation which rendered zeolites economically unattractive for isobutane/alkene alkylation processes. Furthermore, the occurrence of deactivation made proper experimental handling of the reaction difficult, the situation being severely aggravated by complex product distributions.

For the most part these difficulties were circumvented by integral sampling procedures over relatively long periods which, however, can only furnish a time-averaged picture of the catalytic events. Moreover, unduly simple analytical methods were employed in some cases. In an attempt to arrive at a better understanding of isoalkane/alkene alkylation catalyzed by zeolites we developed experimental techniques which we consider to be appropriate to the complexity of the system. They combine instantaneous sampling with high resolution product analysis. The present paper intends to give a detailed description of these methods along with some pertinent results obtained with three cerium exchanged zeolites of the faujasite family.

2   EXPERIMENTAL

The pressure apparatus for conversion of liquid hydrocarbon mixtures on solid catalysts is depicted in Fig. 1. Essentially it consists of a vessel equipped with a magnetically driven stirrer for preparation and storage of the liquid feed mixture, a piston-type pump for pulsation-free operation at very small throughputs, and the fixed bed reactor made from stainless steel. During a run the reactor is immerged into a circulating oil bath which allows a thorough control of temperature and an effective removal of the heat of the exothermic reaction. For in-situ pretreatment of the catalyst at higher temperatures the reactor can alternatively be housed in an electric furnace.

A six-port and a three-port valve are arranged in such a way that prior to an experiment the desired amount of each feed component can be pumped into the storage vessel. An automatically actuated four-way valve controls the functions of the two cylinders of the pump. By means of another four-way valve the reactor can be by-passed, e.g. for analysis of the feed mixture. In order to define the beginning of a run as accurately as possible the following procedure was applied: The feed mixture was pumped while the reactor filled with nitrogen under reaction pressure was by-passed. When the four-way valve was switched the liquid hydrocarbons replaced the nitrogen cushion. Downstream movement of the phase boundary towards the needle valve was visible through pressure resistant teflon tubing.

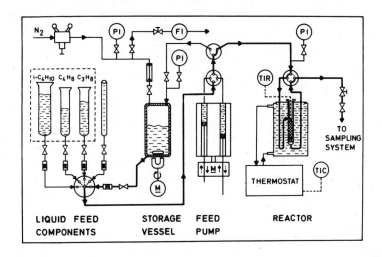

Fig. 1.   Scheme of the apparatus.

As shown in Fig. 2 the liquid effluent from the reactor is depressurized in a needle valve, vaporized, and diluted with nitrogen. The subsequent sampling system embraces a device for instantaneous or differential sampling with glass ampules, a cooling trap for collecting a liquid sample, and the system for on-line GC analyses.

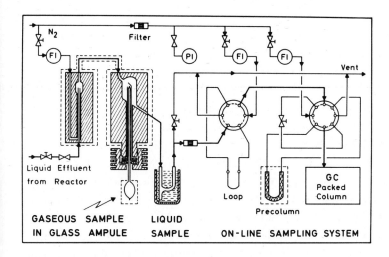

Fig. 2.   Scheme of the sampling system.

The sampling system with glass ampules is an adaptation of the method first applied by Pichler and Gärtner in a study on catalytic cracking (ref. 20). The top of a glass capillary connected to an evacuated and heated ampule is immerged into the gaseous product stream. The sealing

system resembles a GC injection port. At a predetermined time the top of
the capillary can be broken, e.g. by rotating, so that the ampule fills up
with gaseous product. Immediately thereafter the ampule is withdrawn and
sealed. In this state the sample inside the ampule can be stored without
loss of any components until analysis. The whole sampling procedure can be
repeated in very short intervals of one minute or less.

The liquid sample is normally collected at ca. 10 $^\circ$C during the period
of one or several hours. Although this sample is not needed for quantitative evaluation of a run it is useful for qualitative assignment of peaks
in the chromatograms. Moreover, collecting hydrocarbons at subambient temperatures avoids any condensation in the downstream system for on-line
sampling. The latter contains a six-port valve with the loop and a precolumn in which remaining C_5^+ hydrocarbons are removed and which can be
backflushed via an eight-port valve. The main purpose of the on-line GC-
analyses is to monitor the degree of butene conversion during a run.

The final evaluation is based on the analyses of the samples in the ampules. For analysis an ampule is crushed in a specially designed apparatus
which is heated and connected to the injection port of a capillary GLC unit
with a flame ionization detector. Stainless steel columns of 100 m length
and 0.25 mm internal diameter with polypropylene glycol, squalane, and
OV-101 as stationary phases are used. For satisfactory resolution a temperature program starting at subambient temperature, e.g. -15 $^\circ$C along with
a low heating rate of 1.0 $^\circ$C/min or even less must be applied. Typically,
the time required for one analysis amounts to 4 h.

The products contain isobutane in a very large excess. Since the size of
our samples was chosen such as to obtain maximum sensitivities for the rest
of the hydrocarbons, the isobutane peak inevitably was beyond the linear
range of the detector. Hence, isobutane could not be included in the
material balances.

Assignment of the paraffinic peaks up to C_9 is based on commercially
available reference substances, retention index data compiled in the literature (ref. 21), and our earlier work on hydrocracking (ref. 22). No
attempt was made to identify C_{10} or higher hydrocarbons. In order to recognize olefinic peaks additional analyses were carried out with the liquid
sample in the same GLC equipment with and without application of precolumns.
Both an olefin subtracting precolumn filled with a H_2SO_4/H_3PO_4 coated
carrier and a hydrogenation precolumn filled with a Pd/Al_2O_3 catalyst were
utilized. These methods revealed that even with the high resolution capillary GLC procedures some overlap between certain octane and octene peaks
occurred.

Starting from NaX and NaY from Union Carbide the catalysts were prepared
by conventional ion exchange cycles with an aqueous solution of 0.03 mol-%
$Ce(NO_3)_3$ at 80 $^\circ$C. In the preparation of the highly exchanged CeY sample
two intermediate calcination steps were involved as described elsewhere
(ref. 23). The molar ratios Al:Ce:Na in the final catalysts were

3.30 : 1.00 : 0.13, 6.85 : 1.00 : 3.55, and 2.20 : 1.00 : 0.078 for CeX-96, CeY-46, and CeY-98, respectively. The figures denote the formal degree of cerium exchange as calculated simply from the cerium and sodium contents. Whereas the aluminum/cation balance is fulfilled within ca. 5 % for CeX-96 and CeY-46 the CeY-98 sample shows an overall cation excess which, according to literature data (ref. 24), is due to the calcination steps. Each catalyst was pressed binder-free and ground to 0.25 - 0.50 mm. 2.00 cm^3 of this particle size fraction were employed in each run. After in-situ pretreatment at 350 $^{\circ}$C in a purge of dried nitrogen the mass of CeX-96, CeY-46, and CeY-98 were 1.52, 1.09, and 1.42 g, respectively.

The molar ratios of the components in the feed mixtures were as follows: i-butane : 1-butene : propane 11.1 : 1.00 : 0.42 and 11.6 : 1.00 : 0.37 for CeX-96 and Ce-Y-46, respectively, and i-butane : cis-2-butene : propane 11.0 : 1.00 : 0.38 for CeY-98. Propane served as an internal standard for GC analyses. In preliminary experiments it was ascertained that propane was neither consumed nor formed during the reaction. The liquid feed rate at ambient temperature, the pressure, and reaction temperature amounted to 7.5 cm^3/h, 3.1 MPa, and 80 $^{\circ}$C, respectively.

3   RESULTS

3.1   Qualitative description of isobutane/butene conversion on faujasites

When a zeolite of sufficiently high acidity and sufficiently large pore size, e.g. CeX or CeY, is exposed to an isobutane/butene mixture the following results can be revealed by instantaneous sampling:

During an initial stage lasting typically 20-40 min under our conditions the butene is completely removed. Immediately after start-up of the reaction a complex mixture of C_5^+ isoalkanes, i.e. alkylate, can be detected. Moreover, some small amounts of n-butane are formed (ref. 23). It is particularly noteworthy that the C_5^+ product is entirely free from olefins or any type of cyclic hydrocarbons during this initial alkylation stage.

After a certain time on stream butenes appear as a mixture of 1-butene, cis- and trans-2-butene. Almost simultaneously olefins are detected among the C_5^+ hydrocarbons, especially isooctenes. The amount of butenes and the content of olefins in the C_5^+ product increase rapidly with time on stream. In the following sections appropriate ways are outlined to describe such a highly non-stationary behaviour in a quantitative manner.

3.2   Differential and integrated yields

For each ampule drawn the yield of individual products or groups of products on butene charged can be evaluated due to the use of an internal standard. These yields in terms of g/g butene charged will be referred to as differential yields. In Fig. 3 the differential yields of C_5-C_{12} hydrocarbons and butenes as well as their sum are plotted versus time on stream for one experiment. From the liquid feed rate, the composition of the feed, the densities of the components (assuming ideal behaviour of the liquid

mixture), and the mass of catalyst the cumulative butene charge per unit
mass of catalyst which has been referred to as catalyst age (ref. 11) can
be calculated. It is given as a second abscissa in Fig. 3.

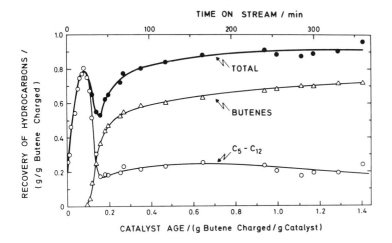

Fig. 3. Conversion of isobutane/cis-2-butene on CeY-98. Differential yields
versus time on stream and catalyst age.

Integrating the differential yields, e.g. graphically, up to a certain catalyst age gives the integrated yields in terms of g/g catalyst. In the above
example integrated yields of the C_5-C_{12} hydrocarbons and of the total hydrocarbons at a catalyst age of 1.4 g butene charged/g catalyst are 0.36 and
1.175 g/g catalyst, respectively. The deficiency of hydrocarbons at the
selected catalyst age is then given by 1.4 - 1.175 = 0.225 g/g catalyst.

To account for this deficiency the following effects have to be considered: 1. Coke was formed from butene and accumulated on the zeolite.
2. Butene was converted into isobutane which could not be determined quantitatively. 3. Quantitative application of our sampling technique with
glass ampules is limited to hydrocarbons up to ca. C_{12}. Any higher products
would not be detected quantitatively even if they were desorbed from the
catalyst. However, in the case of CeY-98 C_{13}^+ products were found to be
absent in the liquid sample where they would have been collected if they
had formed. Hydrocarbon deficiency is then due to the formation of coke and
isobutane. Although for the moment we are unable to separate both effects
we believe that most of the deficiency can be attributed to coke formation.

3.3 Alkylation versus alkene oligomerization. Integrated yields of alkylate

In Fig. 4 the composition of the main carbon number fraction, i.e. C_8,
is given in terms of the mol-fraction of octanes and octenes for the run
with CeY-98. Due to peak overlap the values are less accurate at lower
octane mol-fractions. The dotted part of the curve should be regarded as
an upper limit of the octane mol-fraction.

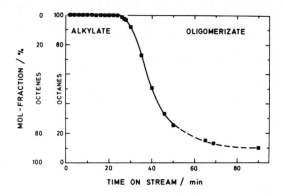

Fig. 4. Conversion of isobutane/cis-2-butene on CeY-98. Composition of the C_8-fraction.

It is evident from Fig. 4 that initially alkylate is formed with an utmost degree of selectivity while late in the run the reaction is better described in terms of butene oligomerization. In order to evaluate the integrated yield of alkylate for a given experiment one must define a criterion which ranks a product as alkylate or otherwise. We consider a product to be an alkylate if its content of alkanes in the C_8-fraction is 90 % or higher. In the case of CeY-98 this requirement is fulfilled up to a time on stream of 30 min.

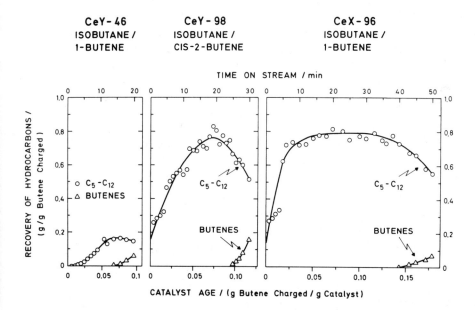

Fig. 5. Integrated yields of alkylate for three cerium exchanged faujasites.

In Fig. 5 the differential yields obtained with the three cerium exchanged zeolites are presented. In each case the end point of the curve has been

fixed according to the above criterion. Note that the catalyst age scales
are somewhat different in relation to the time on stream scales reflecting
mainly small differences in the mass of catalysts. Integration up to the end
points of the C_5-C_{12} curves gives integrated yields of alkylate amounting
to 9, 70, and 125 mg/g catalyst for CeY-46, CeY-98, and CeX-96, respectively.

From these results and Fig. 5 some interesting conclusions can be drawn.
Whereas according to the idealized mechanism the differential yield of alkylate should be slightly above 2 g/g butene charged, the experimental
values even do not reach 1 g/g butene charged. Moreover, our results based
on differential sampling reveal that the differential yields of alkylate
pass through a maximum. Although it is too early to develop a detailed
interpretation, the strong adsorption of the olefin and the accumulation of
coke are considered to be important factors which influence the yield of
alkylate. The integrated yields for CeY-46 and CeY-98 clearly show that a
high degree of cerium exchange is desirable for isobutane/butene alkylation.
Moreover, it is found that CeX-96 is superior to CeY-98 which at least
qualitatively corresponds to a higher aluminum and cation content and hence
a higher number of acid sites per unit mass in the X zeolite.

3.4 Carbon number distributions

Table 1 gives carbon number distributions of the C_5^+ products obtained
with the three catalysts during their alkylation stage. It follows from
these data that although C_8 is the main carbon number fraction alkylation
on the zeolites just as with H_2SO_4 or HF as catalysts (ref. 4-7) always
yields products with other carbon numbers.

TABLE 1. Carbon number distributions (wt.-%) of alkylates

Catalyst	CeY-46	CeY-98		CeX-96	
Time on stream (min)	9	1	30	1	40
C_5	5.7	22.6	6.1	26.0	8.9
C_6	5.9	11.9	5.0	14.4	7.6
C_7	5.9	9.0	5.3	10.4	7.9
C_8	65.9	47.0	53.8	45.8	49.8
C_9	6.1	6.4	8.1	2.3	7.0
C_{10}-C_{12}	10.5	3.1	21.7	1.1	18.8

The data for CeY-98 and CeX-96 reveal a distinct influence of time on
stream on the carbon number distributions. In the fresh state both catalysts
yield large amounts of C_5 - C_7 products while towards the end of the alkylation stage considerably more heavy hydrocarbons are formed. The occurrence
of C_5 - C_7 and C_9 - C_{11} paraffins is generally explained by the so-called
destructive alkylation, i.e. the intermediate formation of higher alkyl
carbenium ions with 12 or 16 carbon atoms followed by one or more steps of
cracking. Obviously, the zeolites loose a great deal of their cracking

activity during on stream use.

3.5 Distributions of individual isomers

Alkylation on the zeolites at 80 $^{\circ}$C selectively leads to i-alkanes with
at least one tertiary carbon atom. The n-alkanes as well as i-alkanes lacking a tertiary carbon atom (e.g. 2,2-dimethylhexane, 2,2,3,3-tetramethyl-
butane, etc.) were found to be absent. The only exceptions of this rule,
which was found to be valid for all carbon numbers, were n-pentane and n-
octane which occasionally occurred in traces. In Table 2 the distributions
of the octane isomers are listed for the samples already selected in
Table 1.

TABLE 2. Distributions of individual C_8 isomers (mol-%)

Catalyst	CeY-46	CeY-98		CeX-96	
Time on stream (min)	9	1	30	1	40
2-M-Heptane	0.1	0.2	0.2	0.1	0.4
3-M-Heptane + 3-E-Hexane	1.7	0.4	0.8	0.2	1.1
4-M-Heptane	0.2	0.1	0.1	-	0.2
2,3-DM-Hexane	9.4	4.8	11.7	4.7	12.9
2,4-DM-Hexane	3.2	6.2	4.9	5.8	7.0
2,5-DM-Hexane	0.3	3.0	2.3	1.6	4.2
3,4-DM-Hexane[a]	45.8	6.9	24.3	6.9	13.4
3-E-2-M-Pentane	2.2	0.6	1.8	0.3	1.8
2,2,3-TM-Pentane	3.3	4.4	2.8	4.4	2.9
2,2,4-TM-Pentane	3.0	22.3	12.2	23.3	18.5
2,3,3-TM-Pentane	16.5	27.9	19.6	30.7	20.8
2,3,4-TM-Pentane	14.3	23.2	19.3	22.0	16.8

[a]Both diastereomers

These data show that monobranched isooctanes are formed in almost negligible amounts. Moreover, they reveal that of the two structures predicted
by the idealized mechanism as primary products of isobutane/butene alkylation, i.e. 2,2-dimethylhexane and 2,2,3-trimethylpentane, one is absent
whereas the other occurs in relatively small concentrations. While these
discrepancies are usually accounted for by postulating rearrangement steps
of i-octyl cations (ref. 4,5) the pronounced dependencies of the experimental isomer distributions upon both the degree of cerium exchange in the
Y zeolites and time on stream are not well understood at present. We believe that beside the idealized mechanism there are principally different
routes leading to i-octanes. One of these can be seen in the so-called self-
alkylation of the olefin, particularly in the alkylation of sec. butyl cation
with 2-butene which leads to the carbon skeleton of 3,4-dimethylhexane and
to a much lesser extent in the alkylation of sec. butyl cation with 1-butene
which results in the formation of the skeleton of 3-methylheptane. The contribution of self-alkylation is obviously more pronounced when the zeolite

is less acid, as in the case of CeY-46, and towards the end of the alkylation stage. A third route to isooctanes is the breakdown of polymeric material. Such a pathway has been claimed by Weeks and Bolton (ref. 25) to produce mainly 2,3- and 2,5-dimethylhexane as well as 2,2,4-trimethylpentane.

4   CONCLUSIONS

Isobutane/alkene alkylation, which is perhaps one of the most complex catalytic reactions, requires special experimental techniques to cope with the simultaneous occurrence of non-stationary behaviour and complex product distributions. Differential sampling in intervals, which are short compared to the characteristic time of non-stationary behaviour, appears to be a powerful tool especially in combination with high resolution analyses. A systematic application of the methods outlined in this paper is under way in our laboratory and will, hopefully, lead to a more detailed picture of the low temperature chemistry of alkyl carbenium ions, including rearrangements and low temperature cleavage. Furthermore, the results can be expected to shed light on special properties of zeolite catalysts, e.g. hydride transfer power, and their response to coke deposition, which are at present very little understood.

ACKNOWLEDGEMENTS

I thank Mr. W. Stober for assistance with the experiments.
Financial support by the German Science Foundation (Deutsche Forschungsgemeinschaft) is gratefully acknowledged.

REFERENCES

1   P.A. Jacobs, Carboniogenic Activity of Zeolites, Elsevier Scientific Publishing Co., Amsterdam, Oxford, New York, 1977, 253 pp.
2   M.L. Poutsma, in J.A. Rabo (Ed.), Zeolite Chemistry and Catalysis, Am. Chem. Soc. Monograph, Vol. 171, Am. Chem. Soc., Washington, D.C., 1976, pp. 437-528.
3   A.P. Bolton, ibid., pp. 714-779.
4   L. Schmerling, in G.A. Olah (Ed.), Friedel-Crafts and Related Reactions, Vol. II, Part 2, Interscience Publ., New York, London, Sidney, 1964, pp. 1075-1131.
5   R.M. Kennedy, in P.H. Emmett (Ed.), Catalysis, Vol. 6, Reinhold Publ. Corp., New York, 1958, pp. 1-41.
6   T. Hutson, Jr. and G.E. Hays, Preprints, Div. Petr. Chem., Am. Chem. Soc., 22 (1977) 325-342.
7   L.F. Albright, ibid., 22 (1977) 391-398.
8   F.W. Kirsch, J.D. Potts and D.S. Barmby, Oil Gas J., 66, No. 29, July 15 (1968) 120-127.
9   F.W. Kirsch, J.D. Potts and D.S. Barmby, Preprints, Div. Petr. Chem., Am. Chem. Soc., 13, No. 1 (1968) 153-164.
10 W.E. Garwood and P.B. Venuto, J. Catal., 11 (1968) 175-177.
11 F.W. Kirsch and J.D. Potts, Preprints, Div. Petr. Chem., Am. Chem. Soc., 15, No. 3 (1970) A-109 - A-121.
12 F.W. Kirsch, J.L. Lauer and J.D. Potts, ibid., 16, No. 2 (1971) B-24 - B-39.
13 F.W. Kirsch, J.D. Potts and D.S. Barmby, J. Catal., 27 (1972) 142-150.
14 E.S. Mortikov, S.M. Zen'kovskii, N.V. Mostovoi, N.F. Kononov, L.I. Golomshtok and Kh. M. Minachev, Izv. Akad. Nauk SSSR, Ser. Khim., 7 (1974) 1551-1554.

15 E.S. Mortikov, S.M. Zen'kovskii, N.V. Mostovoi, N.F. Kononov and
    L.I. Golomshtok, ibid., 10 (1974) 2237-2239.
16 Kh. M. Minachev, E.S. Mortikov, S.M. Zen'kovskii, N.V. Mostovoi and
    N.F. Kononov, Preprints, Div. Petr. Chem., Am. Chem. Soc., 22 (1977)
    1020-1024.
17 R. Schöllner, H. Hölzel and M. Partisch, Wiss. Zeitschr. Karl-Marx-
    Univ. Leipzig, Math.-Naturwiss. R., 23 (1974) 631-641.
18 R. Schöllner and H. Hölzel, Journal f. Prakt. Chemie, 317 (1975)
    694-704.
19 R. Schöllner and H. Hölzel, Zeitschr. Chem., 15 (1975) 469-475.
20. H. Pichler and R. Gärtner, Brennstoff-Chem., 43 (1962) 336-340.
21 A. Matukuma, in C.L.A. Harbourn (Ed.), Gas Chromatography 1968, The
    Institute of Petroleum, London, 1969, pp. 55-75.
22 J. Weitkamp, in J.W. Ward and S.A. Quader (Eds.), Hydrocracking and
    Hydrotreating, Am. Chem. Soc. Symp. Series, Vol. 20, Am. Chem. Soc.,
    Washington, D.C., (1975) pp. 1-27.
23 J. Weitkamp, Proc. 5th Intern. Conf. Zeolites, Naples, June 2-6, 1980,
    in press.
24 A.P. Bolton, in R.B. Anderson and P.T. Dawson (Eds.), Experimental
    Methods in Catalytic Research, Vol. 2, Academic Press, New York,
    San Francisco, London (1976) pp. 1-42.
25 T.J. Weeks, Jr. and A.P. Bolton, J. Chem. Soc., Farad. Trans. I, 70
    (1974) 1676-1684.

B. Imelik *et al.* (Editors), *Catalysis by Zeolites*
© 1980 Elsevier Scientific Publishing Company, Amsterdam — Printed in The Netherlands

PARTICULAR ASPECTS OF ZEOLITE CATALYZED REACTIONS

M. GUISNET [*], N.S. GNEP, C. BEAREZ and F. CHEVALIER
Laboratoire de Chimie 7, ERA CNRS 371, Faculté des Sciences, 40 avenue du Recteur Pineau;
86022 POITIERS CEDEX, FRANCE

Zeolites present for many hydrocarbon reactions a higher activity than that of amorphous catalysts and a selectivity which is often quite different. This particular selectivity can be attributed to two main causes :

(i) the porous structure can either limit the access of certain reactant molecules to the active sites of the zeolite, hinder the desorption of the products or inhibit the formation of reaction intermediates [1].

(ii) according to Rabo "et al" [2], zeolites can be considered as strong solid electrolytes with, as a consequence, a very strong reactant adsorption favourable to bimolecular reactions.

We have chosen two series in order to illustrate these aspects of zeolite selectivity : first, the transformation over a series of protonic zeolites of xylenes, second, the transformation of light paraffins over mordenites.

1. TRANSFORMATION OF XYLENES OVER PROTONIC ZEOLITES

Shape selectivity in the case of reactions of alkylbenzenes on zeolites [3-5] has already been exposed.

The transformations of orthoxylene : isomerization, disproportionation into toluene and trimethylbenzenes and coke formation have been compared in the case of the catalysts described in table 1 ; furthermore, over fluorinated alumina, mordenite and ZSM5 zeolite, a kinetic study of the isomerization of the three xylenes has been carried out.

Table 1 shows that the initial activity of zeolites can be up to thirty times higher than that of silica alumina or of the most active fluorinated alumina [6] ; however, erionite whose pore size is too small to allow orthoxylene to pass through, is practically inactive ; the low activity of the ZSM5 zeolite (four times less active than the mordenite) could probably be explained by the difficulty of access of orthoxylene to the active sites.

The ratio between the disproportionation rate and the isomerization rate (D/I) is equal to zero over ZSM5 and erionite ; it is very low over silica alumina, fluorinated alumina and mordenite but it is very high over Y zeolite (table 1) ; according to Poutsma [7] the high value of the D/I over Y zeolite as compared to that found over amorphous catalysts is explained by a stronger orthoxylene adsorption over Y zeolite than over amorphous catalysts. The values found with mordenite and ZSM5 could be explained either by the difficulties for forming in the pore structure of these zeolites

[*] To whom correspondence should be addressed.

the large 1,1-diphenylalkane-type intermediate for disproportionation or by the limitations to the desorption of the trimethylbenzenes whose kinetic diameters are bigger
than those of the xylenes. This latter hypothesis can be ruled out since over ZSM5 the
formation of toluene, the smaller kinetic diameter product, is not observed and since
over mordenite the formation of toluene and trimethylbenzenes takes place at the same
rate.

The ratio between the amounts of orthoxylene transformed into coke (m_k) and isomerized (m_I) during 7 hour experiments is practically the same for silica alumina and
Y zeolite whereas it is lower for mordenite and very low for ZSM5 (table 1). The low
values found for these two latter catalysts can be explained by the difficulty of forming in the pore structure the intermediates of this reaction which, like the intermediates of the disproportionation, are necessarily bimolecular.

TABLE 1

Orthoxylene transformation at 350°C :

a_I : isomerization activity at time zero ; D/I : ratio between disproportionation rate and
isomerization rate ; m_k : weight of orthoxylene transformed into coke ; m_I : weight
of isomerized orthoxylene.

Catalysts	Origin or preparation technique	$10^4 a_I$ mole $h^{-1}g^{-1}$	D/I	$10^3 m_k/m_I$
Silica alumina	LA3P Ketjen	60	0.06	4.2
Fluorinated alumina	GFS 300 Rhône Progil, 7 wt % F	50	0.05	
Y Zeolite ::	NH$_4$ ion exchange of Union Carbide NaY, wet-air cal-cination, 1.15 wt % Na	1000	0.75	3.8
Mordenite ::	Zeolon 900 H, Norton 0.4 wt % Na	1700	0.05	0.5
ZSM5 Zeolite ::	Preparation according to Mobil procedure, 0.01 wt % Na	450	0	0.15
Erionite-Chabazite ::	NH$_4$ ion exchange of Zeolon 500 Norton, wet-air calcina-tion, 0.23 wt % Na, 0.2 wt % K	3	0	50

:: from Institut Français du Pétrole.

The selectivity of the orthoxylene isomerization is different over zeolites and
over amorphous catalysts : over zeolites, orthoxylene leads directly to metaxylene but
also to paraxylene, while over amorphous catalysts, only metaxylene is formed directly.
This direct transformation of orthoxylene into paraxylene is particularly important
over ZSM5 zeolite and over mordenite (figure 1). This reaction has already been exposed
over Y type zeolites, by Bolton "et al" [8] who explained it by a reactional pathway
involving disproportionation products. This hypothesis can probably explain the results
obtained over mordenite. Indeed, detailed study of the influence of various treatments

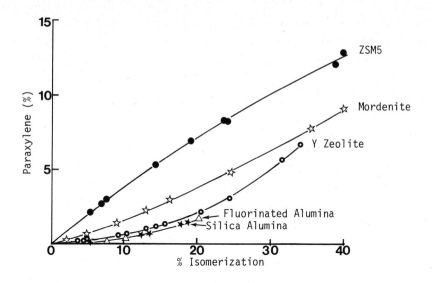

Fig. 1. Orthoxylene isomerization at 350°C.

(dealumination, wet-air calcination, nickel ion exchange) on the selectivity of protonic mordenite shows that the importance of direct isomerization of orthoxylene into paraxylene increases as D/I increases (figure 2). But this proposition cannot account for the results obtained over ZSM5, since this catalyst is completely inactive in disproportionation. The important direct formation of paraxylene observed in this catalyst is probably due to a more rapid diffusion of paraxylene than that of its isomers which have a bigger kinetic diameter. This hypothesis is confirmed by the privileged formation of

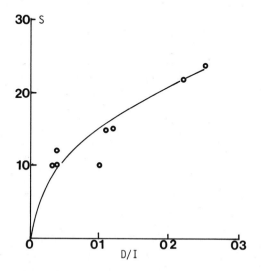

Fig. 2. Orthoxylene transformation at 350°C over variously treated H-mordenites.
S : percent molar fraction of paraxylene in the isomerization products ; D/I : ratio between the disproportionation and the isomerization rates.

paraxylene by the isomerization of metaxylene over ZSM5 and by the fact that paraxylene has been transformed over this catalyst twice faster than the meta and orthoxylenes. However, this hypothesis cannot explain the direct paraxylene formation on mordenite because over this catalyst, the selectivity is the same as over fluorinated alumina : all the xylenes isomerize nearly at the same rate ; metaxylene is transformed into paraxylene as rapidly as into orthoxylene.

2. LIGHT ALKANES TRANSFORMATION ON MORDENITE

2.1 n-Hexane and 2,4-dimethylpentane cracking [9]

At 250°C, the light products resulting from the cracking of these alkanes are not those which were expected from a simple scission of hydrocarbons. Indeed, the mixtures of the hydrocarbons obtained do not contain practically any C_1 nor any C_2 yet do contain a very important quantity of the complementary C_{n-1} and C_{n-2} fragments ; nearly all the products are saturated hydrocarbons. n-Hexane produces a very important quantity of branched hydrocarbons : isobutane and isopentane (table 2). This distribution can only be accounted for by supposing that a bimolecular intermediates is formed either between two molecules of reactant, or between two molecules resulting from a primary cracking of the reactant, or one molecule of reactant and one molecule of product. It is hardly probable that the light products come from a dimer intermediate of the reactant because the presence of products heavier than the reactant is not observed. Fast alkylation reactions between the primary products of cracking, alkenes and carbenium ions, are more probable. These secondary reactions probably result from a very strong adsorption of these compounds on mordenite at 250°C. This hypothesis is corroborated by the fact that the importance of these secondary reactions decreases as the temperature increases : at 400°C, over mordenite, the distribution of light products is practically the one expected from a simple scission of alkanes.

TABLE 2
Cracking of n-hexane and 2,4-dimethylpentane at 250°C on H-mordenite. Product composition (mole %).

Reactant / Products	n-hexane		2,4-dimethylpentane	
Methane	0	0	0.05	0.05
Ethylene	0	0.5	0	0.15
Ethane	0.5		0.15	
Propene	0.2	44.1	2.45	29.85
Propane	43.9		27.4	
Isobutene	0		0.6	
Isobutane	22.1	39.2	47.6	57.85
n-butane	17.1		9.5	
2-butenes	0		0.15	
Isopentane	7	16.2	6.7	7.9
n-pentane	9.2		1.2	
Hexanes			4.2	4.2

2.2 (2-^{13}C)-isobutane transformation

The intervention over hydrogen mordenite of bimolecular processes is clearly demonstrated by the isobutane transformation at 350°C. Indeed, this hydrocarbon isomerizes into n-butane, gives light products (ethane and essentially propane) and also n-pentane and isopentane. At a low conversion rate, the number of pentane molecules is practically equal to that of the n-propane molecules and these hydrocarbons are probably formed by isobutane disproportionation. The formation of n-butane and of pentanes is a second order reaction in respect to isobutane. This seems to show that the isomerization of isobutane like its disproportionation occurs through bimolecular intermediates.

The (2-^{13}C)-isobutane transformation has been studied in a recycling reactor at conversion rates in the 10 to 70 % range. Figure 3 shows the ^{13}C content of the products : isobutane, n-butane ^{13}C contents are practically equal to that of the starting material, whereas the ^{13}C content of isopentane is higher and that of the propane is lower. It can be remarked that the ^{13}C content of propane molecule added to that of a pentane molecule is more or less equal to that of two butane molecules. This confirms the fact that these products are obtained by isobutane disproportionation.

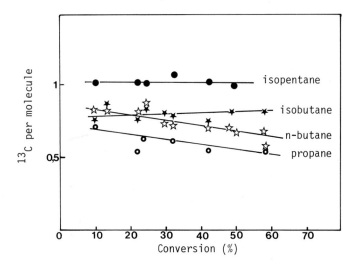

Fig. 3. (2-^{13}C)-isobutane transformation over mordenite at 350°C.

Table 3 gives the ^{13}C distribution for isobutane, n-butane, isopentane and propane at a 20 % conversion rate. The n-butane produced and the isobutane recovered have approximatively the same composition which is quite different from the composition of the reactant. Thus, the recovered isobutane would be, like n-butane, a product of the (2-^{13}C)-isobutane transformation. In these products, one finds a great quantity of molecules containing a) no ^{13}C and b) two ^{13}C ; 5 % of the isopentane molecules also contain three ^{13}C. At this conversion rate the ^{13}C distribution is close to that obtained by a mechanism involving dimer intermediates in which the ^{13}C would be statistically distributed (table 3).

TABLE 3

Transformation products of $(2\text{-}^{13}C)$-isobutane at conversion rate of 20 %. Numbers in brackets : statistical distribution.

Products	^{13}C per molecule			
	0	1	2	3
Isobutane	30(29.5)	58(53.4)	12(16.8)	(0.3)
n-butane	30(29.5)	55(53.4)	15(16.8)	(0.3)
Isopentane	25(19.2)	48(52.7)	22(27.2)	5(0.9)
Propane	41(42.9)	47(48.1)	12(8.9)	(0.1)
Starting Material	15	81.5	3.5	

To account for these results, it is necessary to invoke the fast equilibrium between several dimer intermediates ; these intermediates could be cyclohexanic as proposed by Bolton and Lanewala [10], although the mechanism as to how cyclization and scission of the resulting dimethylcyclohexanic intermediates might occur is not as yet quite clear. However if one supposes that such intermediates do exist it is necessary then to assume that their rearrangement for example through cyclopentanic compounds should be very fast (figure 4) in order to account for the nearly statistical distribution of ^{13}C.

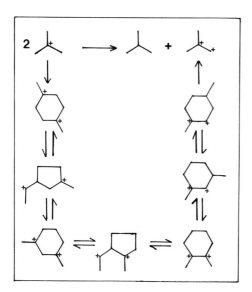

Fig. 4. Example of $(2\text{-}^{13}C)$-isobutane transformation into isobutane with zero and two ^{13}C by cyclohexanic and cyclopentanic intermediates.

CONCLUSION

The selectivity of zeolites for the transformation of saturated or aromatic hydro-carbons is different from the selectivity of amorphous catalysts. Several particular mechanisms have been exposed, namely the intervention of bimolecular processes in xylenes or isobutane isomerizations.

The different types of shape selectivity defined by Csicsery [1] were observed on some zeolites and especially on the ZSM5 zeolite. On this catalyst, paraxylene which has a smaller size than its isomers reacts twice as rapidly as they do, while on the other catalysts all three xylenes have the same reactivity (reactant selectivity) ; metaxylene isomerizes into paraxylene almost three times faster than into orthoxylene (product selectivity). Bimolecular reactions (disproportionation or coke formation) are highly inhibited on ZSM5 (restricted transition state selectivity). When there is no shape selectivity, the bimolecular transformations of aromatics and alkanes on zeolites are strongly privileged. These reactions occur between molecules of the reactant itself and also between the molecules of the primary products which frequently undergo numerous secondary transformations.

Therefore, it may be concluded that the particular selectivity of zeolites is largely a consequence of the fact that with these catalysts the rates of surface reactions are often higher than either the diffusion rates of the reactants or of the products or of their adsorption-desorption rates.

REFERENCES

1   S.M. Csicsery, A.C.S. Monograph, 171 (1976) 680-713.
2   J.A. Rabo, R.D. Bezman and M.L. Poutsma, Acta Physica et Chemica, 24 (1978) 39-52.
3   S.M. Csicsery, J. Catal., 19 (1970) 394-397 ; 23 (1971) 124-130.
4   T. Yashima and N. Hara, J. Catal., 27 (1972) 329-333.
5   S. Namba, O. Iwase, N. Takahashima, T. Yashima and N. Hara, J. Catal., 56 (1979) 445-452.
6   D. Marsicobètre, N.S. Gnep, M. Guisnet et R. Maurel, Rev. Port. Quim., 18 (1976) 313-316.
7   M.L. Poutsma, A.C.S. Monograph, 171 (1976) 437-528.
8   M.A. Lanewala and A.P. Bolton, J. Org. Chem., 34 (1969) 3107-3112.
9   G. Lopez, G. Perot, C. Gueguen and M. Guisnet, Acta Physica et Chemica, 24 (1978) 207-213.
10 A.P. Bolton and M.A. Lanewala, J. Catal., 18 (1970) 1-11.

B. Imelik *et al.* (Editors), *Catalysis by Zeolites*
© 1980 Elsevier Scientific Publishing Company, Amsterdam — Printed in The Netherlands

LOW TEMPERATURE REACTIONS OF OLEFINS ON PARTIALLY HYDRATED ZEOLITE H-ZSM-5

J.P. Wolthuizen, J.P. van den Berg, J.H.C. van Hooff

Laboratory for Inorganic Chemistry

Eindhoven University of Technology

P.O. Box 513

5600 MB Eindhoven

The Netherlands

SUMMARY

   Thermogravimetry is used to study the adsorption and reaction of ethene, propene and i-butene on H-ZSM-5. It is shown that already at roomtemperature ethene, although much slower than propene and i-butene, reacts on dehydrated H-ZSM-5 to form strongly adsorbed products. The inhibiting effect of preadsorbed water on this reaction has been investigated. By solid-state ^{13}C-NMR the reaction product of ethene could be identified as an aliphatic hydrocarbon with 8-10 carbon atoms.

INTRODUCTION

   Zeolite H-ZSM-5, a few years ago introduced by the Mobil R & D Corporation (ref. 1) shows very good properties for the quantitative conversion of methanol to hydrocarbons and water (ref. 2). Several authors have described the overall reaction sequence (ref. 2-8) which according to us (ref. 3, 6, 8) can be represented by the scheme

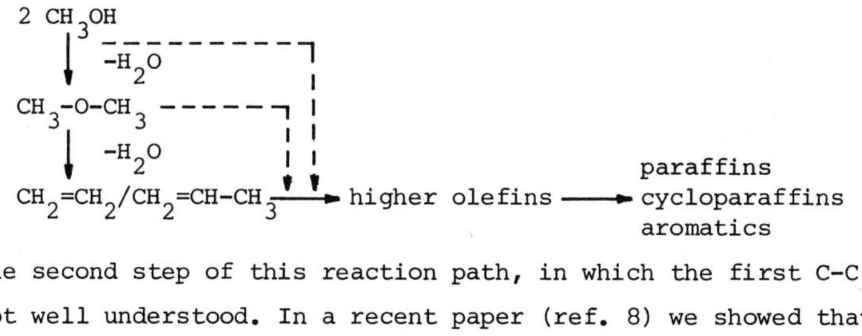

The second step of this reaction path, in which the first C-C bonds are formed is still not well understood. In a recent paper (ref. 8) we showed that both ethene and propene are the primary formed olefins and explained the formation of these compounds by a carbenium ion mechanism. Also Keading and Butter (ref. 7) have shown that ethene is the initial hydrocarbon product when methanol is converted over ZSM-5 class zeolite catalysts modified with phosphorus compounds, and they propose a carbenium ion mechanism, slightly different from ours, to explain the formation of this compound. A prerequisite for this model is that ethene itself can be readily converted to higher hydrocarbons over zeolite H-ZSM-5. It is for this reason that other investigators (ref. 4, 9, 10) have doubted our model, because in their experiments ethene showed a relatively low reactivity. Also Vedrine et al. (ref. 11) have reported a poor reactivity of ethene compared to methanol or propene when reacting over H-ZSM-5. However in our experiments (ref. 8) starting at relatively low temperature (~ 500 K) ethene is readily converted to higher olefins, paraffins and aromatics. IR, ^{13}C-NMR and TG studies, recently reported by us (ref. 12), concerning the adsorption, activation and reaction of ethene near roomtemperature on H-ZSM-5 have shown that under these conditions ethene can be readily protonated on the Brønsted acid sites, forming carbo-cations, that initiate a polymerization reaction. We have also shown that at these low temperatures water interferes with this reaction. Analogous results have been reported by Novakova et al. (ref. 13) based upon IR experiments. In this paper we will present new experimental data on the adsorption and reaction of small olefins at roomtemperature on H-ZSM-5 as measured by thermogravimetry; A comparison will be made between the reactivity of ethene, propene and iso-butene. At that special attention will be given to the influence of preadsorbed water on the reaction of ethene. Furthermore ^{13}C-NMR experiments will be reported to support the statement that already at roomtemperature ethene polymerizes on H-ZSM-5.

EXPERIMENTAL

Materials

   Two H-ZSM-5 samples have been prepared according to the previously described procedure (ref. 1, 3) and were characterized by chemical analysis and X-ray diffraction. The results are given in table 1.

Table 1.
Chemical composition of the H-ZSM-5 samples

Sample	Weight percentage			Mole ratio		
	Na_2O	Al_2O_3	SiO_2	Na_2O	Al_2O_3	SiO_2
BII	0.03	3.17	94.2	0.02	1.00	50.5
GII	0.15	4.10	92.4	0.06	1.00	38.5

Ethene, Propene and iso-Butene were high-purity reagents (99 + %) and were dried by mol.sieve before use. The vectorgas Helium was purified by passing it successively over a BTS, Carbosorb and Mol.sieve column. For the ^{13}C-NMR experiment we used 90% enriched 1,2 ^{13}C ethene from Stohler Isotope Chemicals.

TG-experiments

   A Cahn-RG-Electrobalance, fitted with an Eurotherm Temperature Programmer was used for these experiments. Prior to each experiment the H-ZSM-5 catalyst (sample BII) was calcined at 600°C, rehydrated in air at roomtemperature and next dehydrated to the desired water content by purging it with He (80 cm^3/min) at the appropriate temperature (see text). The reaction mixtures were made by adding the olefin to the He flow in the ratio of olefin: He = 1:4. Especially in the case of ethene the observed weight increase is caused for a substantial part by physisorbed ethene. To eliminate this contribution the following procedure has been followed. (see figure 1). First the H-ZSM-5 sample is contacted with the ethene/He mixture during a certain time interval Δt_1, then the ethene addition is stopped and the sample is purged with pure He for about 15 minutes to obtain a more or less constant weight. Then again ethene is added to the gas stream for another

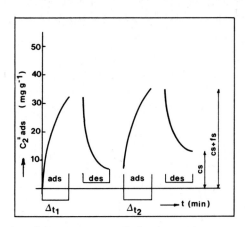

Fig. 1 Measurement of chemisorption
of ethene

time-interval t_2 and so on. Finally the amount of chemisorbed ethene as a function of the total time of adsorption, $\sum_i \Delta t_i$ is obtained.

^{13}C-NMR-experiments

The solid-state ^{13}C-NMR spectrum was recorded on a 180 Mc Double Resonance Spectrometer with dipolar decoupling and magic-angle spinning in a Kel, F sample holder. The H-ZSM-5 zeolite (sample GII) was dehydrated at 573 K and 0.1 Pa. Next 1,2 ^{13}C-ethene was added at 193 K until an equilibrium pressure of 13.3 KPa was reached. The spectrum was recorded at roomtemperature after 24 hours.

RESULTS

TG-experiments

Dehydration of H-ZSM-5. Figure 2 shows the change of weight of a H-ZSM-5 sample when it is purged with He at the indicated temperature. A continuous loss of adsorbed water is measured at temperatures up to about 550 K, then there is a region of constant weight, followed by a small but significant loss of weight above about 700 K. Obviously the first loss of weight represents the dehydration of the hydrated Brønsted acid sites, resulting in a completely dehydrated zeolite after purging at 550-650 K. At still higher temperatures, dehydroxylation occurs generating Lewis-acid sites.

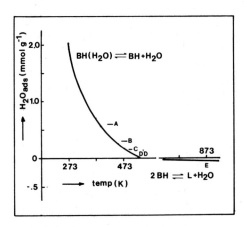

Fig. 2 Dehydration and dehydroxylation
curve of H-ZSM-5

<u>Adsorption and reaction of small olefins on H-ZSM-5 at roomtemperature</u>. The adsorption
curves of ethene, propene, i-butene and n-butane on completely dehydrated H-ZSM-5 recorded
at roomtemperature are presented in figure 3. This figure clearly shows the great
difference between the adsorption rates of ethene on one hand and propene and i-butene
on the other. For propene and
i-butene maximum adsorption is reached
within 20 minutes, while it takes over
60 hours for ethene. By following
desorption experiments it can be
shown that propene and i-butene are
completely strongly chemisorbed
under these conditions, while
even at equilibrium, a part of the ethene
is weakly physisorbed. (see table 2).
If we assume that the number of these
active sites is equal to the number
of present Al-atoms, it is possible
to calculate the average number of
olefin molecules that is chemisorbed
per adsorption site (column 3 of
table 2). The fact that this number
is larger than 1, together with the
shape of the adsorption curves,
supports our earlier assumption
(ref. 12) that the chemisorption
of olefins actually is a
polymerization reaction.
Based on this model it is possible

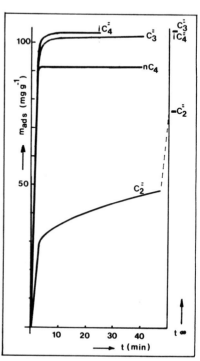

Fig. 3 Adsorption curves of ethene,
propene, i-butene and n-butane
on dehydrated H-ZSM-5
(purged at 573 K).

to calculate the average C-number
of the formed polymers and with the help of an estimated density of 0.75 g/ml also its
volume (column 4 and 5 of table 2). These figures show that the branched hydrocarbons
formed as polymerization products of propene and i-butene better fill the pores than
the linear polymer to be expected in ethene chemisorption.

Table 2 Adsorption of small olefins on H-ZSM-5

	Total adsorbed mg/g zeol.	Chemisorbed mg/g zeol.	Average number of mol. chemisorb. per site	Average C-number of polymer	Volume of polymer ml/g zeol.
C_2H_4	87	76	4.3	8.6	0.11
C_3H_6	102	102	3.8	11.4	0.14
i-C_4H_8	103	103	2.8	11.5	0.14
n-C_4H_{10}	91		pore volume		0.151

<u>The influence of preadsorbed water on the polymerization of ethene</u>. The results of the
measurements of ethene chemisorption on partly dehydrated H-ZSM-5 samples are shown in
figure 4 by the curves A, B, C, D and duplo D'. The corresponding levels of hydration are
indicated in figure 2. As under these conditions no Lewis-acid sites are present, this

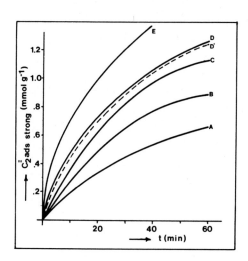

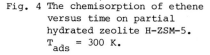

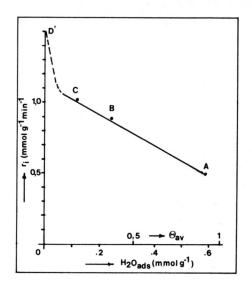

Fig. 4 The chemisorption of ethene
versus time on partial
hydrated zeolite H-ZSM-5.
T_{ads} = 300 K.

Fig. 5 The initial rate of polymerization
on partial hydrated zeolite
H-ZSM-5.
T_{ads} = 300 K.

means that the polymerization of ethene occurs on the Brønsted-acid sites. For comparison
the chemisorption of ethene has also been measured after purging with O_2/He (20-80) at
873 K (curve E), under which condition dehydroxylation can take place. It is clearly
shown that the presence of Lewis-acid sites enhances the rate of polymerization of
ethene. This result is in agreement with that reported by Kubelkova et al. (ref. 16).
They observed no reaction of ethene on the Brønsted-acid sites in HY zeolite at 353 K;
only after dehydroxylation a reaction of ethene was measured. In all experiments (A to E)
at equilibrium the maximum recorded amount of chemisorbed ethene was the same as given
in table 2. This shows that the presence of water only decreases the rate of polymerization. In figure 5 the initial rate of polymerization (r_i) (derived from Fig. 4) is
plotted versus the amount of adsorbed water. If we express this amount in molecules
per active site (θ_{av}) then figure 5 shows a linear relation between r_i and θ_{av} in
the interval $0.1 < \theta_{av} < 0.8$. Assuming a homogeneous distribution of acid strength
and that firstly only one molecule of H_2O is adsorbed per active site (BH) this result
indicates that in the interval $0.1 < \theta_{av} < 0.8$ only the number of dehydrated Brønsted-
acid sites determines the rate of polymerization, i.e. adsorption of one molecule of
water on a brønsted-acid site inhibits this site for ethene polymerization. In the
range $\theta_{av} < 0.1$ obviously stronger acidic sites are involved, while for $\theta_{av} > 0.8$
the assumption that only one molecule of water is adsorbed per active site is not longer
valid. The latter is in agreement with recently reported IR-experiments (ref. 14) by
which the presence of dehydrated Brønsted-acid sites (3605 cm^{-1}) could be detected on a
H-ZSM-5 sample pretreated in vacuum at RT, while after this pretreatment an amount of
water is left on the zeolite corresponding with θ_{av} = 1.3.

^{13}C-NMR experiments

By high resolution ^{13}C-NMR it could be shown (ref. 12) that after adsorption of ^{13}C-enriched ethene at RT on dehydrated H-ZSM-5, initially the characteristic resonance of ethene is recorded. However in course of time also at RT, the signal intensity decreases and finally the signal completely disappears, indicating a conversion of ethene in non-detectable products. Only after thermal desorption above 473 K these products (or decomposition products of it) could be observed in this way (ref. 12, 15). We now present a solid-state ^{13}C-NMR spectrum recorded after the completed reaction of ^{13}C-enriched ethene on dehydrated H-ZSM-5 at RT (see Figure 6). The spectrum shows resonances at 28.8 and 13.5 ppm (to TMS) representative for aliphatic $-CH_2-$ and $-CH_3$ groups respectively. Both signals are strongly broadened because of carbon-carbon coupling interactions but nevertheless a fine-structure can be seen. This fine-structure may be caused by the presence of inequivalent C-atoms in the reaction products or by a variable interaction with the zeolite (e.g. differences in acid strength). From the intensity ratio of the two signals it can be roughly derived that ethene has been converted to straight-chain aliphatic hydrocarbons with a chain length of about 8-10 carbon atoms. And from the fact that these polymers cannot be 'seen' by normal high-resolution NMR we may conclude that they are strongly bonded to the zeolite.

DISCUSSION

By TG and ^{13}C-NMR experiments it is clearly shown that ethene, propene and i-butene are readily polymerized at RT on the Brønsted acid sites of zeolite H-ZSM-5. Comparison of TG-data of ethene and propene adsorption shows two important differences:

i. propene is polymerized much faster than ethene and

ii. the polymerization products of propene better fills the internal pore volume than the product of ethene polymerization.

The higher reactivity of propene can be easily understood because it is more easy to protonate propene to form secundary carbenium ions than to form primary carbenium ions by protonation of ethene.

In propene polymerization only branched hydrocarbons can be formed, which according to the data in table 2, fitt well in the pores of the zeolite (95%). On the contrary after ethene polymerization only about 70% of the pore volume is filled. It indicates that in ethene polymerization linear hydrocarbons are formed. This conclusion is supported by the solid state ^{13}C-NMR spectrum shown in Fig. 6, which proves

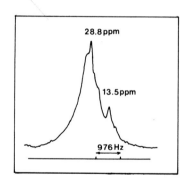

Fig. 6 Solid state ^{13}C-NMR spectrum of the adsorption product of ^{13}C-ethene on H-ZSM-5
freq. 45.267 MHz
MAS > 3 KHz
Kel F sample holder
54.000 fids
waiting time 1 sec
90^0 pulse (8 μsec)

the presence of linear aliphatics strongly adsorbed on the zeolite surface after ethene polymerization.

These experimental data provide new evidence for a reaction mechanism for ethene oligomerization in which ethene reacts with a protonated higher olefin and not vice versa as proposed by Anderson et al (ref. 10).

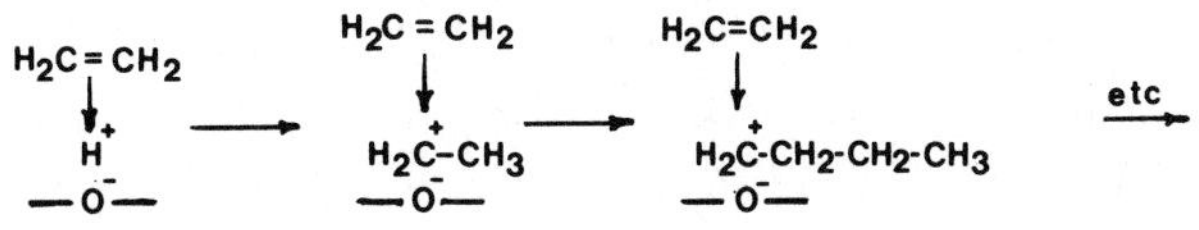

By analogy the polymerization of propene can be represented by the scheme:

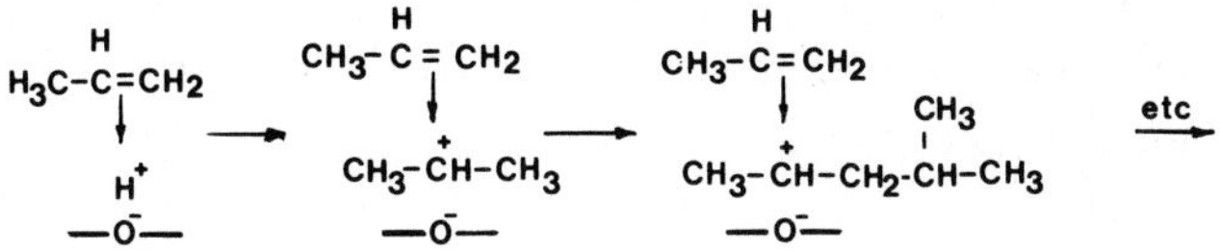

In a recent paper (ref. 8) we proposed that in the methanol and dimethylether conversion on zeolite H-ZSM-5 ethene and propene both are formed as primary olefins. We now have shown that propene as well as ethene can really act as a reaction intermediate.

The adsorption experiments on partially hydrated H-ZSM-5 have shown that at RT water interferes with the adsorption and reaction of ethene. The results (Fig. 5) indicate that the presence of water decreases the number of active sites, i.e. once a water molecule is adsorbed on a Brønsted-acid site this site is blocked for activation of ethene.

At increasing temperatures the residence time of a watermolecule on the site becomes shorter; when ethene and water are co-fed to the zeolite reaction of ethene is measured at temperatures above 400 K (ref. 15). When ethene or propene are converted at temperatures above 573 K the presence of water results in a decrease of the rate of deactivation (ref. 8, 11, 17) and an increase of the selectivity to aromatics (ref. 8, 17). These experimental observations now can be explained by the presupposition that at these temperatures water is in a dynamic adsorption equilibrium with the zeolite surface. As such it interferes with the adsorption and consecutive reactions of the starting olefins and reaction products already formed (higher olefins and aromatics). As a consequence, the residence time of the various hydrocarbons in the pores decreases whereby among other things the rate of coke formation will decrease.

This on its turn will cause:

i.   a decrease of the rate of deactivation and

ii. a decrease of the loss of reaction products, especially aromatics, in coke deposits
     and consequently an increase of the selectivity to aromatics.

ACKNOWLEDGEMENTS

The authors are very grateful to Drs. E.M. Menger and Dr. W.S. Veeman (University of Nijmegen, Department of Molecular Spectroscopy) for recording the ^{13}C-NMR spectra. This work was supported by the Netherlands Foundation of Chemical Research (SON) with financial aid from the Netherlands Foundation for Pure and Scientific Research (ZWO)

REFERENCES

1 R.J. Argauer and G.R. Landolt, US Patent 3.702.886

2 C.D. Chang and Silvestri, J. Catal. <u>47</u> (1977) 249

3 E.G. Derouane, P. Dejaifve, J.B. Nagy, J.H.C. van Hooff, B.P. Spekman, J.C. Vedrine and C. Naccache, J. Catal. <u>53</u> (1978) 40

4 J.R. Anderson, K. Foger, T. Mole, R.A. Rajadhyaksha and J.V. Sanders, J. Catal. <u>58</u> (1979) 114

5 N.Y. Chen and W.J. Reagan, J. Catal. <u>59</u> (1979) 123

6 P. Dejaifve, J.C. Vedrine, V. Bolis, J.H.C. van Hooff and E.G. Derouane in 'Preprints Div. Petr. Chem. - Symp. Recent Adv. Petr. Process - ACS Meeting Honolulu - April 1979' Am. Chem. Soc. <u>24</u> p. 286

7 W.W. Kaeding and S.A. Butter, J. Catal. <u>61</u> (1980) 155

8 J.P. van den Berg, J.P. Wolthuizen, J.H.C. van Hooff, preprints Int. Conf. on Zeolites, Naples 1980, in press

9 B.J. Ahn, J. Armando, G. Perot and M. Guisnet, C.R. Acad. Sci. Paris, <u>C 288</u> (1979) 245

10 J.R. Anderson, T. Mole, and V. Christov, J. Catal. <u>61</u> (1980) 477

11 J.C. Vedrine, P. Dejaifve, C. Naccache and E.G. Derouane, preprints 'The Seventh International Congress on Catalysis', Tokyo, 1980, in press

12 V. Bolis, J.C. Vedrine, J.P. van den Berg, J.P. Wolthuizen and E.G. Derouane, JCS Faraday Trans. I, 1980, in press

13 J. Novakova, L. Kubelkova, Z. Dolejsek and P. Jiru, Coll.Czech Chem. Commun. <u>44</u> (1979) 3341

14 J.C. Vedrine, A. Auroux, V. Bolis, P. Dejaifve, C. Naccache, P. Wierchowski, E.G. Derouane, J.B. Nagy, J.P. Gilson, J.H.C. van Hooff, J.P. van den Berg, and J.P. Wolthuizen, J. Catal. <u>59</u> (1979), 248

15 J.P. Wolthuizen, J.P. van den Berg, J.H.C. van Hooff, to be published

16 L. Kubelková, J. Novakova, V. Bosáček, V. Patzelová and Z. Tvaruzkova, Acta Phys. Chem. <u>24</u> (1-2) (1978) 189

17 W.E. Garwood, P.D. Ceasar and J.A. Brennen, US Patent 4.150.062 (1979)

B. Imelik *et al*. (Editors), *Catalysis by Zeolites*
© 1980 Elsevier Scientific Publishing Company, Amsterdam — Printed in The Netherlands

USE OF ZEOLITE CONTAINING CATALYSTS IN HYDROCRACKING.

C. MARCILLY, J.P. FRANCK
INSTITUT FRANCAIS DU PETROLE - B.P. 311 92506 - RUEIL-MALMAISON CEDEX (FRANCE)

INTRODUCTION

The beginning of hydrocracking may be traced back to the year 1927 when the first BERGIUS plant for hydrogenating brown coal at LEUNA (GERMANY) was put on stream. Hydrocracking only appeared a few years later in the U.S.A. where this process is now more widespread than elsewhere.

At the present time, hydrocracking is used for converting various petroleum cuts, from light naphtas to atmospheric or deasphalted vacuum residua, to lighter and more valuable products like propane, butane, gasoline, jet fuels, diesel or heating oils and, in some cases, lubricating oils (1). It has essentially been developed in the U.S.A. for motor fuel production. In other countries, hydrocracking is not as common and is directed rather to the production of middle distillates (jet fuels and gas oils).

The first zeolite based catalysts were used in 1964. Compared with conventional amorphous catalysts, their advantage is especially apparent when a maximum gasoline is aimed for. Hydrocracking is now the second largest use for zeolite containing catalysts.

I. THE CONVENTIONAL HYDROCRACKING ON AMORPHOUS CATALYSTS.

According to the required products and the treated feedstock, the process scheme involves one or two stages (2, 3). The first one essentially frees the feedstock from the nitrogen and sulfur compounds as NH_3 and SH_2, and hydrogenates most of the aromatics without any serious cracking of the hydrocarbons. In the second one, the main reactions are cracking and hydrogenation of the effluents coming out from the first stage. Each of them contains a specific catalyst specially adapted to the desired transformations.

I.1 The catalysts

They are of bifunctional type combining a hydrogenating and an acidic function.

For the first stage, the hydrogenating function is produced from cobalt or nickel sulfides combined with molybdenum or tungsten sulfides. The atomic ratio of metals from the VIIIth group to those from the VIth group is near 0.25 (4). For the second stage, that function usually comes from nickel or palladium (5).

The acidic function is provided by an acidic carrier, like alumine, silica-alumina, silica-magnesia, etc... It gives the catalyst its cracking activity. The acidity level depends on the feedstock and the required transformation : it must be higher in the second step than in the first one.

The selectivity of the transformation largely depends on the balance between both functions. If the hydrogenation function is high, in comparison to the acidic one, the main reactions, besides deazotation and desulfurization, are hydrogenation and isomerization. If

94

it is low, the main reaction is cracking. That can be seen in figure 1 (6) which compares
selectivities of two catalysts having the same hydrogenation activity for the n heptane
transformation : the most acidic catalyst, with the highest heat of ammonia sorption, has
the highest cracking activity. This explains the difference in the products distribution
obtained from the same feedstock, on the one hand by catalytic cracking, and on the other
hand by hydrocracking. As shown in figure 2 (7), hydrocracking of hexadecane on Pt/SiO_2-
Al_2O_3 gives products with carbon numbers varying between 3 and 12-13 through a maximum between 4 and 11, whereas catalytic cracking on silica-alumina gives products with carbon numbers between 2 and 7 and a greater maximum at about 4-5. The catalytic hexadecene cracking
results in a distribution of products very similar to that obtained from hexadecane hydrocracking : replacing hexadecane by hexadecene is comparable with providing an infinite
hydrogenation function to the catalyst. These results explain why catalytic cracking is
better for gasoline production, whereas hydrocracking is better for middle-distillates
production, (jet fuel, diesel or heating oil). Ref. 8 gives the respective levels of hydrogenation and acidity that are to be reached for various industriel cases.

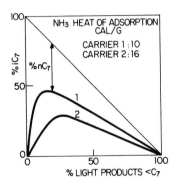

Fig. 1 (6) : hydrocracking of
n-heptane (Ni-Mo sulfides/SiO_2-Al_2O_3)

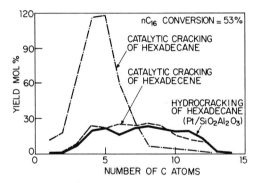

Fig. 2 (7) : comparison between hydrocracking
and catalytic cracking

I.2 The process

Hydrocracking is a very flexible process allowing the treatment of various feedstocks,
from naphtas to residua, and the acquirement of a large diversity of products. Operating
conditions are within the following limits : temperature from 350 to 450°C, partial pressure
of hydrogen from 100 to 150 bar, LHSV from 0.3 to 1.5 h^{-1}, molar ratio H_2/HC near 20. Compared with catalytic cracking, the use of high partial pressures of hydrogen and of relatively low temperatures decreases the kinetics of coke formation and favors the hydrogenation
of aromatic compounds. It results in a fixed bed technology, a very exothermic reaction, an
important chemical consumption of hydrogen and, contrary to catalytic cracking, nearly non-
olefinic and non-aromatic products.

The process outline usually requires only one stage when the desired transformation is
not important : for instance, production of middle-distillates from vacuum distillates or
heavy coking distillates or deasphalted vacuum residua. It often involves two stages when
the required transformation is important : production of a maximum gasoline from vacuum

distillates or deasphalted vacuum residua. Both processes' schemes are respectively shown in figures 3 and 4 (3, 6). In figure 4, it can be noticed that each step possesses its own hydrogen recycle. This results from the necessity of keeping the second step catalyst's environment free of NH_3 and SH_2 and, therefore, of getting rid of both compounds at the outlet of the first step.

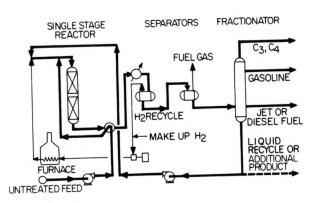

Fig. 3 : single stage conventional hydrocracking
(amorphous catalyst)

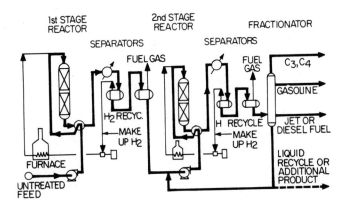

Fig. 4 : two-stage conventional hydrocracking
(amorphous catalysts)

Table 1 (8) gives an example of results performed in a two stage industrial unit with the possibility of liquid recycle, when products heavier than those aimed at are to be avoided. The main products that can be obtained are light C_3-C_4 gases, light gasoline C_5-80°C, (essentially C_5-C_6 hydrocarbons), naphta 80-185°C, jet fuel 185-230°C and gas oil. The following points deserve to be emphasized :

- the light gasoline has a high research octane number due to its high isoparaffinic character,

- the heavy gasoline is a very good feedstock for catalytic reforming because of its high naphtene content,

- the burning qualities of jet fuel and gas oil, respectively measured as smoke point and diesel index, are excellent as a result of their low aromatic content,

- the pour point and freezing point of jet fuel and gas oil are good owing to their low content of linear paraffinic compounds.

TABLE 1 (ref. 8)

Yield and quality of products from hydrocracking middle-east vacuum distillate.

(d_4^{20} = 0,905   ASTM 360-480°C   S = 2,41 %   N = 900 ppm)

Initial point of recycled residual oil (°C)	200	280	350	Without recycle
Yield (wt %)	Gasoline	Jet fuel	Gas oil	Gas oil
Gas.......................................	16.4	10.7	7.7	6.7
Light gasoline........................	22.0	10.3	6.3	5.6
Naphta..................................	64.9	21.7	13.4	13.0
Jet fuel...............................		60.3		
Gas oil................................			74.8	69.0
Residuum...............................				8.0
	103.3	103.0	102.2	102.3

Characteristics :					
Light gasoline....	d_4^{20}	0.670	0.668	0.675	0.665
	NOR	82	75	74	76
	NOR + 0.5 ‰	93	89	88	91
Naphta...........	d_4^{20}	0.742	0.742	0.770	0.765
	PNA	56/34/10	51/37/12	48/39/12	49/38/13
Jet fuel.........	d_4^{20}		0.802		
	S (ppm)		< 10		
	Smoke point (mm)		25		
	Ar (% pds)		15		
	Cristallisation point (°C)		<- 60		
Gas oil..........	d_4^{20}			0.827	0.814
	S (ppm)			30	40
	DI			62	70
	Freeze point (°C)			- 45	- 33
	Flash point (°C)			70	74

II. HYDROCRACKING ON ZEOLITE BASED CATALYSTS.

The development of new highly active zeolite based catalysts has resulted in the modification of the conventional process scheme. The new process which was first used in 1964 at the UNION OIL Co. of California refinery in Los Angeles, is essentially characterized by a first stage with two different catalysts : the first one is for classical hydrotreating, the second one contains a zeolite and promotes a substantial cracking of the feedstock. This new design combines, in one stage, the two stages of the conventional process without any intermediate separation. It was proposed by UNION OIL Co. of California and Esso Research and Eng. Co. as Unicracking-JHC integral hydrotreating-hydrocracking process (3, 9).

II.1 The catalysts

II.1.1 Composition and preparation

The hydrogenating function is provided by metals from the VIIIth group, particularly palladium, or by combining metals from the VIIIth and VIth groups (10, 11). Reaching good noble metal dispersion and macroscopic distribution into the zeolite is of great importance. This aim can be reached by ion exchanging palladium as $Pd(NH_3)_4^{2+}$. Techniques for deposing of non noble metals are not well known. Some of them are described in patents (12, 13, 14).

The acidic function comes from a zeolite mixed with a refractory oxide like alumina, silica-alumina, silica-magnesia, etc... The zeolite content is from about 20 to 80 wt %. Among the most open structures, faujasite and large pore mordenite, the former is more selective, and therefore more suitable, than the latter for producing gasoline (15, 16). That clearly appears in figure 5 (16) which compares the distribution of light products on Pd HY and Pd HMordenite and shows that the former gives more gasoline than the latter. Industrial catalysts contain Y zeolite in H form or exchanged with bivalent metals, like magnesium (17). Contrary to the first catalysts, the new ones which appeared at the end of the 60th (18) seem to contain a hydrothermally stabilized form of Y zeolite. The trend would now be towards increasing use of ultrastable Y (19) like Z14US (20).

TABLE 2 (ref 36)

Comparison between zeolite containing catalyst and amorphous silica-alumina for maximal diesel oil production.

Vol %/feed	Zeolite containing catalyst	Amorphous catalyst
C_4	4.0	2.2
C_5-82°C	8.0	6.4
82-166°C	20.5	6.7
166-371°C	81.5	97.2
Total C_4^+	114.0	112.5
H_2 consumption Kmol/m^3	10.15	10.68
NOR clear (C_5-82°C)	83.0	74.0
Cetane number (166-371°C)	54.0	60.0

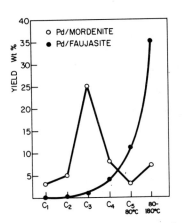

Fig. 5 (16) : hydrocracking of an atm. Gas oil (P.I. = 180°C) conversion 177⁻

The stabilizing treatments leading to ultrastable form of Y zeolite, increase the SiO_2/Al_2O_3 ratio of the framework and modifie the acidity. The main techniques for obtaining such stabilized forms may be classified in 4 groups :

(1) - Self-steaming

M$_c$ Daniel and Maher (21) have proposed 2 techniques from NaY. One of them involves 4 steps : 2 ion-exchanges of sodium with ammonium, each of them preceding a calcination, the first one at 540°C, the second one at 815°C under self-steaming conditions. In these conditions the unit-cell parameter is 24.3 Å instead of 24.65 Å for NaY. The calcination of NH_4Y in deep-bed conditions under static atmosphere, reported by KERR (22), belongs to that group too.

(2) - Controlled steaming

Hansford (23) and Eberly (24, 25) describe other techniques in which NH_4Y or HY are calcined from about 370 to 800°C under dynamic atmosphere containing air and 3 to 100 vol % of water vapor. The unit-cell parameters of the thus obtained stabilized HY is down to about

24.3 Å.

(3) - Thermal treatment by halogen or halogenic acid in a vapor phase

In such a technique, potented by B.A.S.F. (26), dehydrated HY is calcined under chlorine or hydrochloric acid between 400 and 700°C.

(4) - Treatments in liquid phase

One method consists of exchanging sodium of NaY in an autoclave containing an ammonium salt solution, the pH of which is from 2.5 to 5, at a temperature between 150 and 230°C (27). In another method, the zeolite is refluxed in a H_4 EDTA solution (28, 29, 30, 31) or in acetylacetone (30, 31, 32).

II.1.2 The balance between hydrogenating and acidic functions

The performances of the catalyst greatly depend on the balance between both functions. The possibility of modifying the acidity of zeolites by suitable ion exchanges or treatments, and the hydrogenating function, allows the preparation of "tailored" catalysts which are adapted to a refiner's specific need (18). Concerning noble metal, the hydrogenating activity depends on the reduction extent, on the accessibility and the size of metallic particles in zeolite structure. So far as it is known, no thorough study, similar to that of GALLEZOT and al. (33, 34) on PtY, has been published, concerning the positions of palladium atoms or particles in stabilized Y structure according to the calcination and reduction conditions.

Some modifications of the balance between both functions may occur during operation. They seem to be reversible when caused by the presence of poisons like NH_3 or SH_2 in the reacting atmosphere. In some cases, they may be irreversible and require, as it will be seen further, "rejuvenation" of catalysts.

II.1.3 Advantages of zeolitic catalysts over amorphous catalysts

Zeolitic catalysts present 3 major advantages over amorphous catalysts.

(1) They have a greater acidity and so a greater cracking activity.

(2) They are much more selective for producing a maximum gasoline. Fig. 6 (35) presents the true boiling point distillation curve of products from hydrocracking a vacuum gas oil ; it appears that a zeolitic catalyst gives a yield of about 65 % of C_5-180°C gasoline instead of a maximum of 40 % with an amorphous catalyst. As shown in table 2 (36), this advantage disappears when the production of a maximum gas oil is required and, in this particular case, the use of a zeolitic catalyst is not justified. The use of a zeolitic catalyst may be profitable when a great flexibility is looked for, for example with maximal gasoline production in summer and maximal gas oil production in winter.

(3) They have a much better resistance to nitrogen and sulfur compounds.

One of the best advantages of zeolite based catalysts compared with amorphous catalysts is their much higher resistance to nitrogen poisons. That superiority, ascribed to the greater acid sites number in the zeolite, is clearly apparent in figure 7 (37) which presents the evolution of n-heptane conversion as a function of the quinoline content, for respectively a zeolitic and an amorphous catalyst ; in the presence of 2000 ppm of quinoline the decrease of conversion is about 20 % for the zeolitic catalyst instead of 80 % for the amorphous one. In other respects, the greater acidity of the zeolite enhances the resistance of the hydrogenating function to poisoning by sulfur compounds (38, 39, 40).

As it will be seen further, these advantages result in a large saving for the refiner who wants to produce a maximum gasoline.

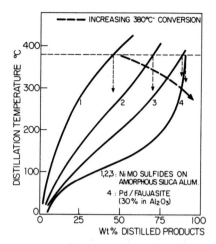

Fig. 6 (35) : hydrocracking of a
vacuum gas oil (P.I. = 380°C)

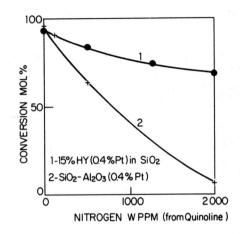

Fig. 7 (37) : influence of N content on
hydrocracking of n-heptane (T = 400°C,
P = 25 bar)

II.1.4 Catalyst regeneration and rejuvenation

In an industrial unit, the activity of the catalyst slowly decreases, in spite of high
hydrogen partial pressures. For palladium containing catalyst, deactivation may have
different reasons :

- classical coke formation on the active surface,
- aging of the hydrogenating function (41),
- inhibition of the acidic function (41).

Coke and simultaneously highly basic nitrogen and sulfur compounds, fixed on the surface,
are burned by classical combustion in the presence of poor oxygen containing air (regeneration).

Aging of the hydrogenation function is essentially due to the metal sintering. This
phenomenon should rapidly occur when the catalyst is put into contact with high water vapor
partial pressures, at temperatures above about 250°C, either during regeneration or, acci-
dentally, during operation (42). Contacting the catalyst with a neutral gas, such as nitrogen, at high temperatures, would also result in decreasing the hydrogenation activity
(43). Different rejuvenation techniques have been devised according to the metallic particle sizes :

(1) particles with diameters < 100-200 Å (macro agglomeration)

Succession of oxidation-reduction (44, 45) or sulfurization-oxidation (46) cycles,
usually performed "in situ" after coke combustion, enables metal redispersion towards particles with diameters slightly smaller than 50 Å.

(2) Particles with diameters < 50 Å.

The regenerated and either oxidized or sulfided catalyst is contacted with a mixture of
water vapor and ammonia gas or with an aqueous ammonia solution, under 150°C, in order to
fill up the micropores. Metallic particles are dissolved as $Pd(NH_3)_4^{2+}$ cations which migrate again in the structure towards ion exchange sites (42, 47 à 50). This technique seems
to be performed "ex-situ" (51).

Causes of the acidic function inhibition are less clear. It would seem that accidents which are responsible for metal sintering, are also responsable for the redistribution of non-reducible metallic ions (Na, Ca, Mg, etc...) on acid sites ; metal ions which initially occupy sites in small cavities (sodalite cage and hexagonal prism) would migrate towards the acidic sites in the supercage (42) and this would decrease the cracking activity. Besides, after use, catalysts are found to contain iron scale and sulfate ions (52). Most of these ions and impurities can be eliminated by controlled ion exchange with ammonium ions, at room temperature, at a pH from 4.5 to 6.5. Such a treatment which enables the elimination of 60-70 % of sodium ions, 90 % of iron and 80 % of sulfate ions (52), would be performed "ex situ" (42, 47, 48, 52). Another technique consists of contacting the regenerated catalyst with H_4 EDTA solution or with an ammonium salt of H_4 EDTA (53).

II.2 The process

The hydrocracking process using zeolite based catalysts involves one or two stages according to the feedstock and the desired products.

II.2.1 Description of one-stage and two-stage processes

As shown in figure 8 (41), the one-stage scheme combines, in one step, the two steps of the conventional process, i.e. hydrotreating and hydrocracking, without any intermediate separation of NH_3 and SH_2. That is made possible by the high resistance of the cracking catalyst to these poisons. In the second reactor, the conversion is about 40 to 70 %. After separation, the unconverted feedstock is recycled to the cracking reactor. In spite of its high ammonia resistance, the presence of large quantities of this compound (up to about 10^4 ppm) results in a definite suppression of the cracking activity which can be compensated by an increase of temperature. In such an "acidity controlled" environment, the total activity is limited by the cracking activity of the catalyst. Therefore, this catalyst must exhibit a very high acidity. This may be done by adjusting the zeolite content in the catalyst (54).

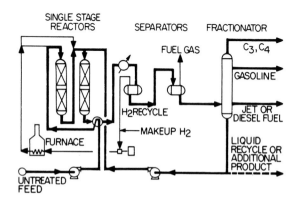

Fig. 8 : single stage Unicracking-JHC unit
(zeolitic catalyst)

The two-stage scheme involves one more hydrocracking reactor. As shown in figure 9 (41), the total effluent from the first stage is combined with that from the second stage and sent to the fractionator where the following products are separated : propane and butane, light and heavy gasolines, jet fuels and gas oils. The heavier residual fraction is recycled to extinction in the second stage at 50-80 % conversion per pass (55, 56). The recycle gas from the combined high pressure separator is stripped of ammonia, so that the second stage catalyst operates under an atmosphere essentially free of ammonia (55, 56). Thus, the environment is essentially "hydrogenation controlling" and the catalyst must exhibit a high hydrogenation activity. Such a catalyst probably consists of palladium on Y zeolite (41).

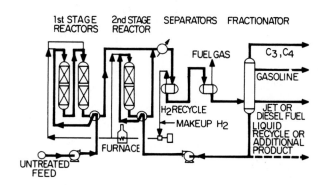

Fig. 9 : two-stage Unicracking-JHC unit
(zeolitic catalysts)

II.2.2 Comparison of the new hydrocracking process with the conventional process

The advantages of zeolitic catalysts (seen in part II.1.3) over amorphous catalysts results in a large saving for the refiner who wants to produce a maximum gasoline. At constant production, a conventional two-stage process gives a C_5-80°C gasoline with NOR of 84 instead of 86 with a first stage integral hydrotreating-hydrocracking process. This allows the increase of the premium/regular ratio and so results in a high profit. Moreover, the first version requires one more fractionment stage and one more furnace. Thus there is important saving of money on the total investment (10 to 20 % for a 10^6 t/year capacity) and on the operating cost (6).

II.2.3 Comparison of one-stage with two-stage processes

The reasons for choosing one version rather than the other are not very simple. They depend upon the required operating flexibility, the characteristics of the feedstock and the yields and qualities of the desired products.

(1) Flexibility : the two-stage scheme offers a better flexibility than the one-stage scheme.

(2) Characteristics of the feedstock : the one-stage scheme is preferally used when the feedstock is not too difficult to transform, such as those containing few highly basic nitrogen compounds and with not a too high final boiling point, or when the required transformation is not very important. The two-stage scheme is more suitable for difficult feedstocks, such as highly aromatics or rich nitrogen ones (9).

(3) Yields and quality of products : to produce gasoline with a high octane number, it is desirable not to hydrogenate completely the unsaturated compounds, in order to keep some aromatics in the heavy gasoline. Table 3 (18) shows that the one-stage integrated process,

where the hydrocracking catalyst operates under atmospheres rich in NH_3 and SH_2 and at high temperature, is suitable for obtaining a good quality gasoline. In two stages, the obtained naphta is free of aromatics, sulfur and nitrogen and is then an excellent catalytic reforming feedstock.

TABLE 3 (ref 18)

Unicracking-JHC of Los Angeles Basin Heavy Virgin Gas Oil for Gasoline and Turbine Fuel.

Feedstock properties		Product yields and properties			
Gravity, °API	20.3	Yields	One Stg Cat A	One Stg Cat B	Two Stg Cat C
Distillation, D-1160, °F		C_1-C_3, scf/b feed	146	50	110
IBP	520	Liquid Products, vol % of feed			
10	641	Butanes	12.1	8.2	8.6
30	689	C_5-C_6 light gasoline	23.5	18.3	16.5
50	728	C_7-plus gasoline	40.6	34.1	34.4
		Turbine fuel	45.0	61.1	61.3
70	762	Total C_4-plus	121.2	121.7	120.8
90	820	H_2 Consumption, scf/b			
EP	890	feed	1750	1950	2110
Sulfur, wt %	1.33	Product properties			
Nitrogen, total, ppm	2270	C_5-C_6 Octane, F-1+3 ml TEL	99.0	98.0	96.5
Aromatics, wt %	35	C_7-plus gasoline			
Heterocyclics, wt %	21	Octane, F-1+3 ml TEL	87.0	84.7	82.8
		Naphtenes + Aromatics	70.1	68.2	68.3
		Turbine Fuel			
		Gravity, °API	37.5	39.7	41.4
		Aromatics, vol %	34	19	2.0
		Smoke point, mm	13.6	20.1	29.7
		Luminometer number	25	44	67
		Flash point, °F	117	126	120
		Freeze point, °F	-51	-58	-66
		ASTM Dist., °F 10 %-95%	352-514	358-510	356-512

By contrast, good quality jet fuels and gas oils must contain as few aromatics (smoke point specifications) and linear paraffins (pour point or freeze points specifications) as possible. The two-stage process, which allows complete hydrogenation of aromatics compounds and obtention of very isoparaffinic products, is particularly well adapted for such a production (Table 3). When operating in one stage, an additional separate unit of aromatics saturation may be necessary (18).

CONCLUSION

Hydrocracking on zeolite-based catalysts is a young process which has chiefley developped in the U.S.A. for gasoline and jet fuel production. In Europe, where the need for such products is less drastic, its development is much more limited and operating units are rather of a conventional type using amorphous catalysts and are oriented towards maximal middle distillates production. Anywhere a great operating flexibility is not aimed at, hydrocracking suffers severe competition from catalytic cracking. The main advantage of hydrocracking is indeed its versatility which allows, on the one hand, the treatment of heavy feedstocks rich in polynuclear aromatics, and particularly resistant to catalytic cracking, and, on the other hand, very varied products of good quality, especially for jet fuels and gas oils. It has the drawback of requiring more expensive investment (high pressure) and operating cost (hydrogen consumption...).

Future prospects seem to be more favorable for hydrocracking in Europe. It is fairly well established that on the horizon of 1985, the need for petroleum products will be characterised by a decrease in heavy fuel consumption and an increase of gasoline consumption. So it will be necessary to transform the excess of heavy black cuts into lighter white products. So, hydrocracking will find its place in refining schemes for producing middle-distillates and gasoline and will continue, in most cases, to compete with catalytic cracking. It may be noticed that constraints coming from the struggle against pollution may, to a certain extent, impede the development of catalytic cracking in Europe. Besides, other way of progress are now open to hydrocracking. Thus, production of 80-160°C naphtas transformable in BTX aromatics for petrochemical use, seems to have a promising future. The development of more active and more stable new catalysts would allow the decrease of the temperature and hydrogen partial pressure, the increase of LHSV and, therefore, the decrease in the reactor-size, the cost of which is about 35 % of the total unit.

REFERENCES

1   A. Billon, J.P. Franck and J.P. Peries, Hydroc. Proc., sept. 1975, p. 139.
2   P. Trambouze, A. Billon and J.P. Peries, Conference on hydrocracking, Institut Belge du Pétrole, Bruxelles, oct. 1968.
3   N. Choudhary and D.N. Saraf, Ind. Eng. Chem., P.R.D., vol 14 (2), 1975, pp. 74-83.
4   S.P. Ahuja, M.L. Derrien and J.F. Le Page, P.R.D., vol. 9 (3), 1970, p. 272.
5   Petroleum Times, july 1970, p. 5.
6   J.P. Franck, Rev. I.F.P., vol XXXIII (4), 1978, 597-612.
7   H.L. Coonradt and W.E. Garwood, Ind. Eng. Chem., P.D.D., vol 3 (1), 1964, p. 38.
8   J.W. Scott and A.G. Bridge, ACS/CIC Conference, Toronto, may 26, 1970.
9   J.W. Ward, Hyd. Process, sept. 1975, 101-106.
10 U.S. 3.132.089, 1964.
11 U.S. 3.159.568, 1964.
12 U.S. 3.835.027, 1974.
13 U.S. 3.890.247, 1975.
14 U.S. 4.120.825, 1978.
15 M. El Malki, thèse Paris 1978.
16 J.P. Franck (I.F.P.), unpublished results.
17 U.S. 3.173.853, 1965.
18 R.P. Vaell, J.L. Lafferty and J. Sosnowski, Proceed. Div. Refin. 1970, 1011-1021.
19 D.E.W. Vaughan, Symp. on Propert. and Applicat. of zeolites, London, April 18-20, 1979.
20 U.S. 3.293.192, 1966.
21 C.V. Mc Daniel and P.K. Maher, 1st Intern. Congress on Molecular Sieves, London, 1967, 186-195.
22 G.T. Kerr, J. Catal., 1969, 15, 200-204.
23 U.S. 3.354.077, 1967.
24 U.S. 3.506.400, 1970.
25 U.S. 3.591.488, 1971.
26 Fr 2.303.764, 1976.
27 U.S. 4.058.484, 1977.
28 G.T. Kerr, J. Phys. Chem., 1968, 72, 2594.
29 G.T. Kerr, J. Phys. Chem., 1969, 73, 2780.
30 R. Beaumont and D. Barthomeuf, J. Catal., 1972, 26, 218.
31 R. Beaumont and D. Barthomeuf, J. Catal., 1972, 27, 45.
32 D.W. Breck, "Zeolite Molecular Sieves", John Wiley and Sons, 1974, pp. 505-506.

33 P. Gallezot, A. Alarcon-Diaz, J.A. Dalmon and B. Imelik, C.R. Acad. Sci., Paris, 1974, t. 278, 1073-1075.
34 P. Gallezot, P.A. Alarcon-Diaz, J.A. Dalmon, A.J. Renouprez and B. Imelik, J. Catal., 1975, 39, 334.
35 J.P. Franck and A. Billon (I.F.P.), unpublished results.
36 Iranian Petroleum Institute, Bulletin 66, 1977, 17.
37 J. Gonzalez Salas, thèse Paris, 1974.
38 P. Gallezot, J. Datka, J. Massardier, M. Primet and B. Imelik, VIth Congress on Catalysis, London, 1976, vol. 2, 696-704.
39 G.D. Chukin, M.L. Landau, V. Ya. Kruglikov, D.A. Agievskii, B.V. Smirnov, A.L. Belozerov, V.S. Asrieva, N.V. Goncharova, E.D. Radchenko, O.D. Konolvachikov and A.V. Agafonov, VIth Congress on Catalysis, London, 1976, vol. 2, 668-676.
40 M.V. Landau, V. Ya. Kruglikov, N.V. Goncharova, O.D. Konoval'Chikov, G.D. Chukin, B.V. Smirnov and V.I. Malevich, Kin. Cat., 1976, 17, 1281-1287.
41 A.D. Reichle, L.A. Pine, J.W. Ward and R.C. Hansford, Oil G.J., july 29, 1977, 137-143.
42 U.S. 4.139.433, 1979.
43 U.S. 4.116.867, 1978.
44 U.S. 3.197.399, 1965.
45 U.S. 3.357.915, 1967.
46 U.S. 3.287.257, 1966.
47 U.S. 3.899.441, 1975.
48 U.S. 3.835.028, 1974.
49 U.S. 3.849.293, 1974.
50 U.S. 3.692.692, 1972.
51 Oil G.J., july 29, 1974.
52 U.S. 4.055.482, 1977.
53 U.S. 4.148.750, 1979.
54 J.W. Ward, R.C. Hansford, A.D. Reichle and J. Sosnowski, Oil G.J., may 28, 1973, 69-73.
55 J.H. Duir, Oil G.J., august 21, 1967, 74-77.
56 J.A. Rabo, R.D. Bezman and M.L. Poutsma, Proc. on the Symp. on Zeolites, Szeged (Hungary), sept. 11-14, 1978, 39-52.

B. Imelik *et al.* (Editors), *Catalysis by Zeolites*
© 1980 Elsevier Scientific Publishing Company, Amsterdam — Printed in The Netherlands

SELECTIVE FORMATION OF p-CRESOL BY ALKYLATION OF PHENOL WITH METHANOL OVER Y TYPE ZEOLITE

SEITARO NAMBA, TATSUAKI YASHIMA, YASUSHI ITABA and NOBUYOSHI HARA
Department of Chemistry, Tokyo Institute of Technology
Ookayama, Meguro-ku, Tokyo, 152, Japan

1.   INTRODUCTION

Many researchers have reported that the alkylations of various aromatics are effectively promoted by zeolite catalysts [1,2]. In the alkylation of toluene with methanol over the usual acid catalysts, the primary products are thought to be o- and p-xylenes. However, an equilibrium composition of xylene isomers is obtained at a considerably high yield of xylene, because the isomerization of the xylene produced is also promoted by the same acid catalysts under the alkylation conditions. However, we have found that p-xylene can be produced at a high selectivity using Y or ZSM-5 type zeolite catalysts [3,4]. We have also reported that the selective formation of p-xylene demands not only the Brønsted acid sites with the suitable strength but also the narrow space of the zeolites for the shape selective reaction [4,5].

In the acid-catalyzed phenol alkylation, para/ortho orientation generally predominates, too. So, to suppress the formation of m-cresol, which is hardly separated from p-cresol by distillation, the catalytic activity for the isomerization of cresol must be reduced. The selective formation of p-cresol as well as that of p-xylene in the alkylation of toluene may demand the Brønsted acid sites with the suitable strength and the narrow space, which may prefer the alkylation rather than the isomerization.

In this paper, we are reporting the effects of the acid strength of Y zeolites and the reaction conditions on the activity and the selectivity.

2.   EXPERIMENTAL

2.1   Materials

Catalysts: A series of H-Na-Y zeolite catalysts was prepared by a con-
ventional cation-exchange procedure at 348K using 0.5-1N aqueous solution of ammonium chloride and Na-Y zeolite(SK40), followed by the calcination of NH_4-Na-Y zeolites to remove ammonia. The degree of cation-exchange was varied from 26 to 78%.

A series of H-K-Y zeolite catalysts was prepared as follows; the Na-Y zeolite was transformed into K-Y zeolite by a conventional cation-exchange procedure with 1N queous solution of potassium chloride at 348K, and then

Table 1. Fraction of cations in H-K-Y zeolites

zeolite	cation		
	H^+	K^+	Na^+
H-K-Y-1	84	11	5
H-K-Y-2	76	16	8
H-K-Y-3	70	23	7
H-K-Y-4	60	31	9
H-K-Y-5	47	43	10

the K-Y zeolite was transformed into H-K-Y zeolite in a similar manner as H-Na-Y zeolites. The fraction of cations in various H-K-Y zeolites are shown in Table 1.

The degree of cation-exchange was determined by flame photometry.

Reactants: Phenol and methanol were obtained from a commercial source with a purity of 99.5%.

2.2 Apparatus and Procedure

The alkylation of phenol with methanol was carried out in a fixed bed-type apparatus with a continuous flow system at atmospheric pressure. The catalyst was placed in an electrically heated quartz reactor and calcined at a desired temperature for 2h in a nitrogen stream, and then brought to the reaction temperature (200-300°C) in situ. The mixture of phenol and methanol was fed by a micro-feeder to start the reaction. Nitrogen was used as a carrier gas and the molar ratio of nitrogen to the reactants was 5. The products were cooled with a dry ice-methanol trap (-25 — -20°C), and the samples for gas chromatographic analysis were collected every 20 min.

3. RESULTS AND DISCUSSION

In the alkylation of phenol with methanol over the Y zeolite catalysts, three cresol isomers, anisol($C_6H_5OCH_3$), xylenols, methoxytoluenes and unknown compounds were found as the alkylation products.

Fig.1 shows the activity change with process time on the H-K-Y-1 catalyst. The yields changed with process time and the maximum yield of p-cresol was observed at 60min of process time. In the subsequent studies, the yields of the alkylation products at the process time when the yield of p-cresol was maximumized were chosen to represent the catalytic activity and selectivity of the catalysts.

In the series of the H-Na-Y zeolite catalysts, the effects of the degree of cation-exchange on the yields of the alkylation products and on the acidities determined by n-butylamine titration were examined and the results are shown in Fig.2. The weak acid sites ($+1.5 \geq H_0 > -3.0$) are helpful to produce anisol and the strong acid sites ($-8.2 \geq H_0$) are effective for the secondary alkylation to produce xylenol and for the isomerization of p- and o-cresols to produce m-cresol. On the other hand, the moderate acid sites ($-3.0 \geq H_0 > -8.2$) are advantageous for the production of p- and o-cresols. The yield of p-cresol was maximumized at 70% of the degree of cation-exchange. The yield of p-cresol was 19% and the selectivity for p-cresol based on phenol converted was 29%.

In the case of the H-K-Y zeolite catalysts, the optimum degree of cation-exchange may be shifted to higher degree, because potassium is more electronegative than sodium. The activities of the series of the H-K-Y zeolites

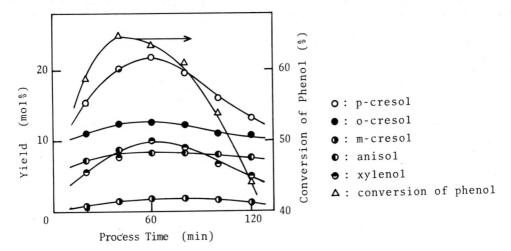

Fig. 1. Activity change with process time

        Catalyst: H-K-Y-1
        Reaction conditions: calcination temperature, 400°C;
        reaction temperature, 250°C; W/F, 90g·h/mol-reactants;
        methanol/phenol, 2mol/mol

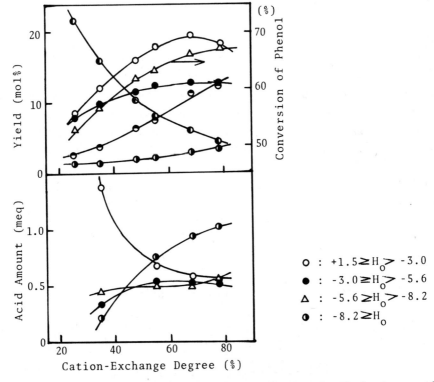

Fig. 2. Effects of cation-exchange degree on yields of alkylation products
        and acidities

        Symbols in upper figure are defined in Fig. 1.
        Catalyst: H-Na-Y zeolite
        Reaction conditions except calcination temperature are
        described in Fig. 1. Calcination temperature; 500°C

were also examined and the results are shown in Fig.3. At the highest
degree of cation-exchange (84%) in the series, the maximum yield of p-cresol
(22%), which was more than that in the H-Na-Y zeolites, was observed.
Moreover, the formations of m-cresol and xylenol by the secondary reactions
were more suppressed than those on the H-Na-Y zeolites. The selectivity
reached to 35%. Accordingly, potassium cation in H-Y zeolite is more
effective to reduce the strong acid sites $(-8.2 \geq H_0)$ without reducing the
moderate acid sites $(-3.0 \geq H_0 > -8.2)$ than sodium cation. Thus, it may be
possible to control the acidities of H-Y zeolites by choosing the second
cation introduced in H-Y zeolites.

The optimum reaction conditions on the best catalyst (H-K-Y-1) were
examined. The effect of the reaction temperature on the activity is shown
in Fig.4. At 250°C, the maximum yield of p-cresol was observed. The higher
reaction temperature provided the higher yields of o-cresol, m-cresol and
xylenol. While, the lower reaction temperature provided the higher yield of
anisol. Thus, the higher reaction temperature than 250°C is helpful for the
secondary reactions.

Fig.5 shows the effect of contact time (W/F) on the activity. The yield
of p-cresol was maximumized in the region of W/F=90-150g·h/mol-reactants.
The longer contact time provided the higher yields of m-cresol and xylenol
and the lower yields of anisol. Too long contact time is helpful especially
for the isomerization.

Fig.6 shows the effect of the molar ratio of the reactants on the
activity. At the molar ratio methanol/phenol=2, the yield of p-cresol was
maximumized. The higher methanol/phenol molar ratio provided the higher
yields of m-cresol and xylenol and the lower yields of anisol and o-cresol.
At the high molar ratio, a considerable amount of unknown products, which
were not polyalkylphenols, were detected.

Fig.7 shows the effect of calcination temperature on the activity. At

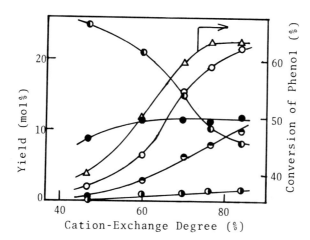

Fig. 3. Effect of cation-exchange degree on yields of alkylation products
       Symbols are defined in Fig. 1. Catalyst: H-K-Y zeolite
       Reaction conditions are described in Fig. 1.

400°C of the calcination temperature, the yield of p-cresol was maximumized. At the calcination temperature higher than 500°C, the yields of m-cresol and xylenol decreased, while the yield of anisol increased. The yield of o-cresol scarecely changed with the calcination temperature.

From above results, the optimum reaction conditions on the best catalyst (H-K-Y-1) were found as follows; calcination temperature=400°C, reaction temperature=250°C, W/F=90-150g·h/mol-reactants, reactants molar ratio methanol/phenol=2. Under such conditions, the yield of p-cresol was 22%, the selectivity for p-cresol based on phenol converted was 35% and the fraction of p-cresol in cresols was 61% and was higher than 23% of equilibrium value.

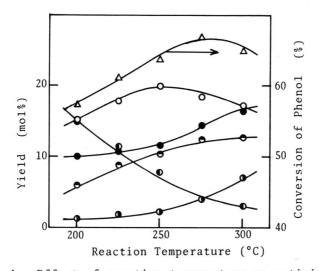

Fig. 4. Effect of reaction temperature on activity
Symbols are defined in Fig. 1. Catalyst: H-K-Y-1
Reaction conditions except calcination temperature
are described in Fig. 1.
Calcination temperature; 500°C

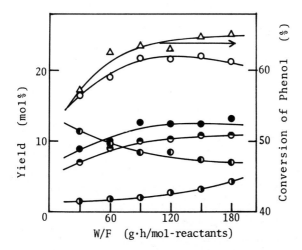

Fig. 5. Effect of contact time on activity
Symbols are defined in Fig. 1. Catalyst: H-K-Y-1
Reaction conditions except W/F are described in Fig. 1.

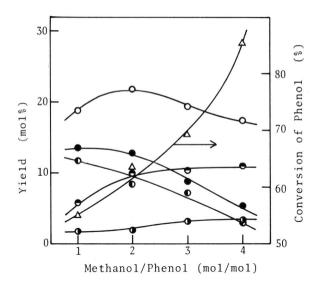

Fig. 6.   Effect of reactants molar ratio on activity
          Symbols are defined in Fig. 1.   Catalyst: H-K-Y-1
          Reaction conditions except reactants molar ratio
          are described in Fig. 1.

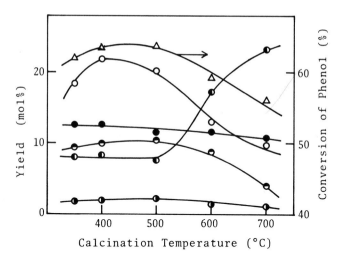

Fig. 7.   Effect of calcination temperature on activity
          Symbols are defined in Fig. 1.   Catalyst: H-K-Y-1
          Reaction conditions except calcination temperature
          are described in Fig. 1.

From the results in Figs. 2, 3, 4 and 5, the mild reaction conditions were
advantageous for the formation of anisol.   So, it was suggested that anisol
might be a intermediate to produce p-cresol in the alkylation.   The conversion of anisol over the H-Na-Y zeolite, whose degree of cation-exchange was
36%, was studied and the results are shown in Table 2.   The yields of cresols
in the conversion of anisol were lower than those in the alkylation of phenol
with methanol.   The same phenomena were also observed when the conversion of
anisol was carried out in the presence of methanol or phenol-methanol.

Table 2. Product distribution in anisol conversion*

reactant	anisol	methoxy-toluene	phenol	cresol o-	p-	m-	xylenol
phenol/methanol (1/2)	15.7	3.5	46.4	10.7	12.8	2.4	3.8
anisol	68.3	5.0	19.5	3.1	2.6	0.2	0.8
anisol (W/F = 180)	59.6	5.2	26.3	4.5	4.5	3.0	0.7
anisol/methanol (1/1)	64.0	6.5	15.7	4.0	4.0	1.3	1.8
anisol/methanol (1/2)	73.9	7.5	9.4	2.7	2.5	1.0	1.1
anisol/phenol/methanol (1/1/2)	41.5	4.6	32.5	6.0	7.6	1.8	2.0
anisol/phenol/methanol (1/1/4)	49.8	4.5	22.9	5.8	6.5	2.3	3.3

* Reaction conditions are described in Fig. 2.

Accordingly, it is concluded that anisol is not a main intermediate in the formation of cresol by the alkylation of phenol with methanol.

4.   CONCLUSION

   In the alkylation of phenol with methanol over H-Y zeolite catalysts, the moderate acid sites ($-3.0 \geqq H_0 > -8.2$) on H-Y zeolite are advantageous to produce p-cresol selectively. While, the weaker acid sites are effective to produce anisol and the stronger acid sites are helpful for the secondary reactions.

   K^+ in H-Y zeolites is more effective for the selective formation of p-cresol than Na^+ in H-Y zeolites. Especially, the H-K-Y zeolite, whose degree of cation-exchange is 84%, is best in the catalysts examined here. The yield of p-cresol is 22%, the selectivity for p-cresol based on phenol converted is 35% and the fraction of p-isomer in cresols is 61% on the best H-K-Y zeolite.

   Anisol is not a main intermediate in the formation of cresol by the alkylation of phenol with methanol over H-Y zeolite catalysts.

REFERENCES
1. P.B. Venuto, L.A. Hamilton, P.S. Landis and J.J. Wise, J. Catal., 5, 81 (1966); ibid., 5, 484 (1966).
2. Kh.M. Minachev and Ya.I. Isakov, Intern. Chem. Eng., 7, 18 (1967); ibid., 7, 91 (1967).
3. T. Yashima, H. Ahmad, K. Yamazaki, M. Katsuta and N. Hara, J. Catal., 16, 273 (1970).
4. T. Yashima, Y. Sakaguchi and S. Namba, 7th Intern. Congr. Catalysis, Tokyo (1980), in press.
5. T. Yashima, K. Yamazaki, H. Ahmad, M. Katsuta and N. Hara, J. Catal., 17, 151 (1970).

B. Imelik *et al.* (Editors), *Catalysis by Zeolites*
© 1980 Elsevier Scientific Publishing Company, Amsterdam — Printed in The Netherlands

DEHYDRATION OF PROPAN-2-OL ON HY ZEOLITES

ROBERT RUDHAM AND ALAN STOCKWELL
Department of Chemistry, University of Nottingham, Nottingham, England.

ABSTRACT
    The dehydration of propan-2-ol on HY zeolites has been studied in a flow
system operating at a total pressure of one atmosphere. The effects of
temperature, extent of exchange and base poisoning suggest that both propene
and di-isopropyl ether are formed at a single Brønsted acid site. A
mechanism involving carbonium and oxonium ion intermediates is discussed.

INTRODUCTION
    Variations in activity with alcohol structure and with the nature and
extent of cation exchange show that the catalytic dehydration of alcohols
on zeolites proceeds by ionic intermediates [1-3]. The source of the
activity is the Brønsted acidity of the structural OH groups which are
generated by deammoniation following NH_4^+ exchange, by cation hydrolysis or
by cation reduction with hydrogen or compounds containing hydrogen. From a
study of propan-2-ol dehydration on HX zeolites [4], an ionic mechanism
involving one OH group for propene formation and two OH groups for di-
isopropyl ether was proposed. This view was supported by Jacobs et. al. [5]
from a study of propan-2-ol dehydration on alkali cation exchanged X and Y
zeolites, where the OH groups were formed by hydrolysis of the alkali
cations or polyvalent cation impurities. For NiY and CuY zeolites the
activity for both propene and di-isopropyl ether formation increased linearly
with the absorbance of the 3650 cm^{-1} infra-red band associated with α-cage
OH groups [6]. Since this suggests that di-isopropyl ether formation is not
invariably associated with two OH groups, propan-2-ol dehydration merits
further investigation.

EXPERIMENTAL
    Catalytic dehydration was studied in a continuous-flow system operating
at atmospheric pressure with a 0.5 cm^3 s^{-1} flow of propan-2-ol and N_2
diluent [4]. The normal propan-2-ol pressure was 1.11×10^3 Pa, but this was
varied in pressure dependence experiments. Infra-red measurements were made
using a cell which took the place of the normal reaction vessel.
    Catalysts were prepared from binder-free NaX (Linde 13X) or NaY (Linde
SK40) by cation exchange with aqueous NH_4Cl at room temperature. A single
highly exchanged sample of NaY was prepared by refluxing with a five-fold
excess of NH_4Cl for 2h. Extents of exchange determined from the amount of

NH_4^+ introduced were in excellent agreement with those determined from Na^+ released. All catalysts were deammoniated by heating in a 1.67 cm^3 s^{-1} flow of pure cylinder N_2 for 16h at 623 K; they are designated HX-n or HY-n, where n represents the nominal concentration of structural OH groups generated per unit cell.

RESULTS

Figure 1 shows the i.r. spectrum of a 50 mg, 2.5 cm diameter disc of HY-14 subjected to the identical activation given catalyst samples. From the elimination of the NH bending band at 1450 cm^{-1} it is evident that complete deammoniation occurs. The disc was active for propan-2-ol dehydration at the i.r. beam temperature of ~ 325 K, with rates of both di-isopropyl ether and propene formation of ~ 3×10^{14} molecules g^{-1} s^{-1}. The spectrum shows a considerable sorption of both water and propan-2-ol under these conditions.

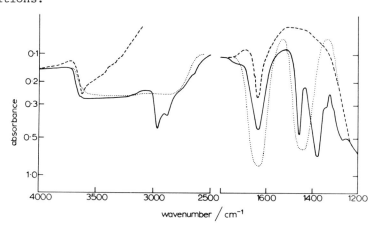

Fig.1. Infra-red spectrum of HY-14: .... dried in N_2 at 325 K; --- following deammoniation at 623 K; —— in flowing propan-2-ol/N_2 at 325 K.

The effect of propan-2-ol pressure on the rates of formation of di-isopropyl ether (r_e, molecules g^{-1} s^{-1}) and propene (r_p, molecules g^{-1} s^{-1}) was determined with 10 mg samples of HX-8.6 and HY-8.4 at a number of temperatures. Stable and reproducible activities were rapidly achieved on changing either pressure or temperature, and there was no evidence for any dehydrogenation to acetone. The results for HX-8.6 confirmed the small pressure dependencies previously observed [4], while those for HY-8.4 indicate departures from zero-order at extremes of temperature and pressure. At low temperatures and high pressures both products show a tendency to inverse order, suggesting that pore filling leads to a diminution in rate. However, at high temperatures and low pressures both products exhibit a positive order, which is more pronounced for ether than for propene.

The effect of temperature on activity was investigated with 100 mg

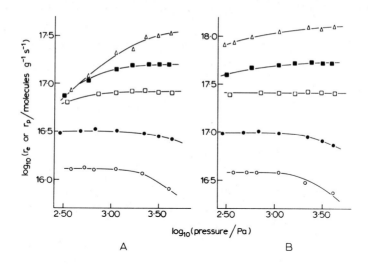

Fig.2. The effect of propan-2-ol pressure on r_e (A) and r_p (B) on HY-8.4:
O, 372 K; ●, 381 K; □, 390 K; ■, 400 K; △, 410 K.

catalyst samples under conditions where minimal pressure dependence was
exhibited and the reactor operated in the differential mode. Under these
conditions r_e and r_p become the zero-order rate constants k_e and k_p, and
measurements covering a range of ~ 35 degrees yielded excellent Arrhenius
plots. Values of the activation energies E_e and E_p, and k_e and k_p at a
common reaction temperature of 370 K, are given as a function of the extent
of exchange for HY catalysts in Fig.3. Selectivities for ether formation
at 370 K, defined as $S_e = 2 k_e/(2k_e+k_p)$, are also plotted in Fig.3. For
HX-8.6 ($E_e = 113$ kJ mol^{-1}, $E_p = 121$ kJ mol^{-1}) and unexchanged NaY
($E_e = 76$ kJ mol^{-1}, $E_p = 106$ kJ mol^{-1}) the observed activation energies
compare favourably with published values [4,5].

Activity measurements extended to considerably higher temperatures, made
with 10 mg samples of HX-8.6, HY-8.6 and HY-35.5, showed that k_p
consistently increased with temperature, whereas k_e passed through a maximum.
The results are presented as plots of S_e against temperature in Fig.4, where
S_e deviates from values calculated from low temperature Arrhenius parameters
at ~ 6% propan-2-ol dehydration.

Progressive poisoning plots were constructed by determining the activity
of 100 mg catalyst samples, between 364 and 409 K, following the adsorption
of successive aliquots of organic base. Fig.5 shows that the plots for k_e
and k_p intersect at zero activity, while S_e remains substantially constant
until high degrees of poisoning are achieved. The concentrations of
adsorbed pyridine required to completely poison the four catalysts studied,
expressed in molecules p.u.c., were: HX-8.6, 8.1; HY-8.4, 8.4; HY-16.6,11.4;
HY-35.5, 46.2. For piperidine the concentrations were: HX-8.6, 8.7;
HY-8.4, 8.5; HY-16.6, 19.3; HY-35.5, 47.5.

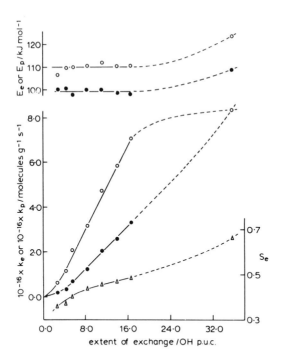

Fig.3. The dependence of activation energy, and of activity and selectivity at 370 K on the extent of exchange in HY catalysts: ●, E_e and k_e; O, E_p and k_p; Δ, S_e.

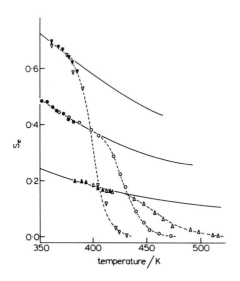

Fig.4. Variation in selectivity for ether formation S_e with temperature: Δ, ▲, HX-8.6; O, ●, HY-8.4; ▽, ▼, HY-35.5; ——, calculated; (low temperature measurements in closed symbols).

DISCUSSION

The infra-red spectra in Fig.1 show the retention of molecular water after deammoniation which results in a broad and poorly resolved spectrum in the OH stretching region. There is no clear evidence for hydroxonium ion

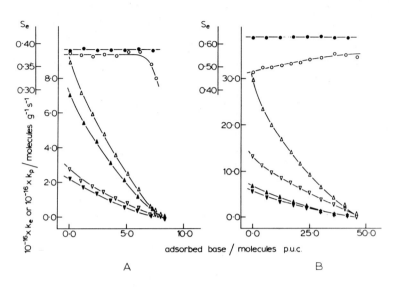

Fig.5. Progressive poisoning plots: (A) HY-8.4; pyridine at 381 K ∇-k_e, $\triangle$-k_p, O-S_e; piperidine at 379 K $\blacktriangledown$-k_e, $\blacktriangle$-k_p, $\bullet$-S_e. (B) HY-35.5; pyridine at 381 K ∇-k_e, $\triangle$-k_p, O-S_e; piperidine at 364 K $\blacktriangledown$-k_e, $\blacktriangle$-k_p, $\bullet$-S_e.

formation, but sufficient H_2O is available for this to occur during propan-2-ol dehydration. Additional effects of water include suppression of Lewis acidity arising from dehydroxylation, and the favouring of structural dealumination during deammoniation of highly exchanged NH4Y. In the absence of Lewis acidity and multivalent exchange cations which coordinate with organic bases, the quantity of base necessary for complete catalyst poisoning may be equated with the concentration of accessible Brønsted acid sites. With HX-8.6 and HY-8.4 there is excellent agreement between the nominal concentration of acid sites and those determined by pyridine and piperidine. The similarity in the concentrations determined by these weak and strong bases suggests that, at low levels of ion exchange, all the Brønsted acidity is accessible within the α-cages and may contribute to catalysis. With HY-16.6 a proportion of the OH groups is inaccessible to both pyridine and propan-2-ol, although piperidine is sufficiently basic to induce migration of protons to accessible positions. The excess acid content detected in HY-35.5 suggests that limited structural dealumination occurred during activation [7], with adsorption of both bases at OH groups and hydroxy-aluminium cations. Progressive poisoning thus shows that the majority of the OH groups formed in HY are active in propan-2-ol dehydration, with an effective concentration which approaches the nominal concentration at low exchange levels.

With the exception of HY-35.5, the activity of HY increases with the extent of exchange without significant changes in the activation energies for product formation. The similar form of the plots of k_e and k_p against extent of exchange up to HY-16.6 (Fig.3) and the insensitivity of S_e to changes in the concentration of acidity (Figs. 3 and 5) suggest a mechanism involving a single OH group for the formation of both products. For HX-8.6, E_e = 113 kJ mol^{-1} and E_p = 121 kJ mol^{-1}, with k_e = 5.33 x 10^{14} molecules g^{-1} s^{-1} and k_p = 3.86 x 10^{15} molecules g^{-1} s^{-1} at 380 K; while for HY-8.4, E_e = 100 kJ mol^{-1} and E_p = 111 kJ mol^{-1}, with k_e = 2.90 x 10^{16} molecules g^{-1} s^{-1} and k_p = 8.16 x 10^{16} molecules g^{-1} s^{-1} at 380 K. Since all OH groups are accessible in these catalysts the differences in activity are controlled by energetic factors, where the increase in acid strength of individual OH groups with an increase in Si/Al ratio [8,9] leads to a decrease in activation energy.

To account for our observations we propose the following reaction sequence. Since molecular water is present after deammoniation and during reaction, accessible OH groups are ionised to form H_3O^+ associated with structural O^- ions:

$$H + H_2O \rightarrow H_3O^+. \tag{1}$$
$$\overset{|}{O}_z \qquad \overset{..}{O^-_z}$$

At high propan-2-ol concentrations proton transfer occurs giving an oxonium ion which yields a carbonium ion by the E_1 elimination of H_2O:

$$H_3O^+ + (Me)_2CHOH \rightarrow (Me)_2CH\overset{+}{O}H_2 + H_2O, \tag{2}$$

$$(Me)_2CHO\overset{+}{H}_2 \rightarrow (Me)_2\overset{+}{C}H + H_2O. \tag{3}$$

Subsequent proton transfer to water (4a) or propan-2-ol (4b) yields propene and regenerates the hydroxonium or oxonium ion:

$$(Me)_2\overset{+}{C}H + H_2O \rightarrow H_3\overset{+}{O} + MeCH = CH_2, \tag{4a}$$

or

$$(Me)_2\overset{+}{C}H + (Me)_2CHOH \rightarrow (Me)_2C\overset{+}{H}OH_2 + MeCH = CH_2. \tag{4b}$$

Alternatively, interaction of the carbonium ion with propan-2-ol gives an oxonium ion, which on proton transfer to water (6a) or propan-2-ol (6b) yields di-isopropyl ether whilst regenerating an ionic species:

$$(Me)_2\overset{+}{C}H + (Me)_2CHOH \rightarrow (Me)_2CH\overset{+}{O}HCH(Me)_2, \tag{5}$$

followed by

$$(Me)_2CH\overset{+}{O}HCH(Me)_2 + H_2O \rightarrow H_3\overset{+}{O} + (Me)_2CHOCH(Me)_2, \tag{6a}$$

or

$$(Me)_2\overset{+}{\underset{\underset{Z}{\overset{\cdot\cdot}{O^-}}}{CHOHCH}}(Me)_2 + (Me)_2CHOH \rightarrow (Me)_2\overset{+}{\underset{\underset{Z}{\overset{\cdot\cdot}{O^-}}}{CHOH}}_2 + (Me)_2CHOCH(Me)_2. \qquad (6b)$$

The proposed mechanism, although basically similar to those previously suggested [4,5] differs in that only one acid site is involved in di-isopropyl ether production. Furthermore, it obviates the unlikely inter-action between similarly charged carbonium and oxonium ions [4]. For reaction at low temperatures, where the flow reactor operated in the differential mode and zero-order kinetics were obeyed, the partial pressure of propan-2-ol was very high relative to that of the water produced. It follows that direct regeneration of the $(Me)_2C\overset{+}{H}OH_2$ oxonium ion by (4b) and (6b) can be regarded as more likely than (4a) followed by (2) and (6a) followed by (2). Presumably the E_1 elimination of water (3) makes the largest contribution to the overall activation energy, and thus accounts for the observed similarity in E_e and E_p. The rapid decrease in the selectivity for di-isopropyl ether formation at high temperatures occurs when the reactor no longer operates in the differential mode. Under steady reaction conditions the molecular contents of the α-cages will reflect the fall in the average partial pressure of propan-2-ol, and the considerable rise in the partial pressure of water. A fall in the concentration of propan-2-ol within the α-cages is unfavourable to reactions (4b) and (6b), but predominantly so to reaction (5) which is essential for ether production. Since propene is produced by (4a), operation in the integral reactor mode tends to a limiting condition in which propene is the sole reaction product, as observed in pulsed microreactors and recirculatory flow reactors [10,11].

REFERENCES

1   P.A. Venuto and P.S. Landis, Advances in Catalysis, 18 (1968) 259-371.
2   P.A. Jacobs, Carboniogenic Activity of Zeolites, Elsevier, Amsterdam, 1977, pp. 99-107.
3   R. Rudham and A. Stockwell, Catalysis, Vol. 1, Specialist Periodical Reports, The Chemical Society, London, 1977, Ch.3, pp. 87-135.
4   S.J. Gentry and R. Rudham, J.C.S., Faraday I, 70 (1974) 1685-1692.
5   P.A. Jacobs, M. Tielen and J.B. Uytterhoeven, J. Catalysis, 50 (1977) 98-108.
6   R.A. Schoonheydt, L.J. Vandamme, P.A. Jacobs and J.B. Uytterhoeven, J. Catalysis, 43 (1976) 292-303.
7   D.W. Breck and G.W. Skeels, Proc. 6th Int. Congr. Catalysis, The Chemical Society, London, 1977, pp. 645-653.
8   R. Beaumont and D. Barthomeuf, J. Catalysis, 26 (1972) 218-225.
9   P.A. Jacobs, W.J. Mortier and J.B. Uytterhoeven, J. Inorg. Nucl. Chem., 40 (1978) 1919-1923.
10 K.V. Topchieva and H.S. Tkhoang, Kinetika i Kataliz, 14 (1973) 398-402.
11 V.S. Levchuk and K.G. Ione, Kinetika i Kataliz, 13 (1972) 949-953.

B. Imelik *et al.* (Editors), *Catalysis by Zeolites*
© 1980 Elsevier Scientific Publishing Company, Amsterdam — Printed in The Netherlands

TRANSFORMATIONS OF BUTENES ON NaHY ZEOLITES OF VARIOUS EXCHANGE DEGREES AS STUDIED BY IR SPECTROSCOPY

J. DATKA
Institute of Chemistry, Jagiellonian University, Cracow, Poland

ABSTRACT

The transformations of but-1-ene adsorbed on NaHY zeolites were studied by the IR spectroscopy at various temperatures which enabled the discrimination of particular steps of the reaction. The kinetics of the isomerization of but-1-ene adsorbed on the zeolites was also studied by the IR spectroscopy.

INTRODUCTION

The interaction of butenes with NaHY zeolites is complex and comprises several stages such as : hydrogen bond formation, isomerization, and oligomerization of butene. In the present research the transformations of butenes adsorbed on NaHY zeolites were studied by the IR spectroscopy at various temperatures (from 220 K to 450 K) which enabled the discrimination of the particular processes overlapping when butenes interact with NaHY zeolite. The NaHY zeolites with various exchange degree were studied in order to characterize the influence of the concentration and the acid strength of OH groups on the butene transformation.

EXPERIMENTAL

The $NaNH_4Y$ zeolites were obtained from a NaY preparation ($SiO_2/Al_2O_3 = 5.22$) prepared at the Institute of Industrial Chemistry, Warsaw. Three samples exchanged to 21.0, 39.7, and 77.1 % (denoted as NaHY/21, NaHY/40, and NaHY/77 resp.) were prepared. The wafers of $NaNH_4Y$ (3-6 mg/cm^2) were pretreated in vacuo at 723 K for 3 hours. Under these conditions the complete decomposition of the ammonium form and transformation into hydrogen one took place. The adsorbates : but-1-ene (Fluka), and pyridine (POCh - Gliwice) were used without any additional treatment. The IR spectra were recorded with "Specord 75 IR" grating spectrometer (Carl Zeiss, Jena).

RESULTS AND DISCUSSION

The concentration of HF OH groups in NaHY zeolites studied in the present research is given in [1]. It is equal to 7.0, 15.5, and 16.7 per unit cell in NaHY/21, NaHY/40, and NaHY/77 zeolites resp. The experiments with the thermodesorption of pyridine [1], adsorption of aromatic hydrocarbons [2], and adsorption of alkanes [3] lead to the conclusion that the acid strength of the HF OH groups increases with the exchange degree of zeolite.

But-1-ene and but-2-ene-trans were adsorbed at 223 K in NaHY zeolite. Butene isomerization was too slow to be observed at this temperature. But-1-ene, and but-2-ene-trans formed hydrogen bond with HF OH groups. The corresponding IR band was shifted from 3640 cm^{-1} to 3160 and 3050 cm^{-1} for but-1-ene and but-2-ene-trans resp. in the case of NaHY/77 zeolite, and to 3200, and 3100 cm^{-1} in the case of NaHY/40 zeolite. The adsorption of 19.5 molecules of but-1-ene, and 18.0 molecules of but-2-ene-trans per unit cell resulted in the disappearence the HF OH band at 3640 cm^{-1}. Both values are close to the number of the HF OH groups in unit cell. It may be therefore concluded'that in the average each HF OH groups forms hydrogen bond with one butene molecule. In the case of NaHY/21 zeolite hydrogen bond formation was not observed. It seems that butene molecules adsorbed on NaY, and NaHY/21 zeolites are mostly bonded to Na$^+$ ions, but in the case of zeolites with higher cation exchange degrees (higher concentration of HF OH groups) butene molecules hydrogen bonded with the HF OH groups predominates.

At 243 K and higher temperatures isomerization of adsorbed but-1-ene molecules is also observed. The bands of hydrogen bonded but-1-ene are substituted by the bands of but-2-ene cis, and but-2-ene-trans also hydrogen bonded. The kinetics of this isomerization was studied by the IR spectroscopy at temperatures 253 - 313 K. The intensity of the 1631 cm^{-1} band characteristic for C=C- vibration in the but-1-ene molecule changed in accordance with the first order reaction law. Assuming that HF OH groups are active centres of the catalytic reaction the turnover numbers were calculated from the known values of rate constants and concentrations of HF OH. The kinetic experiments were made at various temperatures and the activation energy of but-1-ene isomerization was determined. It should be observed here that when investigating the kinetics of a reaction occurring in the adsorbed layer without any participation of the molecules from gas phase one eliminates transport processes. On the other hand if but-1-ene isomerization is studied at temperatures as low as in the present investigation it is not disturbed by the oligomerization of butenes which reaction is very slow in the temperature region 253 - 313 K. One can therefore conclude that the energy of activation determined at such conditions is so called true energy of activation.

In the series of NaHY zeolites the catalytic activity of HF OH groups increases (the activation energy decreases and the turnover number increases) as the exchange degree of zeolite increases. There are therefore the correlations between the acid stregth of the HF OH groups and their catalytic activity. It seems that the increase of the acid strength of HF OH groups and their catalytic activity is the result of induction effect in the zeolite lattice. The replacing of the Na$^+$ ions by protons (which have much higher electrical charge density than Na⁺ ions) result in the increase of the polarization of neighbouring O-H bonds and hence in the increase of their acid strength.

TABLE 1

The catalytic properties of NaHY zeolites

Zeolite	Activation energy KJ mol^{-1}	Turnover at 293 K 10^{-4} s^{-1}
NaHY/21	65.7	0.80
NaHY/40	51.4	11.0
NaHY/77	38.8	62.5

At room temperature and higher ones, the butene molecules undergoe further reaction - oligomerization. Fig. 1 shows the spectra recorded 5 min. (spectrum b), and 20 hours (spectrum c) after but-1-ene adsorption at the room temperature on the NaHY/77 zeolite. The spectrum b is the spectrum of but-2-ene-trans, and but-2-ene-cis hydrogen bonded to the HF OH groups. They are the products of fast isomerization of but-1-ene. Spectrum c is the spectrum of oligomer. The broad band at 3050 cm^{-1} of HF OH groups hydrogen bonded to the but-2-ene molecules disappeared. The intensity and half-width of LF OH band apparently increased. This latter effect may be explained as follows : the HF OH groups which were initially hydrogen bonded to butene molecules are not completely free but interact with oligomers. Such interactions due to the weak intermolecular forces were also observed in the case of adsorption of alkanes [3]. They result in the shift of HF OH band of about 100 cm^{-1} and its overlapping with LF OH band. When measuring the decrease in the intensity of 3050 cm^{-1} (proportional to the concentration of but-1-ene [4]) it has been found that butene oligomerization is a third order reaction.

Weeks at al [5] suggested that the Lewis acid sites (being the result of partial dehydroxylation of NaHY zeolite during its activation) are active sites in butene oligomerization. In the present research the experiments with pyridine adsorption on NaHY zeolites proved that the Lewis acid sites were absent in our samples. NaY zeolite is inactive in butene oligomerization. It can be therefore concluded that HF OH groups may also be active in butene oligomerization on NaHY zeolites. The rate of butene oligomerization increases distinctly with the degree of cation exchange of zeolite. In the case of NaHY/21, NaHY/40, and NaHY/77 zeolites : 20 hours, 2 hours and 15 min. were necessary to obtain the spectrum which did not undergoe any further transformation. It may be concluded that the concentration and acid strength of HF OH groups influence the oligomerization rate.

The product of butene oligomerization is characterized by the absence of unsaturated C=C, and =C-H stretching vibrations. These bands were

124

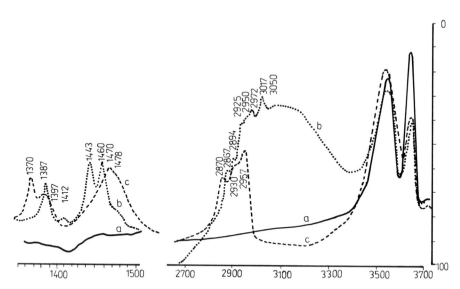

Fig. 1. Butene oligomerization on NaHY/77 zeolite.
a - freshly activated zeolite, b - but-1-ene adsorbed at room
temperature, spectrum recorded 5 min. after adsorption,
c - spectrum recorded 20 h after the adsorption

also absent in the product of propene [6,7] , and hexene [8] polimeri-
zation on NaHY zeolites. The dublet 1370, and 1387 cm^{-1} is characteristic

for isopropyl $\diagup C \diagdown$ $\begin{smallmatrix} CH_3 \\ CH_3 \end{smallmatrix}$ or isobutyl $-C \diagdown$ $\begin{smallmatrix} CH_3 \\ CH_3 \\ CH_3 \end{smallmatrix}$ group.

In the series of NaHY zeolites (investigated in this paper the same
bands are present in butene oligomer. The ratio of the intensity of
1370, and 1380 cm^{-1} bands increases with the exchange degree. According
to [9] this ratio increases as the degree of branching hydrocarbon molecule increases. It may be concluded that in the higher exchanged zeolites the butene oligomer is more branched.

The experiments with the adsorption of consecutive measured portions
of but-1-ene at 353 K were also carried out. In the case of NaHY/40,
and NaHY/77 zeolites resp. the portions of butene corresponding to
22.0 and 25.5 molecules per unit cell were introduced. Butene was
quickly transformed into oligomer, the band of HF OH groups at 3640
disappeared. The next adsorbed portions of but-1-ene isomerized and
but-2-enes appeared in the gas phase. They were very slowly transformed
into oligomer. It is therefore suggested that butene trimers are formed
at first in zeolite supercages. Finch and Clark [10] observed the formation of oligomer containing 2-5 butene molecules on silica-alumina
surfaces.

At temperatures higher than 370 K the product of butene oligomerization decomposes. Gaseous alkanes (mainly isobutane and isopentane) are formed and some polyene species remain on the zeolite indicating that oligomer decomposition is a disproportionation reaction. The band characteristic of polyene compounds (1585 cm^{-1}) was also observed when propene [12] , butene [11] , and hexene [13] remained in contact with NaHY zeolite.

Butene oligomerization is slow at room temperature, but it is fast at higher ones. When but-1-ene isomerization is studied in a dynamic reactor at the temperatures above 350 K zeolite supercage contain the oligomer species. The problem was discussed [11] whether the product of butene oligomerization may play the role of active sites in butene isomerization. In the present research this problem was studied by the IR spectroscopy. At first the butene portions were adsorbed up to the complete disappearing the HF OH band at 3640 cm^{-1} . Then large portion of but-1-ene was introduced and the rate constants of but-1-ene isomerization in the gas phase were calculated on the basis at the decrease of band of C=C- vibration of gaseous but-1-ene at 1645 cm^{-1} (Table 2).

TABLE 2

The catalytic properties of oligomer containing zeolites

Temperature K	Rate constant s^{-1} g^{-1}	
293	14.0	freshly activated zeolite
330	$3.0 \cdot 10^{-3}$	oligomer containing zeolite
354	$3.8 \cdot 10^{-2}$	oligomer containing zeolite
393	<50	oligomer partially decomposed

The same procedure has been also performed in the case of freshly activated NaHY/77 zeolite. In the case of oligomer containing zeolite at 330 K and 354 K the rate constants are much lower than that of non containing oligomer zeolite at 293 K. In order to elucidate why the catalytic activity of oligomer containing zeolite is so low, the but-1-ene was adsorbed at the room temperature on the oligomer containing zeolite. The remaining HF OH groups did not form the hydrogen bonding with butene molecules, indicating that they are inaccessible for adsorbed molecules. It may be the reason of very low activity of oligomer containing zeolite. At 393 K some partial decomposition of oligomer take place, some HF OH groups are already accessible for but-1-ene molecules and may catalyse the isomerization reaction. It should be mentioned that if butene isomerization is studied in traditional reactors above 350 K some zeolite supercages are blocked by oligomer and only a part of the HF OH groups may participate at the reaction.

ACKNOWLEDGEMENTS

The author wishes to express his thanks to Professor A. Bielański for helpful discussion.

REFERENCES

1. J. Datka, J.C.S. Faraday I, paper 9/1218.
2. J. Datka, J.C.S. Faraday I, paper 0/237.
3. J. Datka, J.C.S. Faraday I, paper 8/1476.
4. J. Datka, Bull. Acad. Polon. Sci., Ser. sci. chim., 24 (1976), 173.
5. T.J. Weeks, C.L. Angell, I.R. Ladd, A.B. Bolton, J. Catalysis, 33 (1974), 256.
6. L. Kubelková, J. Novaková, P. Jírů, Reaction Kinetics Catalysis Letters 4 (1976), 151.
7. J. Novaková, L. Kubelková, Z. Dolejsek, P. Jírů, Collection, in press.
8. P.E. Eberly Jr., J. Phys. Chem., 17 (1967), 1717.
9. L.J. Bellamy, "The Infra-red Spectra of Complex Molecules", London: Mathuen and COLTD, New York : John Willey and Sons INC.
10. A. Clark, J.N. Finch, Proc. Int. Congr. Catal. 4th Moscow, 2 (1971), 361.
11. P.A. Jacobs, L.J. Declerck, L.J. Vandamme, J.B. Uytterhoeven, J.C.S. Faraday I, 71 (1975), 1445.
12. P.A. Jacobs, H.E. Leeman, J.B. Uytterhoeven, J. Catalysis, 33 (1974), 17.
13. D. Eisenbach, E. Gallei, J. Catalysis, 56 (1979), 377.

B. Imelik *et al.* (Editors), *Catalysis by Zeolites*
© 1980 Elsevier Scientific Publishing Company, Amsterdam — Printed in The Netherlands

THE ISOMERISATION OF CYCLOPROPANE OVER MODIFIED ZEOLITE CATALYSTS

S.H. ABBAS, T.K. Al-DAWOOD, J.DWYER, F.R. FITCH, A. GEORGOPOULOS,
F.J. MACHADO AND S.M. SMYTH.
CHEMISTRY DEPARTMENT, UMIST, P.O. BOX 88, MANCHESTER U.K.

INTRODUCTION

In order to understand the function of zeolite catalysts in hydrocarbon transformations consideration has been given to several factors including (i) the structure and composition of the zeolite (ii) the acidity of catalytic sites (iii) the influence of cations and promoters, and (iv) the nature of the carbocation intermediates. Several approaches have also been made to the provision of a theoretical basis for carboniogenic activity. Recent extensive reviews cover much of this work (1).

Currently there is considerable interest in the properties of modified zeolites and in this paper we examine the dependence of kinetic parameters for the isomerisation of cyclopropane on zeolites modified by dealumination or ion-exchange. Results are examined in the context of theoretical models for acidity and catalytic activity.

EXPERIMENTAL

Faujasite - like zeolites were NaX (ex. BDH), NaY (ex. Crosfield Chemicals Ltd.) NaY and USY (ex. W.R. Grace Ltd.) Clinoptilolite was kindly provided by Dr. A. Dyer (Univ. Salford). NaY was dealuminated using EDTA(2), USY by steaming and acid leaching(3) and clinoptilolite by acid leaching (4). Ion exchange of NaX with Co^{2+} or Mn^{2+} was made from analar solutions of their chlorides. Nitrogen sorption was determined volumetrically and pyridine chemisorption by microbalance (Cahn). X-ray powder photographs were obtained with a Philips spectrometer. Kinetic measurements were made using a pulsed microreactor (5) modified to allow measurement of retention times on the catalyst. The catalysts were activated in a stream of pure dry helium and temperature was raised stepwise over a few hours to the final activation temperature (usually $400^{\circ}C$) where it was maintained for 16 hours. Results were based on runs with good carbon balances (98±2%). Experimental temperatures were selected randomly and the activation energies derived were found to be independent of the carrier gas flow rate. A surface rate constant (k) was calculated from

$$\ln(1/1-X) = k\tau$$

where X is the fractional conversion and τ the appropriate contact time. Sorption enthalpies ($\overline{\Delta H}$) were calculated from plots of $\ln V_g^{\ominus}$ against reciprocal temperature ($V_g^{\ominus}$ = corrected retention volume). Activation energy and sorption enthalpy are related by

$$E = E_a + (\overline{-\Delta H})$$ where E_a is the apparent activation energy.

DISCUSSION

The mechanism for isomerisation of cyclopropane over acidic catalysts is well established (6) (7) (8). Under the conditions of activation used in the present work reaction almost certainly proceeds via a carbocation intermediate.

$$CH_2 \overset{CH_2}{\diagdown} CH_2 + H^+ \overset{(i)}{\rightarrow} CH_2 \overset{CH_2}{\underset{H^+}{\diagdown}} CH_2 \overset{(ii)}{\rightarrow} [\overset{+}{C}H_2 - CH_2 - CH_2]$$

$$\downarrow (iii)$$

$$CH_2 = CHCH_3 + H^+$$

In the scheme all species are sorbed on the surface and step (i) is taken as rate determining (6) (7). The catalytic sites are presumed to be Bronsted sites and infrared evidence suggests that the hydroxyls giving rise to the band around 3650 cm^{-1} (O_1H) are active in faujasites (9).

According to Zhavoronkov (10) the activation energy for a process involving a carbocation intermediate should depend on the factors which stabilise the surface carbocation, the heat of formation of which is represented by

$$\Delta H = D_{OH} - P_A - e^2/r + \sum_i Z_i e^2/R \tag{1}$$

where D_{OH} is the heterolytic dissociation energy of the hydroxyl bond, P_A is the proton affinity of cyclopropane, e^2/r is the coulombic energy of the carbocation-zeolite pair and the summation represents the coulombic interaction between the carbocation and the cations present in the supercage. However, as pointed out by Jacobs (1) it is difficult to separate out effects associated with the parameters in equation (1). Table 1 gives data for NaX, NaHX, NaY, and NaHY. Values of surface activation energy (E) are generally higher for X than for Y and this does not arise solely from changes in sorption enthalpy ($-\overline{\Delta H}$).

Results for the HY zeolites (Fig. 1) prepared by dealumination of USY by steaming followed by acid extraction show that activation energies decrease and then level out and show some tendency to increase at high Si/Al ratio. These materials are of course not strictly equivalent to Y or X because of the dealumination process. The extent of "healing" is not known with certainty and there is a possibility that new active sites are produced during processing.

However, when all the results are considered (Fig. 3) it seems reasonable to accept that E depends on the ratio Si/Al and that E decreases to a value of Si/Al between about 5 and 8. Of course Fig. 3 shows effects other than Si/Al ratio. Different sources of zeolite are included, and the data of Habgood et al refer to a different activation temperature (500°). Moreover cations may be Na$^+$ or H$^+$. In Fig. 3(b) the same results are plotted against fraction of aluminium and show approximately linear dependence decreasing with decrease in Al/(Al+Si) up to a value of 0.13 after which they either do not decrease or start to increase with further decrease in aluminium fraction. Fig. 3(b) shows that there is considerable scatter about any "line" particularly at low values of Al/(Al+Si) (low Si/Al) which reflects both experimental errors and genuine differences in E at given levels of Si/Al. The implication is that such differences are greater at low Si/Al.

Fig. 3(a) and 3(b) may be compared with plots due to Barthomeuf and based on infrared studies (11). Data from Barthomeufs' paper are replotted in Fig. (4). Comparison of Fig. 3(a) and (4) suggest a correspondence but this of course does not prove that E depends on

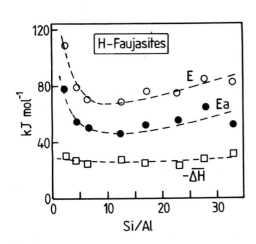

Fig. 1. Isomerisation of cyclopropane over HY zeolites. Kinetic parameters (21)

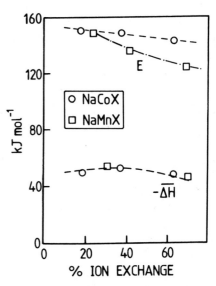

Fig. 2. Isomerisation of cyclopropane over NaMX zeolites

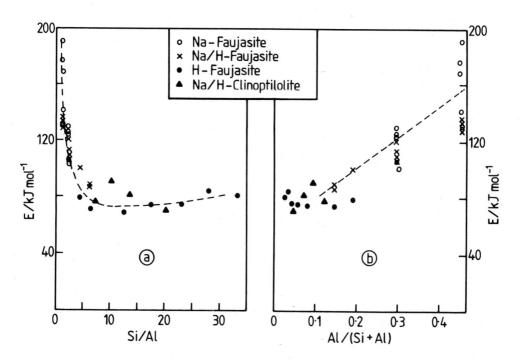

Fig. 3. Isomerisation of cyclopropane. Dependence of activation energy on zeolite composition. Data include results of Habgood et al (6) (7) and Wierzchowski et al (18).

v^2_{OH}. Further work using infrared and kinetic measurements on the same samples is required to confirm this suggestion. A correlation between v^2_{OH} and E would presumably imply that the energy for dissociation of the active hydroxyl is a significant factor in the stabilisation of the protonated cyclopropane carbocation. Incidentally Fig. 4 suggests that v^2_{OH} does not continue to decrease with increase in Si/Al ratio but levels out Si/Al around 5 or 6. Similarly for plots of $\bar{v}$ or $\bar{v}^2$ against aluminium fraction. Interest in the effect of aluminium content on the acid strength of zeolites was stimulated by the important work of Barthomeuf and Beaumont (12) who arrived at the following conclusions.

(i) Strong acidity in NaY zeolites appeared only after about 30% of the Na$^+$ ions were exchanged for H$^+$.

(ii) Dealumination removed only weak acid sites initially. Subsequently strong sites were progressively removed.

(iii) For X, and Y a quantity α, $(0 \leqslant \alpha \leqslant 1)$ defining the effective acidity of the protons was found to be a function only of aluminium content suggesting $\alpha = 1$ at Si/Al = 5.9±0.8. Subsequently Barthomeuf (11) represented H-zeolites as $TO_n(OH)_m$ where (T = Al or Si). By analogy with inorganic oxyacids acid strength was related to 'n' ($n \leqslant 2.0$) which implies a smooth increase in acid strength with dealumination (increasing 'n'). The major change in 'n', and consequently in acid strength takes place over the range of Si/Al from 1 to 6 after which 'n' increases only slowly with Si/Al.

Dempsey (13) suggested an explanation for the results of Barthomeuf and Beaumont based on the distribution of Al atoms in faujasites (accepting Lowenstein's rule). Strong acidity was associated with aluminium atoms having no other aluminiums as diagonal neighbours in 4 rings. The overall probability (1-P) of an aluminium not having diagonal neighbour aluminiums is plotted in Fig. 6 where it is compared with the experimental α values of Barthomeuf et al. The measure of agreement suggests that the average acid strength of protons is indeed related to the geometric distribution of the aluminium atoms. Again the major change occurs over the range of Si/Al 1 to 6 (almost linear in Al/Al+Si). Fig. 7 shows results of calculations using the model of Mikovsky and Marshall (14).

The number of Al atoms with 0, 0 or 1, 0, 1 or 2, or 0, 1,2, or 3 diagonal neighbouring aluminiums in 4- rings is given as a function of Si/Al. The number of aluminiums with no diagonal 4- ring aluminium neighbours increases up to a value of Si/Al ≈6 and thereafter descreases as Si/Al increases. Included in Fig. 7 are some experimental points for the total amount of pyridine retained at 150°C (after outgassing). Chemisorption of ammonia, which can react with protons in both the small pores and in the supercages, involves about 30 molecules in the HY zeolite unit cell (15). This might imply that curve 1 in Fig.7 represents the basis of chemisorption sites in the unit cell. Pyridine, however, is restricted to the large pores and in HY zeolites it seems that about 15-16 molecules are chemisorbed as Py$^+$ions at 150°C (16) (17) and that up to Si/Al = 4.2 there is little change (17). Our own work, on dealuminated USY suggests that as Si/Al ratio is increased beyond 6-7 the amount of chemisorbed pyridine decreases with Si/Al ratio. Apart from one very low point our results, at higher values of Si/Al are a little below those corresponding to curves 0 or 1 (in Fig.7) which coincide at higher Si/Al ratio.

A decrease in activation energy (E) for the isomerisation of butene over NaHY zeolites with increase in Si/Al ratio (from 2.67 to 4.23) was observed by Bielanski et al (17). The effect of Si/Al ratio on E was dependent on the degree of exchange of H$^+$ for Na$^+$. Both

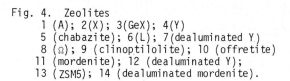

Fig. 4. Zeolites
   1 (A); 2(X); 3(GeX); 4(Y)
   5 (chabazite); 6(L); 7(dealuminated Y)
   8 (Ω); 9 (clinoptilolite); 10 (offretite)
   11 (mordenite); 12 (dealuminated Y);
   13 (ZSM5); 14 (dealuminated mordenite).

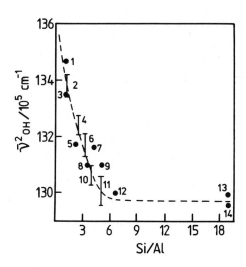

Fig. 4. Vibration frequency of acidic
         hydroxyls in zeolites

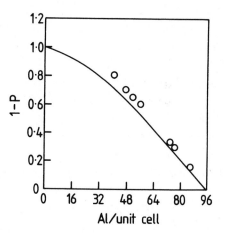

Fig. 6. Smooth curve is the overall
         probability of an Al atom in Faujasite
         not having diagonal neighbour aluminiums
         in 4 rings. Open circles are α values
         of Barthomeuf and Beaumont (12).

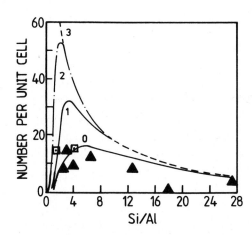

Fig. 7. Distribution of aluminium atoms
         (after Mikovsky and Marshall (14))
         smooth curves give the number of
         aluminium atoms with, 0; 0 or 1;
         0,1 or 2, and 0,1,2 or 3 diagonal
         neighbour aluminiums in 4 rings
   ▲ Pyridine chemisorbed at 150ºC
   ▣ Chemisorbed Pyridine at 150ºC
      on HY zeolites (16) (17)

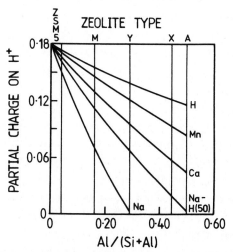

Fig. 5. Partial charges on zeolite protons
         (after Mortier (19)).

increase in Si/Al ratio and in extent of proton exchange led to a decrease in E.

Published work on the effect of exchange of Na^+ for H^+ on E for cyclopropane isomerisation does not support a major effect. Wierzchowski et al (18) reported an increase in E on substitution of H^+ for Na^+ in NaY and no effect in NaX until a level of exchange where structural damage was observed. However we have found for three samples of NaHY (Si/Al = 2.34) which had been 30% ammonium exchanged to remove weak acid sites that E was about 30 kJ mol^{-1} less than for the parent NaY, and a partially exchanged NaHX zeolite also gave E lower than the parent NaX. We have not yet established the detailed pattern of dependence of E on exchange of H^+ for Na^+ and further work will be published elsewhere, but it appears that E is reduced at any rate initially, on exchange of protons into NaX and NaY.

An effect on E due to exchange of H^+ for Na^+ is reasonable if exchange increases the protonic charge (and if this is a significant parameter in equation (1)). In practice it is difficult to calculate protonic charge or to relate it to bulk properties of the zeolite. However, an approach by Mortier (19), based on the electronegativity model of Sanderson is very useful in that relative protonic charges can be expressed quantitatively. Results of calculations, following Mortier are shown in Fig. 5. Protonic charge varies both with Si/Al ratio and with counter cation. The effect of exchange of H^+ for Na^+ is less at higher Si/Al ratio in agreement with the variation of E for butene isomerisation (17), and agrees with our own work on cyclopropane isomerisation.

Fig. 5 implies that the effect of cation exchange is greater for faujasites with low Si/Al ratio. Consequently NaX was used to measure the effect of transition-metal ion exchange on E. Results are shown in Fig. 2. For these particular catalysts E decreases (slightly) with increase in exchange level (as predicted by Zhavaronov (10) from equation (1)). Effects were not large and were assessed within each group (Co(II) or Mn(II)) using two regression models, one assuming that catalysts within a group have the same activation energy and the other that E depends on % ion exchange. Analysis of variance showed that the decrease in E with increase in % M(II) was significant for Mn(II) but not for Co(II). More work is required to clarify the effect of transition-metal ions in relation to type, method of exchange and extent of exchange. For both Mn(II) and Co(II) rates increased rapidly after about 30% exchange, presumably corresponding to the influence of supercage M(II) ions and rates were very dependent upon traces of water. Results were consistent with the production of protons by hydrolysis of coordinated water species.

Based on the few samples of dealuminated clinoptilolite it seems that values for E are similar to those for faujasite in the range of Si/Al from 7 to 20.

The model of Mortier which quantifies protonic charge seems to predict the general pattern for E over faujasites as a function of Al/(Al+Si) for values of this ratio greater than about 0.13. However, the implication that protonic charge should continue to increase as the composition tends to silica does not seem to be borne out by the pattern in E or by that for $\bar{v}_{OH}$. If the quantity (1-P) (Fig. 6) is taken as a measure of average acidity then this quantity does show a linear dependence on Al/Al+Si) at higher values and a decreased dependence at lower values. However experimental results for E or $\bar{v}_{OH}$ suggest a more abrupt departure from linarity than is evident in Fig. 6.

The pattern of variation in activation energy for cyclopropane isomerisation over dealuminated faujasites suggests that the acidity of the faujasites increases until the Si/Al ratio is about 5-7 and thereafter either remains constant or increases slightly.

TABLE 1

	Sample	Si/Al	Activation Temp/$^{\circ}$C	E/kJ/mol^{-1}	$-\overline{\Delta H}$/ kJ mol^{-1}	S.S.A/ m^2/g
a	NaX	1.16	500	192		
	NaX	1.19	500	142		
	NaX	1.22	500	169	44	
	NaY	2.31	500	103	40	
b	NaX	1.18	400	131		
	NaX	1.18	500	130		
	NaXH(20)	1.18	400	136		
	NaHX(20)	1.18	500	154		
	NaHX(39)	1.18	400	128		
	NaY	2.36	400	91		
	NaHY(5.5)	2.36	400	105		
	NaHY(12)	2.36	400	113		
	NaHY(56)	2.36	400	121		
c	NaY	2.43	400	126	40	890
	NaY	2.34	400	134	45	831
	NaHY(17)	2.34	400	100	39	
	NaY	2.46	400	123	48	
	USY	2.54	400	110	31	850
	Dealuminated	4.34	400	82	27	675
	USY	6.50	400	76	26	771
		12.60	400	75	28	625
		17.10	400	76	23	647
		23.0	400	75	21	595
		27.8	400	85	27	832
		33.1	400	81	30	570
d	EDTA	4.33	400	100	33	767
	treated	6.11	400	89	24	840
	NaY					

a) Ref. 6,7, b) Ref. 18, c) this work, d) back exchanged with Na$^+$

REFERENCES

(1) P.A. Jacobs, Carboniogenic Activity of Zeolites, Elsevier, Amsterdam, 1977.
(2) G.T. Kerr, e.g. Adv. Chem. Ser., 121 (1973) 220.
(3) J. Scherzer and J.L. Bass, J. Catalysis, 54 (1978) 285.
(4) R.M. Barrer and M.B. Makki, Can. J. Chem., 42 (1964) 1481.
(5) R.F. Benn, J. Dwyer, A.A.A. Esfahani, N.P. Evmerides and A.K. Szczepura, J. Catalysis, 48 (1977) 60.
(6) D.W. Bassett and H.W. Habgood, J. Phys. Chem., 64 (1960) 769.
(7) B.H. Bartley, H.W. Habgood and Z.M. George, J. Phys. Chem., 72 (1968) 1689.
(8) J.W. Hightower and W.K. Hall, J.A.C.S., 90 (1968) 851.
(9) Nguyen The Tan, R.P. Cooney and G. Curthoys, J. Catalysis, 44 (1976) 81.
(10) M.N. Zhavoronkov, Kin. i Kataliz, 14(2) (1973) 521.
(11) D. Barthomeuf, J. Phys. Chem., 85 (1979) 249.
(12) R. Beaumont, D. Barthomeuf, J. Catalysis, 26 (1972) 218; 27 (1972) 45.
(13) E. Dempsey, J. Catalysis, 33 (1974) 497; 39 (1975) 155; 49 (1977) 115.
(14) R.J. Mikovsky and J.F. Marshall, J. Catalysis, 44 (1976) 170; 49 (1977) 120.
(15) A. Auroux, V.Bolis,P. Wierzchowski, P.C. Gravelle, J.C. Vedrine, J.C.S. Far. I, 75 (1979), 2544.
(16) P.A. Jacobs, L.J. Declerck, L.J. Vandamme and J.B. Uytterhoeven, J. Chem. Soc. Faraday I, 71, (1975) 1545.
(17) A. Bielanski, J. Dakta, A. Drelinkiewicz and A. Malecka, "Proceedings Int. Conf. Zeolites", Szeged (Hungary), (1978) 89.
(18) P.T. Wierzchowski, S. Malinowski, and S. Krzyzanowski, La Chim. e L'Ind., 59 (1977) 612.
(19) W.J. Mortier, J. Catalysis, 55 (1978) 138.
(20) P.A. Jacobs, W.J. Mortier, J.B. Uytterhoven, J. Inorg. Nucl.Chem., 40 (1978) 1919.
(21) S. Smyth, Ph.D. Thesis, U.M.I.S.T., in preparation.

B. Imelik *et al.* (Editors), *Catalysis by Zeolites*
© 1980 Elsevier Scientific Publishing Company, Amsterdam — Printed in The Netherlands

POISONING OF ACIDIC CENTRES IN ZEOLITES WITH SODIUM AZIDE

P. FEJES, I. KIRICSI, I. HANNUS, T. TIHANYI and A. KISS
Department of Applied Chemistry, Jozsef Attila University, Szeged, Hungary

ABSTRACT

Zeolite acid site catalysis was reduced or even eliminated with sodium vapour. Under appropriately selected conditions the remaining carboniogenic activity was zero. The same method could be successfully applied to influence bifunctional catalysis.

1   INTRODUCTION

Owing to their unique properties, zeolites have turned out to be among the most interesting catalysts and catalyst base materials ; they are responsible for many advances that have been achieved in the petroleum refining industry and several other fields of industrial organic syntheses over the past two decades (1, 2, 3) . Metal-containing catalysts, where the metal is deposited upon acidic supports or is reduced from the respective ion in an acidic matrix (e.g. zeolite), exhibit bifunctional activity. While the metal behaves as a hydrogenating-dehydrogenating agent, the acidic component is responsible for the oligomerization-isomerization-cracking activity in hydrocarbon transformations. In the case of zeolites, acidity develops in accordance with stoichiometry simultaneously with the reduction of the metal ions, provided dihydrogen is used for this purpose. Mitigation or in some cases complete elimination of the acid catalysis is a prerequisite of hydrogenation under "clean conditions" over these catalysts. Numerous papers have been devoted to this question.

Acidic action can be greatly reduced, or even eliminated, simply by ion exchange in aqueous media (4) or by adsorption of gaseous nitrogen bases (5, 6, 7). Rabo et al. (8) used anhydrous alkali metal halides for the elimination of zeolitic H^+ at about 573 K. While all these methods are detrimental for the metal component, treatment of the reduced catalyst with Na vapour has been shown to be an efficient procedure which can be selectively applied to the acid sites with full retention of the hydrogenating activity (9) .

This paper is concerned with selective poisoning and its effects on the carboniogenic activities of NaY catalysts in the skeletal isomerization of cyclopropane, and with selectivity changes in the bifunctional action of a nickel-on-zeolite specimen, also caused by selective poisoning with sodium vapour.

2   EXPERIMENTAL

In the experiments two NaY base zeolites were used : a commercial NaY sample[*](Carbon and Carbide Co. Linde Division ; type SK-40 ; lot No. 23606-224 ; this will subsequently be referred to as sample No. 1), and a sample synthesized by L. I. Piguzova[*](sample No. 2).

* Donated by the Carbon and Carbide Co., Tarrytown, USA
* Supplied by L. I. Piguzova (Inst. Neft. Prom., Moscow, USSR),

Catalysts No. 3-8 were obtained from No. 1 by adding various amounts of NaN_3. (A detailed description of the procedure is to be found in (9). Exchange of the Na^+ (degree of exchange : 70 %) for NH_4^+ ions resulted in sample No. 9. Sample No. 10 was obtained by admixing 10 mass % of NaN_3 to sample No. 9. The base material for the bifunctional Ni^0HNaY catalysts was sample No. 2 ; a proportion of the Na^+ ions were exchanged for Ni^{2+} (degree of exchange : 32.5 %) and various amounts of solid NaN_3 were added (samples Nos. 12-16). Catalyst compositions are listed in Table 1.

TABLE 1
Unit cell compositions of the base zeolite samples used

No.	Components					
	Na	NH_4	Ni	Al	Si	O
1	53.5	-	-	60.9	131.1	384
2	58.1	-	-	60.7	131.3	384
9	20.2	39.8	-	60.9	131.1	384
11	39.4	-	9.6	60.7	131.3	384

Sodium azide contents of the modified NaY (No. 1) zeolite samples

No.	3	4	5	6	7	8
NaN_3 content in w%	0.69	1.83	4.17	6.76	13.1	23.4

Sodium azide contents of the modified NiNaY (No. 11) zeolite samples

No.	12	13	14	15	16
$\dfrac{mol\ NaN_3}{mol\ Na}$	0.21	0.86	2.14	4.3	23.8

The nickel-loaded catalyst samples were prepared by reduction with dihydrogen of the samples Nos. 12-16 at 600 K (the decomposition temperature of NaN_3 is 623 K). The degree of reduction of the Ni^{2+} ions after two hours was found to be 0.147, as determined by the method of Bremer (10).

Decomposition of the NaN_3 into Na and N_2 (see later) was accomplished at 723 K by gradually increasing and then (after attainment of 723 K) maintaining the temperature at the same level for a further two hours during continous pumping of the system.

For investigation of the kinetics of the catalytic hydrocarbon reactions, a tank reactor of constant material content was used with intensive gas circulation for GC analysis[*] (see in detail under (9)).

For the IR spectroscopic investigations self-supporting wafers were prepared from

[*] The GC used (Hewlett Packard type : 5710 A) was a gift of the Alexander von Humboldt Foundation.

samples Nos. 9 and 10 (see under (11)). The spectra were recorded using a Pye Unicam
SP-200 IR spectrophotometer with the vacuum infrared cell described by Beyer et al.(12).
A Hungarian-made Q-derivatograph and a Russian Dron-1 X-ray diffractometer were used for
further characterization of the catalyst samples.

3 RESULTS AND THEIR DISCUSSION

3.1 Preparation of catalytically inactive NaY by poisoning the Brönsted centres with sodium vapour

Brönsted acid sites in the zeolite framework can be eliminated by reaction with sodium
vapour. Instead of the use of a sodium source under high vacuum conditions, the reaction
leading to poisoning can be carried out more conveniently by decomposing NaN_3 into Na and
N_2 in more or less intimate mixtures of zeolite and various amounts of NaN_3 at elevated
temperatures. The decomposition temperature depends on the concentration of heavy metal
ions and the physicochemical characteristics of the medium in which the reaction takes
place. In the zeolitic framework decomposition occurs at around 683 K, which is 60 degrees
higher than under ordinary conditions (9). Furthermore, depending on the amount of NaN_3
added, part of the azide does not decompose even at still higher temperatures. (In the case
of sample No. 8 a NaN_3 residue of 0.12 mol kg^{-1} could be found in the zeolite after an
8-hr treatment at 450 K.) It could be shown by IR spectroscopy and X-ray diffraction that
this residual azide content can not be ascribed to solid crystalline NaN_3. It behaves instead like the well-known "salt occlusions" described in detail by Rabo (13).

The catalytic activities of the NaY zeolite and the poisoned samples in the skeletal isomerization of cyclopropane were characterised by first-order rate constants.

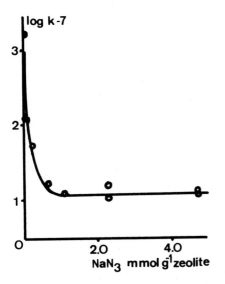

Fig. 1 Catalytic activity of the zeolite as a function of the initial NaN_3 content.

As can be seen in Fig. 1, in the case of the poisoned catalysts the decrease of activity is
considerable, and for sample with an initial NaN_3 content equal to or greater than

1.12 mol kg^{-1} the remaining activity is zero. The very slow residual reaction corresponds to the unimolecular homogeneous decomposition of cyclopropane described in the literature (14, 15).

Even though the duration of interaction between Na atoms and H$^+$ could not have been longer than a few seconds at most, it appeared to be very effective, leading to reduction and finally to complete elimination of the active acid sites, presumably by the reactions

$$\{AlO_2^-\}\ H^+ + Na \rightarrow \{AlO_2^-\}\ Na^+ + H \qquad 1$$

and

$$\{AlO_2^-\}\ H^+ + NaOH \rightarrow \{AlO_2^-\}Na^+ + H_2O \qquad 2$$

Traces of NaOH could be produced in the reaction of Na with the remaining water content of the catalyst sample.

As far as reactions 1 and 2 are concerned, the results of the catalytic investigations are corroborated by direct infrared spectroscopy.

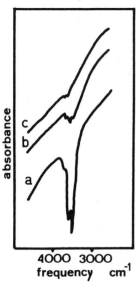

Fig. 2 IR spectra of the zeolite samples No. 9 (HNaY) (a) and No. 10 (containing 10 w % solid NaN$_3$) after heat treatment at 643 K (b) and 773 K (c) in vacuo.

In the spectrum of the HNaY sample (No. 9), which did not contain added NaN$_3$, the LF and HF OH bands are clearly discernible (spectrum a). Spectra b and c were recorded after heat treatment in vacuo at 643 and 773 K, respectively, of the HNaY samples with added NaN$_3$. The reduction of the band intensities corresponds to expectations.

3.2 <u>The effect of sodium poisoning on the selectivity of a Ni-loaded HNaY zeolite catalyst</u>

On the basis of the results obtained in connection with the skeletal isomerization of cyclopropane catalysed by Brönsted acid sites, it was assumed that bifunctional catalyst selectivity could also be influenced by selective poisoning with Na. As a test substance propylene was chosen ; in the presence of dihydrogen over Ni metal, this undergoes hydrogenation-dehydrogenation reactions and complex carboniogenic transformations, which will be called oligomerization-isomerization-cracking (briefly oic reactions) for the sake of convenience.

The kinetic investigations were carried out below 500 K because above this limit the products were too complex and could not be separated and ordered, respectively, into the theoretically possible reaction sequences. Nevertheless, besides hydrogenation and the oic reactions, hydroisomerization and hydrocracking of the paraffinic components could be detected. These latter transformations are practically absent below 500 K.

In the case of poisoned catalysts, only hydrogenation occurs. Under the same conditions, only the complex oic reactions can be observed over NaHY catalysts 16 . The Ni^0HNaY zeolite samples show both activities to varying extents, depending on the experimental conditions.

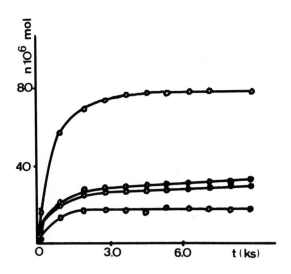

Fig. 3 The products obtained by the bifunctional action of a Ni^0HNaY zeolite catalyst from a 1:1 propylene/dihydrogen mixture ;   propane,   2-methylpropane,   2-methylbutane,   2-methylpentane.

As seen in Fig. 3, the main reaction products are propane (by the hydrogenation of propylene), and various 2-methyl-substituted paraffins (2-methylpropane, 2-methylbutane, 2-methylpentane).

As a first approximation, let the extent of acid catalysis be characterized by the sum of products other than propane,* n_{olig}, produced in the first 6 ks of the experiment, and similarly let the amount of propane, n_{prop}, be regarded as the extent of hydrogenation.

* In separate experiments, which have been omitted here, it has been shown that the amount of propane from oligomeric products does not influence the present deductions.

140

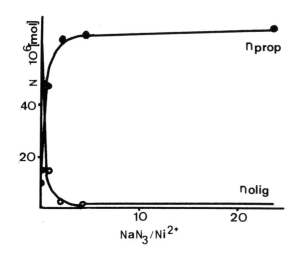

Fig. 4 Change of the acidic (n_{olig}) and hydrogenation (n_{prop}) activities of the poisoned Ni^0HNaY zeolites vs degree of poisoning (initial amount of propylene $584.4 \cdot 10^{-6}$ mol ; reaction temperature 473 K ; reactor volume $65 \cdot 10^{-6}$ m^3 ; volume of the whole system $221.9 \cdot 10^{-6}$ m^3 ; propylene/dihydrogen ratio = 1.

In fig. 4, these quantities are shown against the degree of poisoning (measured as the amount of NaN_3 added to the catalyst). The oligomerization activity is reduced to a great extent ; however, the hydrogenation activity remains uninfluenced, demonstrating the selective character of this poisoning, as referred to in the introduction.

Therefore, it can be concluded that sodium vapour can be selectively applied to poison the carboniogenic activity of a bifunctional catalyst.

REFERENCES

1   P.A. Jacobs, "Carboniogenic Activity of Zeolites", Elsevier Sci. Publ. Co., Amsterdam, 1977
2   J.A. Rabo, "Zeolite Chemistry and Catalysis", A.C.S. Washington, 1976
3   G.K. Boreskov, Kh. M. Minachev, "Application of Zeolites in Catalysis" Akadémiai Kiado, Budapest, 1979
4   C.P. Huang, J.T. Richardson, J. Catal., 57, (1977), 332
5   M.S. Goldstein, in "Experimental Methods in Catalysis Research" (Editor : R.B. Anderson Academic Press, New York, 1968, p. 361
6   K. Tanabe, "Solid Acids and Bases" Academic Press, New York, 1970
7   P. Forni, Catal. Rev., 3, (1973), 69
8   J.A. Rabo, M.L. Poutsma, G.W. Skeels, Proc. Inter. Congr. Catal. 5th, North Holland, Publ. Co., Amsterdam, 1973, p. 1353.
9   P. Fejes, I. Hannus, I. Kiricsi, K. Varga, Symposium on Zeolites, zeged, Hungary, 1978 Acta Phys. Chem. Szeged, 24, (1178), 119
10 H. Bremer, K.H. Bager, F. Vogt, Z. Chem., 14 (1974), 199
11 I. Hannus, I. Kiricsi, K. Varga, P. Fejes, React. Kinet. Catal. Lett., 12, (1979), 309
12 H.K. Beyer, R. Krisch, J. Mihalyfi, React. Kinet. Catal. Lett., 8, (1978), 317
13 J.A. Rabo, "Zeolite Chemistry and Catalysis", (Edited by J.A. Rabo), ACS, Washington, 1976, Ch. 5, p. 332
14 T.S. Chambers, G.K. Kistiakowaky, J. Am. Chem. Soc., 56, (1974), 399
L5 I. Pritchard, P..Sowden, A.F. Trotman-Dickenson, Proc. Roy Soc.,,217, (1953), 563
16 I. Kiricsi, Magy. Kém. Lap., 32, (1977), 605

B. Imelik *et al.* (Editors), *Catalysis by Zeolites*
© 1980 Elsevier Scientific Publishing Company, Amsterdam — Printed in The Netherlands

ISOMERIZATION REACTIONS OF SUBSTITUTED CYCLOPROPANES OVER ZEOLITE OMEGA

by

H. FRANK LEACH AND CHRISTINE E. MARSDEN[*]

(Chemistry Department, Edinburgh University, West Mains Road, Edinburgh, Scotland, EH9 3JJ)

ABSTRACT. The isomerization reactions of methylcyclopropane,
1,1-dimethylcyclopropane, and 1,2-dimethylcyclopropane have been
studied in a static system over an omega zeolite pretreated at
various temperatures. The maximum catalytic activity of the zeolite
occurred at a pretreatment temperature which corresponded to the
generation of maximum surface area, and was closely related to the
conditions under which tetramethylammonium species encapsulated
within the initial zeolite during synthesis were decomposed.

   The kinetics of the reactions of the substituted cyclopropanes
were first order in the temperature range 230-330 K. Initial product
distributions indicated kinetic rather than thermodynamic control, and
have been rationalized in terms of a proton addition mechanism.

INTRODUCTION

   Cyclopropane and its methyl-substituted derivatives are stable molecules which can
thermally and catalytically isomerise with opening of the cyclopropane ring to give a
range of products. The ring opening of cyclopropane itself yields the single product
(propene) - the rate at which this reaction occurs over a particular catalyst can be used
as a measure of catalytic activity. With the methyl- and dimethyl-substituted molecules
however ring opening will produce a distribution of products. The nature of this
distribution is characteristic of the operative reaction mechanism, and consequently a
study of such molecules can provide important evidence to substantiate mechanistic theories.

   Roberts [1] examined cyclopropane isomerization over a range of acidic catalysts. The
reaction was poisoned by basic molecules and it was concluded that the principal reaction
intermediate was an n-propyl carbonium ion formed by attachment of a catalyst proton to a
ring carbon atom. Over silica-alumina tracer techniques were used to establish that intermolecular exchange of one hydrogen atom took place during cyclopropane isomerization [2].
This is consistent with either (a) the protonic mechanism proposed by Roberts [1],
requiring a $C_3H_7^+$ (Bronsted site) surface intermediate, or (b) a bimolecular hydride
transfer mechanism involving a $C_3H_5^+$ (Lewis type) surface intermediate. In further
studies Hall et al. [3] [4] examined the isomerization of methyl-, ethyl- and dimethyl-
cyclopropane over silica-alumina. From the observed product selectivities they were able
to conclude that a protonic type of mechanism was operative, and they invoked the non-
classical cyclopropyl carbonium ion as the surface complex. To explain the product
selectivity observed in the hydrogen chloride catalysed isomerization of 1,1-dimethyl-
cyclopropane Bullivant et al. [5] postulated the simultaneous occurrence of two processes
i.e. (i) a unimolecular isomerization giving 2-methylbut-2-ene and 3-methylbut-1-ene, and

* Now at I.C.I. (Mond Division) Ltd., Runcorn, Cheshire, England.

(ii) a bimolecular HCl catalysed process involving a six-centred transition state and yielding 2-methylbut-1-ene. More recently [6] it was reported that the major product of cyclopropane isomerisation over hydrogen Y zeolite was isobutene - the formation of a non-classical carbonium ion was invoked to rationalize this product distribution. This non-classical species was also suggested as the intermediate in exchange and isomerisation of cyclopropane over deuterated X and Y zeolites [7] [8].

Zeolite omega is a synthetic material with, as yet, no known natural analogue. The synthesis was first patented by Flanigen et al. [9] and an essential feature is the presence of tetramethylammonium (TMA) cations. The amount of TMA cations present in the parent gel is believed to control both the silica/alumina ratio and the proportion of TMA contained in the resultant zeolite. The structure of the aluminosilicate framework of zeolite omega was determined by Barrer and Villiger [10], and consists of an hexagonal type structure composed of gmelinite cages (14-hedra) linked both laterally and longitudinally. The longitudinal linkage occurs via the sharing of six-membered ring faces and produces long columns of gmelinite cages parallel to the c-axis. The lateral joining of six such columns generates the main channels of the zeolite - these run parallel to the c-axis, are circumscribed by 12-membered ring windows, and have a free diameter ($\sim$0.75 nm) which allows the ready sorption of large molecules such as cyclohexane.

A typical zeolite omega (Ω) sample contains approximately eight cations per unit cell - six being sodium and two TMA. A pyrolytic study [11] produced evidence of two different TMA cationic sites. i.e. (i) in unrestricted main channel locations, and (ii) within the restricted gmelinite cages. The bulk of the TMA was believed to be in the latter type of location, and thermal treatment at 773-873 K (in both oxidising and inert environments) was necessary in order to remove such species. Weeks et al. [12] examined the calcination of ammonium exchanged Ω - and followed the decomposition of both the ammonium and TMA cations. They observed that TMA cation decomposition generated hydroxyl groups directly, a similar observation to that reported by Wu et al. [13] for TMA - offrelite.

EXPERIMENTAL

The zeolite omega (Ω) sample used in this study was kindly provided by Shell Development Company, and was similar to the starting material used in the work reported by Cole et al. [11]. X-ray fluorescence spectroscopy and thermogravimetric analysis suggested that the unit cell composition approximated to $Na_{6.3}$ $TMA_{1.7}$ $(AlO_2)_8(SiO_2)_{28}.21$ H_2O with a Si/Al ratio of 3.53, within the range indicated by Breck [14].

Methylcyclopropane (99% pure) was supplied by K. and K. Laboratories, and 1,1-dimethyl-cyclopropane (99% pure) and trans-1,2-dimethylcyclopropane (99% pure) by Pfaltz and Bauer. The reagents were all purified and outgassed by vacuum distillation and repeated cycles of freezing, pumping and thawing. They were stored in cold fingers fitted with Rotaflo teflon taps until required.

Isomerization reactions were followed in a static system with an appropriate pressure of reactant (typically 10^{20} molecules) admitted to the reaction vessel containing the zeolite omega (approximately 0.1 g) after outgassing. The gas phase composition was analysed periodically by passage of a sample through a gas chromatographic analytical arrangement, and the extent of reaction thus measured as a function of time.

Surface area measurements (using nitrogen at liquid nitrogen temperature) and adsorption experiments were undertaken with a Cahn RG Automatic Electrobalance - with a maximum sensitivity of $\pm 1 \times 10^{-6}$ g.

RESULTS AND DISCUSSION

Preliminary. In a preliminary study to establish the pretreatment conditions that would generate optimum catalytic activity the surface area of omega was measured as a function of outgassing temperature. As shown in Figure 1 there is a maximum surface area in the outgassing temperature range 830-870 K. The data shown are for a period of 16 hours outgassing - if the period is extended to 40 hours then further small increases in surface area (to ∿295 m^2 g^{-1}) were noted. These surface area changes can be related to the fact that zeolite dehydration required temperatures of 623-673 K, and the decomposition of TMA has been reported [12] to require temperatures greater than 783 K. It seems clear that the

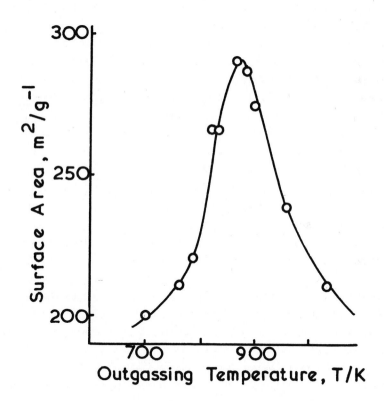

FIGURE 1. Variation of surface area of omega zeolite with outgassing temperature.

decomposition and removal of the bulky TMA cations from within the gmelinite cages is a necessary prerequisite of obtaining the maximum surface area.

The pretreated omega samples displayed no significant catalytic activity for the reactions under study at pretreatment temperatures below 723 K. However overnight outgassing (for 16 hours) at temperatures above 800 K did produce active catalysts, and extended outgassing up to 40 hours enhanced this activity. The activity versus pretreatment temperature displayed a maximum at circa. 850 K in a similar manner to the surface area behaviour, and at higher temperatures the activity began to decrease again. The omega samples that were outgassed under vacuum became grey in appearance (cf. to their initial white colour) which was attributed to some carbonaceous residue from the TMA decomposition. If an oxygen treatment was incorporated into the pretreatment (with exposure to two doses of oxygen – 5×10^{20} molecules – for 30 minute intervals at outgassing temperature) the catalysts retained their initial white colour. Furthermore the maximum catalytic activity of such pretreated samples was achieved after only 20 hours, and this activity was similar to that for the samples only given thermal pretreatment.

The catalytic results over the omega samples pretreated over the range of temperatures gave similar product ratios and activation energies which suggest that the same reaction mechanism is operative – with the variation in activity being a direct consequence of a change in the number of catalytically active sites. Adsorption experiments with the n-butenes and methylcyclopropane showed that equilibrium was readily established, and indicated that the molecules did not enter the gmelinite cages. Thus the internal surface area available to these molecules is not dependent upon the presence or absence of TMA ions within such cages. If total zeolite surface area was the important factor then oxygen treated samples might be expected to exhibit more catalytic activity than thermally treated samples with definite carbon deposition. It seems probable that the variation in activity resulting from the TMA cation removal is caused by the generation of catalytically active sites during the decomposition of these ions. Wu et al. [13] reported evidence for the generation of acidic structural hydroxyl groups from the decomposition of TMA-offretite, and Weeks et al. [12] have reported the appearance of similar species from the calcination of omega. In the present study the observed relationship between catalytic activity and TMA removal strongly suggests that a significant number of such active sites must of necessity be generated in positions accessible to the reactant molecules i.e. in the main channels of the structure rather than within the gmelinite cages.

Methylcyclopropane. The isomerization reactions of methylcyclopropane were studied over omega zeolites pretreated at temperatures above 800 K – and occurred at conveniently measurable rates within the temperature range 295 K → 335 K. Figure 2 shows a typical reaction profile for the reaction at 325 K. There is no iso-butene present in the products and but-1-ene is also present in excess of cis-but-2-ene. i.e. the distribution of products is kinetically, rather than thermodynamically, controlled. The disappearance of methylcyclopropane closely follows first order kinetics from the beginning of the reaction and indeed good first order kinetics are observed in all experiments up to 90% conversion of the initial reactant. Table 1 lists the observed rate constants and Arrhenius

TABLE 1. Rate Constants and Arrhenius Parameters for MCP Disappearance over Zeolite Omega

Outgassing Temp/K	Outgassing Time/hrs	Activation Energy/kJ mol^{-1}	lnA	Reaction Temp/K	$k \times 10^{-4}$ s^{-1} g^{-1}
823	13	50.9	12.5	295	2.77
				308	5.85
				325	17.48
				328	20.75
				335	33.42
823	>40	51.7	14.35	306	24.17
				318	59.67
				328	95.0
				332	119.17

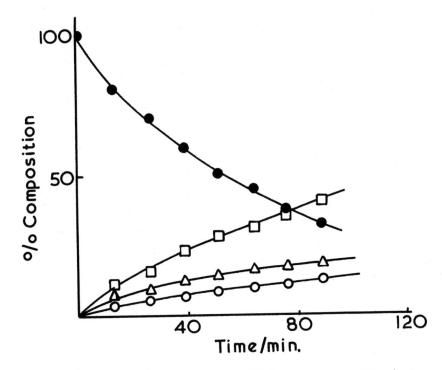

FIGURE 2. Reaction of methylcyclopropane at 325 K over omega zeolite (outgassed at
823 K). (● ≡ methylcyclopropane, □ ≡ trans but-2-ene,
△ ≡ but-1-ene and   O ≡ cis but-2-ene).

parameters for omega samples outgassed at 823 K for (i) 13 hours and (ii) more than
40 hours. The activation energy was the same in both cases with the enhanced activity of
the latter series resulting from the increase in the pre-exponential factor.

The initial product distribution for the MCP reaction was in all instances in the
order trans but-2-ene>but-1-ene>cis but-2-ene, with initial product ratios very different
from equilibrium thermodynamic values. If experiments were carried on for an extended
period the subsequent isomerisation of but-1-ene, giving cis but-2-ene and trans but-2-ene
as products, could be observed. This secondary rate of disappearance of but-1-ene
followed first order kinetics, and comparison of the rate constants showed the MCP
isomerization reaction to be some five times faster than the but-1-ene isomerization
reaction within the observed temperature range. The rates of but-1-ene conversion after
the initial MCP reaction were found to be identical to those reported for reactions in
which but-1-ene was the initial reactant [15].

The absence of isobutene (and cyclobutane) from the observed products of MCP
isomerization is indicative of a proton addition mechanism. A process of this type
involving the secondary butyl carbonium ion as intermediate has previously been postulated
for the interconversion of the n-butenes over zeolite Ω. Such an intermediate seems less
likely in the MCP case however as the product distribution (trans but-2-ene>but-1-ene>
cis but-2-ene) is different in this instance. Differences in product distribution between
n-butene and MCP isomerization reactions of this type have previously been reported [4]

over silica-alumina. Hightower and Hall postulated the non-classical cyclopropyl carbonium
ion as the reaction intermediate for the MCP isomerization. In the present work over
zeolite omega the product distributions are compatible with the involvement of this non
classical ion, and Figure 3 illustrates a possible reaction mechanism.

But-2-enes   But-l-ene

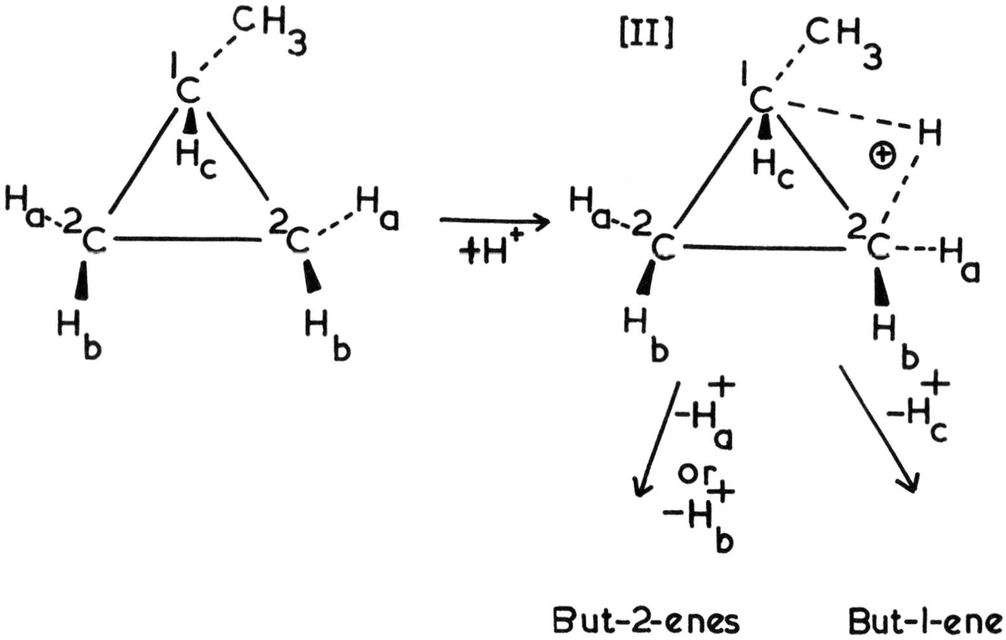

FIGURE 3.   Possible reaction mechanism for isomerization of methylcyclopropane – with
            proton addition giving the non classical carbonium ion (II) as reaction
            intermediate.

It can be seen that cleavage of the C_1-C_2 carbon-carbon bond in the non-classical ion
(II) can occur in two ways. i.e. (i) resulting in the formation of a ^{1}C-H bond, the
cleavage of a ^{1}C-H$_c$ bond and the production of but-1-ene. (ii) resulting in the formation
of a ^{2}C-H bond, the cleavage of either ^{2}C-H$_a$ or ^{2}C-H$_b$, and the production of cis but-2-ene
and trans but-2-ene respectively. If the non classical ion is adsorbed in a plane
parallel to the cyclopropane ring then the production of trans but-2-ene and but-1-ene
would be favoured from steric considerations. Their formation is dependent upon the
abstraction of a proton (H$_b$ or H$_c$) orientated towards the surface. In contrast the
formation of cis but-2-ene is dependent upon the loss of H$_a$ which is orientated away from
the surface – abstraction by the surface is clearly more difficult in this case.

Dimethylcyclopropanes. The isomerization of trans 1,2-dimethylcyclopropane over omega
was followed over a rather lower temperature range (229 K – 293 K) than that required for
the MCP study. The reaction profile is illustrated in Figure 4, from which it can be seen
that all six possible C_6 alkenes are formed although only 3-methyl-but-1-ene and 2-methyl-
but-2-ene in very significant amounts. The pentene products were only formed in small
amounts and could not be accurately determined at low conversions – even after all the
1,2-dimethylcyclopropane had been converted the products contained methyl butenes and
pentenes in the approximate ratio of 9:1. After an initial acceleratory period the rate

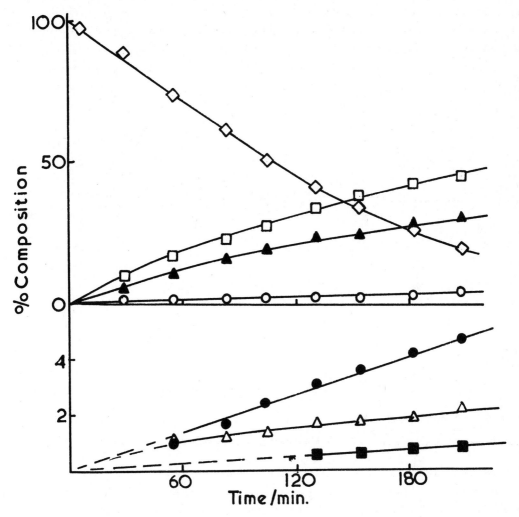

FIGURE 4. Reaction of 1,2-dimethylcyclopropane at 280 K over omega zeolite.
($\Diamond$ ≡ 1,2-dimethylcyclopropane, $\square$ ≡ 3-methylbut-1-ene,
$\blacktriangle$ ≡ 2-methylbut-2-ene, $\bigcirc$ ≡ 2-methylbut-1-ene, $\bullet$ ≡ trans pent-2-ene,
$\triangle$ ≡ pent-1-ene and $\blacksquare$ ≡ cis pent-2-ene).

of 1,2-dimethylcyclopropane disappearance followed first order kinetics closely. Table 2 lists the concise experimental results and Arrhenius parameters.

TABLE 2. Rate Constants and Arrhenius Parameters for 1,2DMCP Isomerization over Omega

Reaction Temp/K	$k \times 10^{-4}$ $s^{-1} g^{-1}$	lnA	Activation Energy/kJ mol^{-1}
229	0.05		
242	0.25		
247	0.40	19.4	60.5
261	2.3		
280	9.61		
293	63.8		

As in the MCP case the product distribution was kinetically, rather than thermodynamically, controlled. If the reaction was followed for an extended period the subsequent and secondary isomerization reaction of 2-methylbut-2-ene (to yield 3-methylbut-1-ene and 2-methylbut-1-ene) could be observed. The product distribution observed was far removed from that predicted for a hydride abstraction mechanism and in fact is compatible with a proton addition process. It seems likely therefore that a similar reaction mechanism to that invoked for the MCP reaction i.e. with the non-classical carbonium ion intermediate, is operative in the 1,2-dimethylcyclopropane case. However in this instance the actual product composition can be well correlated with that predicted from a classical proton addition ring-opening reaction. The relative amounts of the methyl butenes and pentenes can be rationalized in terms of steric and energetic considerations. The trace of 2-methylbut-1-ene formation can be attributed to the involvement of a primary carbonium ion as intermediate whereas the other methyl butene species would be formed via secondary carbonium ion intermediates.

The isomerization reaction of 1,1-dimethylcyclopropane was the fastest of the three reactions examined (at any fixed temperature). It was too fast to accurately measure at room temperature, and results were obtained over the temperature range 213-280 K. The major product of the reaction was in this case 2-methylbut-2-ene, with substantial amounts also of 3-methylbut-1-ene and a small (less than 5% after two hours) but significant amount of 2-methylbut-1-ene. There were no detectable traces of pentene formation in any of the experiments. The rate of 1,1-dimethylcyclopropane disappearance followed first order kinetics, and Table 3 lists the rate constants and Arrhenius parameters.

TABLE 3. Rate Constants and Arrhenius Parameters for Isomerization of 1,1-Dimethyl-cyclopropane over Zeolite Omega

Reaction Temperature/K	$k \times 10^{-4}$ $s^{-1} g^{-1}$	Activation Energy/kJ mol^{-1}	lnA
213	1.07		
228	3.76		
238	10.23	35.1	10.7
250	19.03		
273	85.2		
280	107.3		

As in the previous reactions the product distribution is not thermodynamically controlled, and if followed for an extended period further isomerization of 2-methylbut-2-ene (in this case the major product) can be observed. Similar product distributions have been observed in our laboratories over CeX [16], and by Hightower and Hall [4] over silica-alumina. The latter workers explained their results qualitatively in terms of a proton addition mechanism but the predictions of a classical ring-opening reaction do not correlate with the observed relative amounts of the three products. i.e. 3-methylbut-1-ene would be formed from a primary carbonium ion intermediate and hence not expected in substantial amounts, and statistical and energetic considerations would predict significantly more 2-methylbut-1-ene than 2-methylbut-2-ene. However if the methyl butene species are produced directly from the non-classical cyclic carbonium ion in a more or less concerted process not involving a ring opening to the classical carbonium ion then it is possible to rationalize the observed product distribution.

In conclusion attention is drawn to the work of John et al. [17] who made a microwave investigation of the double bond shift process of $CD_2=CH.CH_3$ over the omega zeolite used in the present study. As they have discussed in detail a significantly different product composition (propenes containing one, two or three deuterium atoms in different positions) results depending on whether an associative mechanism (with proton addition) or a dissociative mechanism (as in hydride abstraction) is operating. Table 4 gives their essential results.

TABLE 4. Composition of d_1- and d_3- Propenes Resulting from the Isomerization of $CD_2=CH.CH_3$ over Omega Zeolite

Normalized Deuterium Distribution/%

$CDH=CH.CH_3$	$CH_2=CH.CH_2D$	Total d_2-Propene/%
87	13	88
89	11	78
75	25	65
100	0	100(a)
50	50	100(b)
40	60	35(c)

$CH_2=CH.CD_3$	$CD_2=CH.CH_2D$	$CHD=CH.CHD_2$	Total d_2-Propene/%
91	9	0	88
87	6	8	78
73	18	9	65
100	0	0	100(a)
50	50	0	100(b)
10	30	60	35(c)

(a) Product distribution predicted from associative mechanism
(b) Product distribution predicted from dissociative mechanism
(c) Equilibrium distribution from 5 exchangeable hydrogens.

The data for the d_1 product species illustrates an initially very high selectivity for $CDH=CH.CH_3$ formation (relative to that of $CH_2=CH.CH_2D$) which is entirely compatible with an associative mechanism. Even after 35% reaction of the $CD_2=CH.CH_3$ starting material the d_1 propene distribution is very different to that predicted statistically at equilibrium or by the dissociative mechanism. Similar behaviour is shown by the d_3-propenes with a high initial preferential formation of the $CH_2=CH.CD_3$ molecule and little $CD_2=CH.CH_2D$ as would be expected if a dissociative process was taking place.

Thus in the double bond shift reaction of the labelled propene over omega there is clear spectroscopic evidence that an associative (proton addition) reaction mechanism is operative. In the present investigation of the isomerization reactions of the methyl substituted cyclopropane species over omega the product distributions observed are also indicative of a similar proton addition process – with the non classical carbonium ion being the most likely reaction intermediate. The Bronsted acidity responsible for this catalytic activity of the omega zeolite is believed to result from hydroxyl groups generated during the decomposition of encapsulated TMA cations, with a significant fraction of these sites being situated in the main channels of the zeolite and thus readily accessible to reactant molecules.

REFERENCES
1 R.M. Roberts, J. Phys. Chem., 63 (1959) 1400.
2 J.G. Larson, H.R. Gerberich and W.K. Hall, J. Amer. Chem. Soc., 87 (1965) 1880.
3 W.K. Hall, F.E. Lutinski and H.R. Gerberich, J. Catalysis, 8 (1967) 391.
4 J.W. Hightower and W.K. Hall, J. Amer. Chem. Soc., 90 (1968) 851.
5 J. Bullivant, J.S. Shapiro and E.S. Swinbourne, J. Amer. Chem. Soc., 91 (1969) 7703.
6 T.T. Nguyen, R.P. Cooney and G. Curthoys, J. Catalysis, 44 (1976) 81.
7 B.H. Bartley, H.W. Habgood and Z.M. George, J. Phys. Chem., 72 (1968) 1689.
8 Z.M. George and H.W. Habgood, J. Phys. Chem., 74 (1970) 1502.
9 E.M. Flanigen and E.R. Kellberg, Netherlands Patent, 6 710 729 (1967).
10 R.M. Barrer and H. Villiger, Chem. Com., 12 (1969) 659.
11 J.S. Cole and H.W. Kouwenhoven, Adv. Chem. Series, 121 (1973) 583.
12 T.J. Weeks, Jr., D.G. Kimak, R.L. Bujalski and A.P. Bolton, Trans. Farad. Soc., I, 72 (1976) 575.
13 E.L. Wu, T.E. Whyte, Jr., and P.B. Venuto, J. Catalysis, 21 (1971) 384.
14 D.W. Breck, Zeolite Molecular Sieves, 570 (1973).
15 C.E. Marsden, Ph. D. Thesis, Edinburgh University (1978).
16 N. McC. Coutts, Ph. D. Thesis, Edinburgh University (1973).
17 C.S. John, C.E. Marsden and R. Dickinson, J. Chem. Soc., Far. Trans. I, 72 (1976) 2923.

B. Imelik *et al.* (Editors), *Catalysis by Zeolites*
© 1980 Elsevier Scientific Publishing Company, Amsterdam — Printed in The Netherlands

HYDROCARBON REACTIONS CATALYSED BY MORDENITES

Hellmut G. Karge and Jürgen Ladebeck
Fritz-Haber-Institut der Max-Planck-Gesellschaft, Faradayweg 4-6, 1000 Berlin 33, Germany

1. INTRODUCTION

The diameters of the orifices and cross-sections of the main channels penetrating
a mordenite crystal [1, 2] are very similar to the dimensions of benzene, pyridine and
similar molecules. In reactions of such molecules over mordenite catalysts one also
should expect preferential formation of products without significantly larger dimensions. For instance, alkylation at the benzene ring should yield mainly monoalkylben-
zenes and among the dialkylbenzenes the para-dialkylbenzenes should be preferentially
formed; meta-dialkylbenzenes, even though thermodynamically favoured, should be re-
pressed for steric reasons; ortho-dialkylbenzenes should be almost entirely absent.
The idea of this so-called "shape selectivity" of mordenite catalysts was one of the
sources of our early interest in this zeolite. Indeed, such a shape selectivity has already been observed, e. g. for the alkylation of benzene by ethylene or propylene [3].
Besides monoethylbenzene and cumene only para- and meta-dialkylbenzenes were formed.
Similar shape selectivities have been reported by Csicsery [4, 5] for transalkylation
of 1-methyl-2-ethylbenzene over hydrogen mordenite catalysts.

In an extensive IR study it could be shown that the Brønsted acid sites of hydrogen
mordenite were catalytically active centers in hydrocarbon reactions such as benzene
alkylation, proceeding via carbenium ion mechanisms [3, 6]. The hydrogen mordenite catalysts used, however, possessed at the same time Brønsted and Lewis acidity, even at
low activation temperatures, i. e. before dehydroxylation starts to occur [6 - 8]. It
therefore remained unclear as to whether the Lewis sites are actually unimportant or
whether they might enhance the acid strength of adjacent OH groups, thus inducing the
extraordinarily high activity of mordenite catalysts [9].

In many investigations of hydrocarbon reactions over mordenite catalysts it was observed, however, that the high catalytic activity of the zeolite was reduced more or
less rapidly with time-on-stream [3, 10 - 12]. Origin and mechanism of the deactivation process accompanying, for example, the benzene alkylation is not yet fully understood. There are open questions concerning, for instance, the role of the different
sites as well as the part played by olefinic and aromatic hydrocarbons in catalyst
aging [3, 13].

We have therefore investigated whether the presence of Lewis sites in mordenite is
a prerequisite for its high activity both in alkylation of benzene and in deactivation reactions. Furthermore, we have attempted to prolong the catalyst life-time by
appropriately modifying the zeolite. Since the decrease of activity might be partially caused by a loss of acidic OH groups during time-on-stream, a detailed study of
rehydroxylation was carried out [14, 15]. Finally, we have attempted to elucidate the
reactions leading to deactivation by separately studying the behaviour of olefin and
aromatic compounds over mordenite catalysts.

2. EXPERIMENTAL

The reactions were conducted in a fixed-bed flow reactor, generally working in the differential mode. The experimental set-up was similar to that in ref. [3]. In the reactor 0.1 - 1.0 g of the catalyst could be prepared under nearly the same conditions as were used in the IR experiments. The reactants were deluted by helium (total pressure 1×10^5 Pa). Feed and product stream (30 ml/min.) were analysed at short intervals by a Perkin Elmer gas chromatograph type 3920 equipped with a benton 34+di-n-decylphthalate and a Porapak R column for separating the aromatics and the light hydrocarbons, respectively. The IR spectra were run with a spectrophotometer 325 (Perkin Elmer), using a modification [16] of the cell designed by Knözinger et al. [17]. The procedure has been described elsewhere [6].

The following materials were used in the catalytic experiments (dry composition in brackets):

(i) commercial hydrogen mordenite, HM ($H_{6.8}Na_{0.4}Al_{7.2}Si_{42.6}O_{96}$);

(ii) commercial sodium mordenite, NaM ($Na_{7.1}Al_{7.9}Si_{36.5}O_{96}$);

(iii) ammonium mordenite, NH_4M ((NH_4)$_{8.5}Na_{0.02}Al_{8.5}Si_{39.5}O_{96}$);

(iv) beryllium mordenite, BeNaM ($Be_{2.1}Na_{3.1}$ $Al_{8.65}Si_{39.69}O_{96}$);

(v) dealuminated beryllium mordenite, BeM-D ($Be_{2.0}Na_{0.16}Al_{4.2}Si_{43.8}O_{96}$).

HM and NaM were supplied by Norton Company, Mass. The cation-containing mordenites were prepared from NaM using standard ion-exchange and acid leaching procedures. BeNaM, MgNaM, and CaNaM, applied in the colour tests for acid strength, were exchanged to 58, 59, and 67 %, respectively.

3. RESULTS AND DISCUSSION

3.1 Brønsted acid hydrogen mordenite

A pure Brønsted acid form of hydrogen mordenite without intrinsic Lewis sites could be prepared via ammonium exchange of sodium mordenite and subsequent deammoniation under mild conditions, i. e. at temperatures not exceeding 650 K and under good high vacuum. Completion of deammoniation was controlled by monitoring a) constancy of the OH band intensity, b) absence of the NH_4^+ band at 1435 cm^{-1}, and c) NH_3 partial pressure by means of mass spectrometry. After pyridine adsorption no band indicating Lewis acidity was detected (see Fig. 1). At the same time this material was also a very active catalyst in alkylation of benzene (see Fig. 2). Consequently, the presence of Brønsted sites, but not of Lewis sites in hydrogen mordenite is a prerequisite for the catalysis of such hydrocarbon reactions. It was confirmed, on the other hand, that fully dehydroxylated hydrogen mordenite, having only Lewis but no Brønsted centers, is practically inactive in alkylation of benzene as well as in dealkylation, transalkylation, benzene cracking, and ethylene reaction (see below).

3.2 Benzene cracking over hydrogen mordenite

When the benzene alkylation with ethylene was carried out at high reaction temperatures (723 K), significant amounts of methylbenzenes were also formed via secondary reactions [12]. The high acidity of HM seems to be sufficient to catalyse even the destruction of the benzene ring. Hence, we reacted pure benzene over HM. Table 1 shows the product distribution over fully hydroxylated HM ($T_{act.}$ = 723 K) at various tempe-

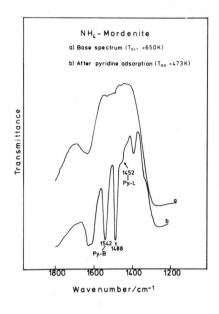

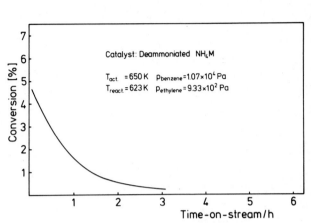

Fig. 1   Spectra of deammoniated NH₄M before   Fig. 2   Alkylation over deammoniated NH₄M
and after pyridine adsorption

atures. High fractions of toluene were detected above 623 K. The fraction of light
hydrocarbons included methane, ethylene, ethane, propene, propane, butene, and butane;
15.5 mol-% of this fraction was ethylene and 64 mol-% propane. Most of the benzene
fragments, however, were involved in coke formation, and benzene cracking activity of
HM dropped very quickly.

Table 1

Conversion of benzene over HM in percent of the feed. (hc: light hydrocarbons, see text;
T: toluene; EB: ethylbenzene; ET: 1.4-ethyl toluene)

$T_{react.}$	hc	T	E B	E T	Coke	Total (%)
523	0.04	--	0.004	--	0.63	0.7
623	0.05	1.34	0.44	--	3.79	5.6
723	0.17	1.64	0.36	0.03	4.60	6.8

This is demonstrated by Fig. 3 which illustrates the rapid decrease of benzene conversion and the concomitant decrease of toluene formation with time-on-stream.

3.3 Reaction of ethylene over hydrogen mordenite

We have found, however, that HM rapidly deactivates during alkylation of benzene
with an olefin [3], even at low temperatures (375 K) when benzene cracking does not occur. We therefore assumed that the main cause of catalyst aging is the reaction of olefin with itself, i. e. polymerisation. Passing an ethylene/helium mixture (partial pressure of ethylene: 3.33 x 10³ Pa) over HM ($T_{act.}$ = 723 K, $T_{react.}$ = 373 and 723 K) resulted in an instantaneous onset of the reaction at high conversion levels (70 %),
followed by a rapid (exponential) decrease of the conversion with time-on-stream [15].
After 40 min., two thirds of the ethylene feed was converted into a polymeric form

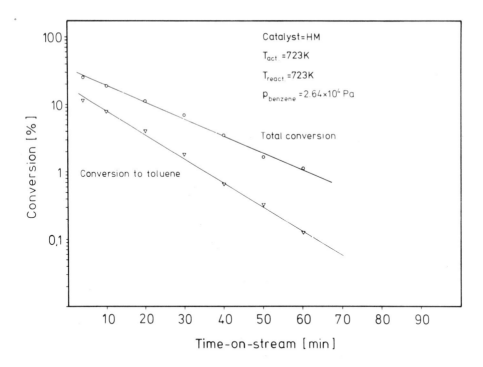

Fig. 3   Toluene formation during benzene cracking over HM

(and carbonaceous deposits), the remainder to light hydrocarbons (CH_4, C_2H_6, C_3H_6, C_3H_8, C_4H_8, and C_4H_{10}). The IR spectra of HM showed after adsorption under 1.33×10^4 Pa ethylene the bands only of saturated hydrocarbons and a remarkable weakening of the band of acid OH groups (3605 cm^{-1}). The bands could not be removed from the spectrum by evacuation at the same temperature; that means formation of the saturated species was practically irreversible. At higher adsorption temperatures (573 K) a broad and intense band around 1585 cm^{-1} developed, indicating coke formation. In the case of dehydroxylated HM ($T_{act.}$ = 973 K) no CH bands whatsoever appeared. The sample remained white.

3.4 Transalkylation and dealkylation of ethylbenzene over hydrogen mordenite

If polymerisation of olefins is the basic step in coke formation on HM, then catalyst aging will consequently occur to a lesser extent for reactions which do not involve free olefin. As an example, we tried transalkylation or disproportionation of ethylbenzene. Whether or not pure transalkylation occurs depends, of course, on the reaction conditions, in particular on the reaction temperature. At higher temperatures dealkylation becomes possible, since it requires higher activation energies than transalkylation. Indeed, we recognize from table 2 that above 500 K obviously no pure transalkylation occurs, since benzene (B) and diethylbenzene (DEB) do not form with the ratio 1 : 1, the fraction of diethylbenzene being significantly lower. Instead, side products appear, viz. toluene (T), xylene (X), styrene (St), and light hydrocarbons (hc). In sharp contrast, at lower temperatures pure transalkylation is observed. Below 500 K benzene and ethylbenzene are the only products, forming in the ratio 1 : 1. The ratio of the 1.4-diethylbenzene to 1.3-diethylbenzene is 1 : 2. The yield of 1.2-diethylbenzene is zero, its dimensions (7.0 x 8.5 x 6.0 Å) being already too large compared with the channel diameter. Thus we encounter the same effect of shape selectivity as in the case of alkylation of benzene. In the high temperature regime where dealkylation is also possible and olefin is evolved, the HM catalyst is rapidly deactivated. In contrast, in the region of pure transalkylation a steady state is quickly reached and the activity (conversion) remains nearly constant for a long time, i. e. more than 5 hrs.

Table 2
Conversion of ethylbenzene over HM to light hydrocarbons (hc), aromatics (ar), and coke in percent of the feed; product distribution of the aromatics in mol-% (details see text).

$T_{react.}$/K	hc	ar	Coke	Total	B	T	X	St	DEB
373	--	11.8	--	11.8	48.3	--	--	--	51.7
473	--	31.5	--	33.2	50.2	--	--	--	49.8
573	5.8	38.5	3.6	47.9	70.2	11.9	2.5	4.0	11.4

3.5 Dehydration of cyclohexanol over hydrogen mordenite

In this reaction free olefin, viz. cyclohexene, actually forms; however, under mild conditions the ring remains intact and no polymerisation occurs. Thus, if a mixture of 266 Pa cyclohexanol and 10^5 Pa helium flows over HM at 410 K, the catalyst does not suffer from aging due to irreversible polymer or coke deposition. A slight loss of activity is caused by adsorption of the second product, i. e. water. This can be easily removed by purging with helium, resulting in complete restoration of the initial activity [18].

3.6 Cation-exchanged mordenites

We suspected that OH groups of extreme acid strength present in HM were responsible for both the high activity and the rapid deactivation. Cation-exchanged mordenites were prepared, with the expectation that these catalysts will have weaker acidity than hydrogen mordenite and, therefore, a significantly longer life-time combined with equally good selectivity. The acid strength of HM, BeNaM, and MgNaM was compared in a colour test, using Hammett and arylmethanol indicators. As can be seen from table 3, their acid strength shows the sequence HM > BeNaM ∿ MgNaM > CaNaM [19]. All three catalysts were active in benzene alkylation, but the life-time of BeNaM and MgNaM was much longer than that of HM. At variance with HM, BeNaM did not catalyse benzene cracking [20] and showed reversible ethylene adsorption at low temperatures (373 K). For the ethylene reaction, BeNaM exhibited much longer life-times than HM and less pronounced coke formation [15]. Pure transalkylation of ethylbenzene turned out to be impossible over BeNaM under our reaction conditions, since a relatively high reaction temperature (723 K) had to be chosen, in the same way as for the alkylation of benzene by an olefin. The reason was that BeNaM did not adsorb benzene sufficiently well below 573 K [20] and the conversion was correspondingly low at these temperatures. Obviously, the effective pore diameter of BeNaM is slightly smaller than that of HM [21].

156

Table 3
Acid strength of mordenites by colour tests

Mordenite	pK$_a$					pK$_R$	
	3.3	1.5	-3.0	-5.6	-8.2	-6.6	-13.3
HM	+	+	+	+	+	+	+
BeNaM	+	+	+	+		+	
MgNaM	+	+	+	+		+	
CaNaM	+	+	+				

3.7 Dealuminated cation-containing mordenite

It is well known that the effective pore diameter of mordenite might be changed by dealumination [22 - 24]. This prompted us first to leach sodium mordenite by moderate acid treatment and subsequently to carry out the ion exchange using beryllium oxalate solution [12]. By this procedure we hoped to obtain catalysts which (i) render possible reactions of benzene and benzene derivates already at low temperatures and (ii) exhibit the good stability of BeNaM. The thus modified beryllium mordenite (BeM-D) indeed showed adsorption properties which were quite similar to those of HM. Furthermore, it is evident from Fig. 4 that with this modified BeM-D, pure transalkylation of ethylbenzene takes place quite efficiently at just the same low temperature as over HM. The only products formed are benzene and diethylbenzene, their molar ratio being 1 : 1.

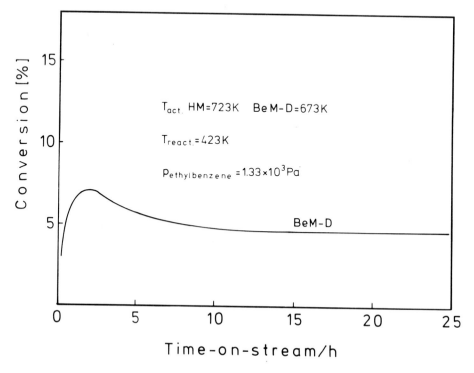

Fig. 4   Transalkylation of ethylbenzene over BeM-D

At 423 K reaction temperature the order of the reaction with respect to ethylbenzene was nearly zero. This means that the catalyst was working under saturation conditions. The plot of conversion vs. the reciprocal temperature which is shown in Fig. 5 exhibits

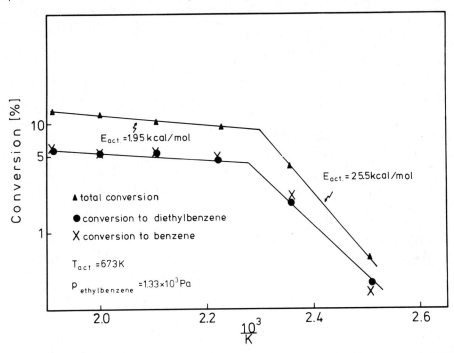

Fig. 5   Arrhenius plot for the transalkylation of ethylbenzene over BeM-D

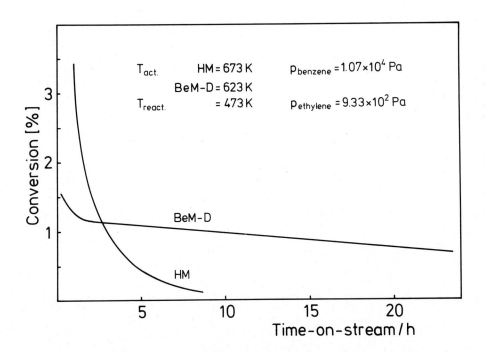

Fig. 6   Alkylation of benzene with ethylene over BeM-D

two regions. At low temperatures, the apparent activation energy is about 100 kJ mole^{-1}, but at $T_{react.}$ > 440 K only about 8 kJ mole^{-1}. Most probably the reaction is diffusion-controlled in the second region.

It was tempting, of course, to check whether or not the modified, dealuminated beryllium mordenite is also able to catalyse the alkylation reaction at low temperatures (compare 3.4 and 3.6). The result of an alkylation run is shown in Fig. 6. In contrast to the non-dealuminated BeNaM, the modified BeM-D catalysed the alkylation of benzene with olefin already at 473 K quite efficiently. In the product stream only ethylbenzene and diethylbenzene were detected. The life-time is - in contrast to HM - remarkably long.

3.8 Rehydroxylation of hydrogen and beryllium mordenite

Only a small part of the original acid OH's of HM could be restored on treating a fully dehydroxylated HM with H_2O vapour at various temperatures. Exposure of fully dehydroxylated BeNaM to H_2O vapour at 673 K, however, completely recovered the acid OH's, i. e. th active centers, and the catalyst regained most of its original activity.

CONCLUSIONS

The experiments have demonstrated the decisive role of olefin compounds in the deactivation of mordenite catalysts, when the olefins are able to polymerise under reaction conditions. The catalyst life-time can be markedly prolonged by zeolite modification, e. g. by ion exchange with beryllium cations. Dealumination by acid treatment and subsequent ion exchange seems to be a promising method of preparing beryllium mordenite and similar catalysts which have improved properties and are able to catalyse aromatics reactions even at low temperatures.

ACKNOWLEDGEMENT

This paper is based on results of Drs. H. Karge, J. Ladebeck, Z. Sarbak, and Dipl.-Chem. H. Kösters, and J. Schweckendiek. The authors gratefully acknowledge the contributions of their three colleagues. They are further indebted to Mrs. Erika Popovič for excellent technical assistance and to Prof. Dr. J. H. Block for his interest in this work as well as for his critical reading of the manuscript. The authors also thank the Deutsche Forschungsgemeinschaft for financial support.

REFERENCES

1   W.M. Meier, Z. Kristallographie 115 (1961) 439
2   W.J. Mortier, J.J. Pluth, and J.V. Smith, Mat. Res. Bull. 10 (1975) 1037
3   K.A. Becker, H.G. Karge, and W.-D. Streubel, J. Catalysis 28 (1973) 403
4   S.M. Csicsery, J. Catalysis 19 (1970) 394
5   S.M. Csicsery, J. Catalysis 23 (1971) 124
6. H.G. Karge, Z. phys. Chem. Neue Folge 76 (1971) 133
7   F.R. Cannings, J. Phys. Chem. 72 (1968) 4691
8   M. Lefrancois and G. Malbois, J. Catalysis 20 (1971) 350
9   J.H. Lunsford, J. Phys. Chem. 72 (1968) 4163
10 T. Yashima, H. Moslehi, and N. Hara, Bull. Japan Petroleum Institute, 12 (1970) 106
11 E.A. Swabb and B. C. Gates, Ind. Eng. Chem. Fundam. 11 (1972) 540
12 H.G. Karge, Proceedings of the Fourth Int. Conf. on Molecular Sieves, Chicago 1977
    (J. Katzer, Ed.) ACS Symposium Series 40 (1977) 584
13 D. Walsh and L.D. Rollmann, J. Catalysis 49 (1977) 369
14 H.G. Karge, Z. phys. Chem. Neue Folge 95 (1975) 241

15   H.G. Karge and J. Ladebeck, Proc. of the Symp. on Zeolites, Szeged, Hungary, 1978
     Acta Physica et Chemica, Nova Series 24 (1978) 161
16   H.G. Karge, submitted to Z. phys. Chem., Neue Folge
17   H. Knözinger, H. Stolz, H. Bühl, G. Clement, and W. Meye, Chemie-Ing.-Technik
     42 (1970) 548
18   H.G. Karge and H. Kösters, to be published
19   H.G. Karge and J. Schweckendiek, to be published
20   H.G. Karge and J. Ladebeck, Proc. of the 6th Canadian Symposium on Catalysis,
     Ottawa, 1979 (C.H. Amberg and J.F. Kelly, Eds.) 1979, 87
21   T. Yashima and N. Hara, J. Catalysis 27 (1972) 329
22   I.M. Belenkaja, M.M. Dubinin, and I.I. Krisztofori, Izv. A. N. SSSR, Ser. chim.
     7 (1971) 1391
23   I.M. Belenkaja, M.M. Dubinin, and I.I. Krisztofori, Izv. A. N. SSSR, Ser. chim.
     12 (1971) 2635
24   P.E. Eberly, jr., Ch.N. Kimberlin, and A. Voorhies, J. Catalysis 22 (1971) 419

B. Imelik *et al.* (Editors), *Catalysis by Zeolites*
© 1980 Elsevier Scientific Publishing Company, Amsterdam — Printed in The Netherlands

PROTON AND OXYGEN MOBILITY IN NEAR-FAUJASITE ZEOLITES

J.J.FRIPIAT

C.N.R.S. - C.R.S.O.C.I. 1B, Rue de la Férollerie, 45045 ORLEANS CEDEX

1. INTRODUCTION

It is well known that the lattices of X and Y zeolites can be easily reversibly distorded, for instance upon adsorbing or desorbing water, without appreciable loss of cristallinity. Such distorsion results in displacing the lattice components to various extent, producing mainly modifications of the intensity of X-ray diffraction lines. There is thus some limited mobility of these lattice components and as the building blocks of these lattices are oxygens tetrahedra centered either on Si^{4+}, Ge^{4+} or Al^{3+} cations, displacements of tetrahedra as a whole in the cuboctahedron and perhaps displacements of oxygens within some tetrahedra may account for the observed modifications.

The distribution of the charge balancing cations, even in the dehydrated state may also be modified by thermal treatment and numerous examples of such redistribution are available in the literature. The aim of this contribution is to review experimental results on protonic motions in decationated HY zeolites obtained by thin bed calcination of the ammonium form (ref.1,2) and on the oxygen exchange in NaY, NaX and NaGeX in contact with CO_2^{18} (ref.3). The preparation and properties of the Germanium Sieve have been published previously (ref.4,6). The other sieves used in these works had the classical composition but their iron contents were lower than 100 ppm. The experimental procedures used for these studies are described in detail in the above references. In summary the proton mobility was studied by NMR whereas the O^{18}-O^{16} exchange was followed by the mass spectrographic analysis of the gas phase. Infrared spectroscopy has been used to study the evolution of the OH and OD bands when the decationated sieve was exposed to CH_3OH and the nature of the surface carbonate species during the isotopic exchange with labeled CO_2.

The comparison of the data obtained on the mobility of proton and oxygen in zeolite should shed some light on the nature of the catalytic sites in this important class of acid catalysts.

2. THERMAL MOBILITY OF PROTONS IN HY ZEOLITE (ref.1)

The measurements of the spin-lattice (T_{1p}) relaxation time by NMR, between - 180 and + 180°C, show that T_{1p} remains constant. T_{1p} is determined in this temperature range by a spin energy diffusion process, the paramagnetic impurities being the sink towards which the proton spin energy moves by a flip-flop mechanism. From 180 to 260°C T_{1p} falls sharply as the temperature is increased. Protons go from oxygen atom to another and therefore they move to oxygen atoms which are alternately inside the hexagonal prism, inside the cuboctahedron and inside the supercage. This motion modulates the proton-paramagnetic interaction which is shown to be the most efficient mechanism for the longitudinal relaxation

The activation energy for the proton jump is 19 kcal mole^{-1} and from the combination of T_{1p} and of the spin-spin relaxation process T_{2p}, the proton jump frequency and diffusion coefficient shown in Table 1 have been obtained.

TABLE 1.
Proton jump frequency ν and diffusion coefficient, assuming an average jump distance of 4.4 Å

T/°C	D/cm^{2}sec^{-1}	ν/s^{-1}
0	2.9 10^{-18}	0.9 10^{-2}
100	3.4 10^{-14}	1.1 10^2
200	7.7 10^{-12}	2.4 10^4
300	2.3 10^{-10}	8.2 10^5
400	3.1 10^{-9}	9.7 10^6

Since each oxygen tetrahedron always has one of its oxygen atoms in the hexagonal prism (see for example fig.2 ref.7), linking the cubo-octahedra, and another in the supercage, the oxygen atoms have an equal probability of occupancy, in agreement with Vedrine et al.(ref.8). R.A.Dalla Betta and Boudart (ref.9) have studied the kinetics of exchange of zeolite hydroxyl groups with D_2 on H-Ca Y. The rate of exchange is first order between 200 and 400°C. The proposed mechanism is a two steps process in which D_2 first exchanges rapidly with a small number of surface species H* and then D* reacts with the surface hydroxyl

$$D_2 + H^* \rightarrow D^* + HD$$
$$D^* + OH \rightarrow OD + H^*$$

If we consider that the diffusion coefficients shown in Table 1 have the same order of magnitude for H-Ca Y as in H Y then the half time for equilibration of hydroxyl groups over a crystallite can be approximated by the Einstein diffusion equation $t = <\ell^2>/2D$, if some reasonable estimate can be made for $<\ell^2>$. The experimentally observed time for half exchange OH-OD at 304°C by Della Betta et al. was 330 s. This leads to $\sqrt{<\ell^2>} = 2.8$ 10^{-4}cm ≈ 0.3 μm. Typical cristallite size for synthetic zeolites is of the order of μm. Thus the proton equilibration over the zeolite surface may well be the rate limiting process in this mechanisms.

3.   PROTON EXCHANGE BETWEEN METHANOL AND H Y (ref.2)

The spin-lattice proton relaxation rate of CD_3OH in H Y is apparently contributed by the proton exchange process between the alcoholic OH and the surface OH. The activation energy for that exchange is practically independent of the filling degree of the cage as shown by the thermal dependence of the correlation time (see Fig.1.).

The H Y-methanol system should be considered as constituted by a pool of protons belonging either to the lattice OH or to CD_3OH and involved in a fast-exchange process relayed by $CD_3OH_2^+$. The methanol molecule may thus be considered as a vehicle that transports the lattice protons within the cage.

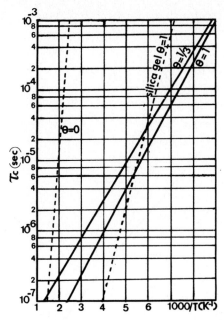

Fig.1 - Variation of the correlation
time ruling the spin-lattice relaxation
in H-Y unloaded and partially (two-thirds)
or totally filled with CD_3OH. For silica
gel data see ref.13.

The infrared spectroscopic study (ref.2) carried out on the same system shows that
CD_3OH is strongly hydrogen bonded to the lattice OH groups, which are strongly acidic
(ref.10) (pKa $\approx$ - 6). Actually when CD_3OH is adsorbed by H Y, the pair of OH stretching
bands at 3670 and 3565 cm^{-1} is replaced by a broad band extending from 3600 cm^{-1} to
3100 cm^{-1} (see Fig.1, ref.2).

The removal of physically adsorbed CH_3OH cannot be achieved completely below 200°C
under vacuum. The isotopic exchange between CH_3OD and the lattice OH occurs also very
rapidly at room temperature. Note also that pKa of $CH_3OH_2^+$ is about - 4.5 (ref.11). Thus
$CH_3OH_2^+$ is almost as a strong acid as the surface OH. The following mechanism for trans-
porting the protons could be then visualized as depicted below

$$CH_3OH_2^+{}_{(surface)} + CH_3OH_{(cage)} \rightarrow CH_3OH_{surface} + CH_3OH_2^+{}_{cage}$$

The OH primarily involved would be of course those directed towards the supercage
(3670 cm^{-1} stretching) because the first step would be the hydrogen bond formation (II)
but since protons are moving within all the oxygens of one tetrahedron, the III $\rightarrow$ I
back reaction may involve OH which were not originally in the supercage. A consequence
of this, is that a molecule such as CH_3OH works as a catalyst for accelerating the protonic
motion. A convincing evidence for that effect results from the comparison of ν in
Table 1 with the value of $1/\tau$ in Figure 1 at the same temperature , as well as by the
decrease of the activation energy from 19 to about 2.5 kcal. Remark that the mechanism
sketched above ressembles that suggested by Stubner et al. (ref.12) in their study of

proton exchange between 2-butanol and η-alumina with the important difference that in η alumina, the alcohol molecule is the proton donor.

4.   OXYGEN ISOTOPIC EXCHANGE IN NaY, NaX AND NaGeX IN CONTACT WITH CO_2^{18} (ref.3)

Isotopic exchange between CO_2^{18} and the lattice oxygen of these sieves is measurable at room temperature. However exchange in NaX and NaGeX is much more rapid under comparable conditions than exchange in NaY. Table 2 shows the time requested to reach 50% exchange assuming that portions of the solid which have previously exchanged with the gas are maintained at equilibrium with the continuously changing isotopic composition of the gas.

TABLE 2.
Time (hrs) at which 50% of the isotopic exchange is achieved and γ, the fraction of lattice oxygen atoms available for isotopic exchange at equilibrium.

Sieve	Temperature/°C	Time (hrs)	γ
NaY	300	>> 300	(a)
	400	200	(a)
	500	69	(a)
NaX	200	8	0.76 (b)
	300	19	0.84 (b)
NaGeX	100	6	0.67 _
	200	1	0.68
	300	< 1	0.82
	300 (c)	< 1	0.97

N.B.   a) the equilibrium value has not been reached even at 500°C.
        b) see text explaining this apparent inversion.
        c) samples pretreated at 300° in CO_2 atmosphere.

The general aspect of the exchange kinetics carried out at 300°C is shown in Fig.2. The exchange process with NaX is characterized by an induction time which is longer at 300°C than at 200°C. This explain the apparent inversion of the experimental results shown in Table 2, N.B. (b). The NaY and NaX sieves were outgassed at 360°C for 12-18 hours and at a dynamical pressure of 10^{-6} torr in order to remove water before the exchange whereas NaGeX was outgassed at 10^{-4} torr at 300°C for the same period of time. The activation energy of the exchange process is in all cases lower than 10 kcal. This is considerably less than those which characterize isotopic exchange in silicate and aluminosilicate minerals (ref.14,15). From these results, it appears clearly than in spite of the similarity of their

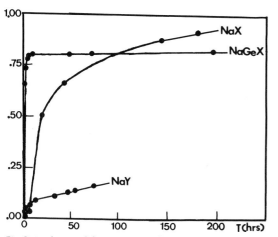
Fig.2: Evolution of the isotopic exchange at 300 C

structures, these sieves have a very different dynamical behavior as far as the oxygen isotopic exchange is concerned. This difference in behavior must be attributed to the effects associated with the substitution of Al^{3+}, Ge^{4+} for Si^{4+} in the tetrahedral sites but also to the nature of the exchangeable cations. Indeed preliminary data suggested that partial replacement of Ca^{2+} for Na^+ increases markedly the rate of isotopic exchange in Y zeolite. In order to understand the exchange mechanism an infrared study of the surface carbonate species in the course of the exchange process was undertaken. The main conclusions of this study can be summarized as follows. The absence of detectable chemisorbed CO_2 species on NaY and its low rate of exchange supports the hypothesis that it is through the surface carbonate species that the oxygen isotopic exchange between CO_2^{18} and the lattice O^{16} takes place. It is actually very difficult to assign to one well specified site the main role as entering gate because in the course of the reaction and depending upon temperature, some surface species can block the reaction whereas at a different temperature this blocking action can be different. For the NaX and NaGeX sieves, a asymmetrical bidentate structure showing bands between 1640 and 1680 cm^{-1} and between 1350 and 1365 cm^{-1} is the favored candidate. There would be more of these sites present in NaGeX than in NaX at similar temperature.

Such a bidentate structure could involved the charge balancing cation and an oxygen linked to a tetrahedral aluminum cation as depicted below.

where M^{4+} = Ge^{4+} or Si^{4+},
and Na^+ is on site III.

This could explain as well the effect of the number of Al substitutions, the effect of substituting Si^{4+} by Ge^{4+} and the influence (to be studied further) of the charge balancing cations. Such a structure has been proposed and discussed by Jacobs (ref.16) in order to explain the difference in behavior of X and Y zeolites with respect to the promotion of CO_2 on the dehydration of alcohol. The bidentate structure is not formed in NaY and, in addition this structure may evoluate into a true carbonate CO_3^- anion (with bands at 1485 and 1425 cm^{-1}) which can block the active sites.

DISCUSSION

The common point between the mechanism reported to explain the proton and oxygen mobility involves ultimately the electronic properties of the oxygen linking an aluminum to a M^{4+} tetrahedron. It seems that the activation of this oxygen by a proton or by a chemisorbed CO_2 molecule is the key of the explanation of the experimental data reviewed in this contribution. Consequently more experimental and theoretical work should be done on the $Al-O-M^{4+}$ linkage to understand these basic features.

REFERENCES

1   M.M. Mestdagh, W.E.E. Stone and J.J. Fripiat, J.C.S. Farad. Trans. I, 72 (1976) pp. 154-162.
2   P. Salvador and J.J. Fripiat, J. Phys. Chem. 79 (1975), pp. 1842-1849.
3   C. Gensse, T.F. Anderson and J.J. Fripiat, submitted to J. Phys. Chem. (1980).
4   L. Lerot, G. Poncelet and J.J. Fripiat, J. Mater. Res. Bull. 9, (1974), 979.
5   L. Lerot, G. Poncelet, M.L. Dubru and J.J. Fripiat, J. Catalys. 37, (1975), 396.
6   M.L. Dubru, L. Lerot and G. Poncelet, J. Catalys. 38 (1977), pp. 205-216.
7   E.G. Derouane, M.M. Mestdagh and J. Vielvoye, J. Catalysis, 33 (1974) pp. 169-177.
8   J.C. Vedrine, D.S. Leniart and J.S. Hyde. Ind. Chim. Belg. 38, (1973) pp. 397-403.
9   R.A. Dalla Betta and M. Boudart, J. Chem. Soc. Farad. Trans. I, 72 (1976) pp. 1723-1732.
10 P.G. Rouxhet and R.E. Sempels, J. Chem. Soc. Farad. Trans. I, 70 (1974) pp. 2021-2031.
11 J.J. Fripiat, A.C.S. Symposium series n°34, Magnetic Resonance in Colloïd and Interface Science (1976), Edited by H.A. Resing and C.G. Wade.
12 B. Stübner, H. Knözinger, J. Conard and J.J. Fripiat, J. Phys. Chem (82) (1978), pp 1311-1817.
13 J.J. Seymour, M.I. Cruz and J.J. FRIPIAT, J. Phys. Chem., 76, (1974) pp. 3078.
14 K. Muchlenbachs and I. Kushito, Carnegie Inst. Wash. Yearb; 73, (1974), pp. 232-236.
15 R.A. Yund and T.F. Anderson, Carnegie Inst. Wash. Publ. 634 (1974), pp. 99- 105.
16 P.A. Jacobs, Carbogenic activity of zeolites, p. 159, Elsevier Scient. Publish. Co. (1977).

B. Imelik *et al.* (Editors), *Catalysis by Zeolites*
© 1980 Elsevier Scientific Publishing Company, Amsterdam — Printed in The Netherlands

CATALYTIC REACTIONS INDUCED BY TRANSITION METAL COMPLEXES SOLVATED IN ZEOLITE MATRICES: OXIDATION, CARBONYLATION AND RELATED REACTIONS

Younès BEN TAARIT

Institut de Recherches sur la Catalyse, CNRS

2, avenue Albert Einstein, 69626 Villeurbanne Cédex

Michel CHE

Université Pierre et Marie Curie

4, place Jussieu

75230 Paris Cédex 05.

1 INTRODUCTION

Early studies on zeolites concentrated almost exclusively on the acidic behaviour of zeolites. Most of the impetus was focused on NH_4^+ form and alkali-, alkaline-earth-, and rare-earth exchanged zeolites. The cations were not studied *per se*. They were solely considered as a means to thermally stabilize the zeolite lattice, while promoting and preserving most of the acidic properties of decationated zeolites. The location of cations within the zeolite framework first drew the attention to the cations themselves. It was initially viewed as an academic exercice and was not primarily intended to account for a particular behaviour of zeolites at least as catalysts. Later it was made use of cristallographers findings with the aim of explaining specific features of zeolites as catalysts. Controversies as to the cations position (including protons) were initiated which unveiled the many parameters which govern the siting of these cations. Consequently, and only then, did the cations location start to be viewed in a more realistic fashion : that is a dynamic phenomenon acted upon by several thermodynamic and chemical parameters.

Although the general prevailing view was that bivalent cations do prefer internal S_I sites, the first evidence to the contrary was brought by EPR studies of copper II Y zeolites (1,2). It was shown that even at the lowest exchange levels, Cu^{2+} ions moved towards different sites, and, at a thourough dehydration state, they were still distributed among at least two different sites. Further, it was evidenced that they could still move backward to yet different sites and eventually to the large cavities upon coordination with adequate, bulky organic molecules (1,2). These findings were further substantiated by X rays analysis of copper location, in samples characterised by higher exchange levels, and direct evidence was again provided as to the initial siting in dehydrated Cu II Y samples and the dynamic behaviour of copper ions within the zeolite porous structure as well (3,4).

Later, investigation of the dynamic behaviour of TMI was extended to other cations and differences inherent to the electronic structure, irrespective of the zeolite effects, were evidenced (5-8).

In the course of these investigations, it was clearly shown that transition metal ions coordinate N-containing ligands, olefins, aromatics, carbon monoxide, oxygen etc... In a number of cases, the transition metal ion was still held to the zeolite framework by a variable number of coordinating lattice oxide ions. Thus the zeolite behaved as a mono-to polydentate macro ligand to the central transition metal ion. Nickel (7), chromium (7) and cobalt nitrosyls (9,10), tetrahedral copper monoammine (2) are illustrative examples of the zeolite behaviour... These complexes were held both coulombically and coordinatively to the zeolite. In other circumstances i.e. hexaaquo and tetraammine and pyridine copper II complexes (2,11), cobalt isocyanate complexes (12) etc...,the central ion is shielded from the zeolite influence by its own ligands field and was only held to the zeolite structure by coulombic forces. The extreme case where only Van der Walls interactions are involved between the complex and the framework was also encountered ; these complexes could be formed in situ (13-15) or sublimed onto the zeolite or extracted from a suitable solution (16).

All these features underline the role of zeolites as a rather common solid matrix (or solid solvent), although cage effects should not be neglected. A more unique feature to zeolites is perhaps their potentialities in view of stabilizing odd oxidation states (whether even or uneven). This ability was examplified by a number of coordinatively unsaturated cations such as Copper I (1,8), Nickel I (7, 17), Rhodium II (18), Palladium I (19,20), Cobalt (o) (10), which can easily activate various molecules : O_2, olefins, CO, NO etc... In this respect i.e. stabilisation of unusual oxidation states, zeolites are singled out from other common solvents and/or plain solid matrices.

In view of the wide span of cationic states in zeolites, the well established mobility of these cations, their extreme reactivity on one hand and the unique stabilizing properties of zeolites on the other hand, one should think TMIZ are doomed to a wide use in catalysis at least in reactions where solution analogues of these transition metal complexes are usually employed. In fact despite the attractive character of well defined complexes which should allow thourough fondamental studies and permit higher dispersion of the active phase and prevent agglomeration of the active species to inert compounds, rather few studies were carried out both in view of basic research or development of new heterogeneous catalysts. Suprisingly polymeric and fonctionnalized common mineral carriers were preferred to zeolites in order to anchor soluble complexes. Perhaps this situation is to be blamed on two types of reasons :(i) the difficulty, where organic substrates are involved, to avoid premature reduction to the metallic state ; this is probably the major drawback (ii) Another inconvenience is related to matter transport or diffusion within the zeolite porous structure. This diffusion might well depress the reaction rate but more inconveniently result in further reaction of the reaction products on standing within the zeolite to produce dimers oligomers etc... Thus TMIZ could only be likely used where at least one of these inconveniences could be easily circumvented or prove to be of no effect. Thus the more likely reaction to be catalyzed over TMIZ are oxidation reactions so that reduction of the cation to the metallic state could be avoided. Other reactions with substrates such as olefins, carbon monoxide and nitrogen oxides and amines which do not seem to result necessarily in a deep reduction to the metallic state could also be contemplated. Hence oxidation, carbonylation and related reactions on the one hand, dimerisation, oligomerisation, cyclisation and addition reactions on the other hand are the most plausible reactions one could reasonably expect to carry out and investigate on transition metal ions or complexes hosted in zeolite matrices. However a compulsory condition would be,for most of these reactions,to be performed at rather mild temperatures. An alternative choice would be the use of quasi-permanent cationic zeolites (Mn^{2+}, Co^{2+}, Zn^{2+}, Fe^{2+}, Cr^{2+}) which seem to resist severe reduction conditions.

2 OXIDATION

The oxidation of hydrocarbons whether alkanes or alkenes to carbon dioxide and water is not within the scope of this review, though informative it might prove in view of understanding mild oxidation processes. In our opinion this reaction is best carried on transition metals and in this respect is not characteristic of transition metal ions catalysis. Therefore we shall concentrate on mild oxidation of hydrocarbons, with or without the incorporation of oxygen onto the organic substrate. Related reactions involving inorganic substrates such as ammonia will also be considered.

2.1 Propelyne oxidation.

Propelyne oxidation on transition metal ions exchanged zeolites (TMIZ) has been the subject of important efforts in the early seventies presumably because the particular suitability of these materials to thourough fondamental studies of the oxidation processes. In that respect, the well defined nature of the complexes formed in zeolite compared with the extreme complexity of real oxide (mixed , supported etc...) catalysts, the isolated state of each transition ion, the feeling that cooperative and intrinsic properties could be discriminated, the belief that the importance of the electronic structure of the cation, irrespective of the otherwise unavoidable differences due to different cirstallization lattices,might be estimated, have certainly originated most of the investigations carried out on these materials. However the hope that TMIZ catalyst might circumvent a number of difficulties specific to oxides, in order to design industrial catalyst has probably played its own part. In particular, an optimal dispersion of the active ion as well as prevention of sintering and relative regeneration ease or at least recovery could reasonably seem better achievable than with usual catalysts.

Though this hope did not fully materialize, the informations gained from these studies certainly deserve to be taken into account to help clarify the oxidation process which is still one of the most mysterious catalytic processes despite the undisputable progress accomplished especially by the use of isotopic labeling (21,22).

Whether because of the failure of this hope to materialize or perhaps based on an independant knowledge, the use of zeolites as mild oxidation catalysts has been considered as both unrealistic and inadequate essentially because of the wide-spread belief that only semiconducting catalysts or at least catalysts supported on semiconductors could reasonably provide for the necessary electron dissipation from the active site to the whole lattice so as to insure rapid reoxidation (that is regeneration) of the active centre (23).

As already outlined, the major difficulty in using TMIZ in catalytic reactions involving organic substrates arises in maintaining the ion at a high oxidation state. In the case of propelyne oxidation a compromise in the feed composition has to be found.: while preservation of the ionic character requires a high O_2/propene ratio, selective oxidation is favoured by excess propene. Another possibility to cirumvent this problem is the use of permanent cation type TMIZ since the Redox pontentials of various TMI vary with the nature of the transition metal ion, the O_2/propene ratio should not affect equally all transition metal ions hosted in zeolites, hence a number of transition metal ions could perhaps be selected in order to perform mild oxidation of olefins provided the O_2/propene ratio is the only determining parameter (which is not probably the case).

Among the first reports on propelyne oxidation over TMIZ, those of Mochida et al. (24)

indicated that the olefin was essentially converted to carbon dioxide and water and trace formaldehyde when Cu II Y was used with an oxygen rich feed. (2 % olefin 48 % O_2 and 50 % N_2). The authors reported that the ionic character of the copper was preserved and that no oxide was formed.

It was anticipated that TMI zeolites might act as bifunctionnal catalysts : that is addition of steam was thought to result in the hydratation of propene to isopropanol via the acidic centres ; isopropanol would in turn subsequently dehydrogenate to acetone (25). The actual result, upon addition of steam, was the formation, besides CO_2, still the main product, of significant amounts of acrolein using Ag(I) and Cu II Y zeolites. Table 1 shows the activities and selectivities achieved under, various experimental conditions of propene conversion over Cu II Y, Ag I Y and Ni II Y zeolites respectively (25)

Table 1

Catalyst	Temperature	Feed rate	Feed composition vol. %				Yield/converted propene		
			C_3H_6	O_2	H_2O	N_2	CO_2	Acrolein	isopropanol
Cu II Y	200	135.3	22.5	37.0	34.0	6.5	1.11	0.52	0.58
"	250	221.4	18.1	28.0	45.2	8.7	19.5	6.32	
"	280	157.0	22.0	36.4	29.3	12.3	23.0	6.67	0
"	300	157.0	22.0	36.4	29.3	12.3	21.9	7.20	0
"	310	288.0	17.4	17.4	0	65.2	0.83	0	0
"	310	300.0	3.0	50.0	0	47.0	3.9	0	0
Ag I Y	300	228.0	19.0	28.4	43.7	8.9	20.1	2.67	0
Ni II Y	300	221.4	18.1	28.0	45.3	8.7	2.1	0	0

Effect of temperature, feed rate, water and O_2/C_3H_6 ratio on the activity and selectivity of TMIZ

Interestingly the best yield in acrolein ($\sim$ 30%) was obtained with Cu II Y, at 250°C for a feed composition of 18 : 28 : 45 in C_8H_6 : O_2 : H_2O, the conversion at higher temperatures increased but the selectivity towards acrolein dropped. It is noteworthy that silver zeolite also produced acrolein besides CO_2 with however a lower yield while nickel zeolite exhibited poor activity and no acrolein was detected. A more extensive investigation of the transition metal ion series was performed (26). The results obtained established that propelyne was converted to CO_2 again; the following activity pattern Pd II>Pt II>Tl II> Cu II>Ag I>Mn II>Ni II>Co II>Cr III>Zu II>V IV? Was observed. It is noteworthy to mention (i) that this series included vanadium which so far has not been proved to be actually exchanged (ii) that Pd II and Pt II Y zeolites were reported to be black which is not so clear an evidence of the cationic character as the authors might have thought, it is rather indicative of the reduction to the metallic state. Furthermore the observed activity pattern, even in the case of a large oxygen excess, showed the same valcano realtionship with respect to the heat of combination per oxygen atom to form the relevant oxide as was observed with supported oxides, which is rather unexpected in the case of TMIZ and at least in sharp contrast with Rudham and coworkers findings (27), and what is more hints to a probable partial reduction of the transition metal ion and formation of a supported oxide. When the O_2/propene ratio was reverted to lower values, the main product was acrolein with CO_2 still produced. The authors however claim that the catalysts were effectively transformed into oxides. The kinetics unfortunatly were investigated irrespective of the nature of the product and only with respect to propene conversion. Nevertheless a partial order of 0.5 was observed with respect to oxygen indicating dissociative adsorption of oxygen as a rate determining step in the prevailing reaction. The order with respect to propene varied considerably indicating a wide range of interaction strength between the transition metal or ion and the olefin. More recently Rudham and coworkers (28) reported the deep oxidation of propelyne over a series TMIZ of type X including Pd,Cu, Co, Zn, Ni, Mn, Cr and Fe which exhibited this activity ordering. Furthermore, the acidity of the matrix, which could be modified via the usual way of sodium exchange with NH_4^+ ions, was shown to have a promoting effect on the deep oxidation activity and conversely pyridine neutraliza- tion of the acidity depressed the reaction rate. It was therefore suggested that propelyne may adsorb as a secondary carbonium ion at a Bronsted site and was later converted by oxygen activated on the transition metal ion. It hinted at the possibility of closely simi- lar intermediates in mild oxidation and deep oxidation. A possible scheme involving the secondary carbonium ion was suggested :

$$-O_L-H + CH_3 - CH = CH_2 \rightarrow CH_3 - \overset{+}{C}H - CH_3 + -O_L-$$

$$CH_3 - \overset{+}{C}H - CH_3 + (Cu - O_2)^+ \rightarrow CH_3 - \overset{+}{C}H - CHO + H_2O + Cu^+$$

$$CH_3 - \overset{+}{C}H = CHO + - O_L- \rightarrow CH_2 = CH - CHO + - O_L H$$

Further oxidation would ultimatly result in CO_2 and H_2O. Thus most reports indicated rather poor selectivities towards aldehydes or ketones. The only contradictory fragmentary report which claimed promising results was authored by Arai and Tominaga (29) but dit not recieve further confirmation. High yields in acetone were obtained using a Cu Pd Y zeolite operating at 120°C with a feed mixture olefin : O_2 : H_2O of 1 : 2 : 2 . 92.8 % of converted propene was acetone at a conversion level of 16.4 %. No mention of the catalyst stability was made.

However in recent experiments Cu II Y exhibited a fair selectivity in carbonyl compounds as a result of propene oxidation. Typically under optimum conditions, (O_2/propene ratio, flow rate and temperature) the yield in acrolein reaches 70 % of converted propene which could be converted to as much as 50 %. CO_2 is the second main product while minor quantities of allylic alcohol, acetaldehyde and propionaldehyde are also detected. Although the colour of the catalyst did change during catalysis, and even simply upon propelyne adsorp- tion at room temperature, there was no hint of formation of either a metallic or oxide lattice. The typical Cu II Y colour was rapidly restored upon heating in flowing O_2 at the reaction temperature. The use of excess propelyne in the feed resulted in a similar behaviour as that reported by Mochida et al. (26).: high yields in acrolein but also in CO_2 were observed and a much higher production of acetaldehyde. Under these circumstances the initial Cu II Y could not be restored and Cu O was detected upon heating at 500°C in flowing O_2. Therefore Cu II Y has been probably reduced to metallic copper and operated possibly as a $Cu(o)/Cu_2O/CuO$ mixture (30). Upon using cobalt II Y the activity as well as the selectivity in acrolein dropped drastically and the only carbonyl compound produced in significant amounts was acetaldehyde (31).

2.1.1 Influence of the kinetic parameters on the activity and selectivity.

. The oxygen to olefin ratio.

Because it primarily determines the overall atmosphere, this parameter is the most critical one. Depending on the range of values taken by this parameter, the nature of the operating catalyst may be drastically affected. EPR measurements showed that upon equili- bration of activated copper II zeolite by an O_2-olefin mixture, a variable degree of copper II reduction was observed (30). This partial reduction level increased with increa- sed olefin/O_2 ratio. Similar reduction has also been reported by Leach et al. (32) in the presence of pure olefin and suspected or evidenced in a number of cases upon adsorption of organic substrates on a variety of TMIZ (1, 33). On the other hand, this ratio probably affects the coverage in both oxygen and propelyne even though the adsorption centres might be different since it is expected that olefins may adsorb on both high and low oxidation state transition metal ions, while O_2 should be primarily coordinated by low valent cations. It was often reported that olefins were more strongly coordinated to TMI than O_2, hence it is of the utmost importance that the O_2/olefin ratio should be well in favour of oxygen, which indeed will unfortunatly enhance deep oxidation rates.

. Reaction temperature : This parameter will influence the relative concentration of partially reduced to oxidized forms of the TMI at least as strongly as the O_2/olefin ratio and therefore would modify the kinetic as well as thermodynamic parameters of the reaction. While at high temperatures effective reoxidation of the TMI should be better achieved, again overheating and kinetic parameters should favour deep oxidation. At lower temperature, where mild oxidation rates are thermodynamically enhanced, oxidation of the TMI is not so well achieved and metal formation as well as other side reactions might either clog and poison the surface or favour deep oxidation. Hence again a difficult compromise especially in view of overheating effects is to be achieved in the choice of the reaction temperature depending strongly on the Redox potential of the TMI. One usually is helped in seeking the best conditions by the possibility to adjust the flow rate so that thermal effects could be minimized and various rates optimized. Also dilution of the reagents mixture should improve selectives. Water vapour proved also to improve selectivities (25). It seems that its high thermal capacity helped achieve a more effective removal of excess heat from the catalyst bed. This latter effect is probably responsible of the depressed deep oxida- tion rate observed by Mochida et al (25) with respect to acrolein formation rate since no acetone was formed which would have been expected, should a bifunctionnal reaction invol- ving an intermediate hydration proceed.

2.1.2 Acidity and electronic structure.

Rudham and coworkers (28) reported that increased acidity favoured deep oxidation. The scheme put forward involved a secondary carbonium ion which subsequently reacted with dioxygen to give rise to a γ protonated acrolein. Thereafter this species could give up a proton and desorb or further reacts to degradation products. Whether such a species might exist is not certain, (usually the double bond is less reactive than the carbonyl towards further oxidation of acrolein i.e. acrylic acid and eventually CO_2), however it is well conceivable that acidity favours condensation of both the olefin and the possible mild oxydation products. The adsorbed polymerised species could then hardly desorb and presu- mably undergo various dehydrogenation reactions leading to carbonaceous materials and, provided temperature is high enough and excess O_2 available. CO_2 is evolved, otherwise the whole surface is simply clogged. It is quite often that oxidation at low or moderate tem- perature on TMIZ exhibits an important carbon unbalance (31). Thus acidity is presumably bound to depress selective oxidation of olefins (except perhaps C_2H_4), even though acidity would be reasonably expected to promote hydration (H_2O always present at least because of the deep oxidation reaction) and lead subsequently to dehydrogenation to form ketones. Unfortunately at temperatures where dehydrogenation aquires a decent rate, the trend is usually to dehydration. Perhaps, extremely high water pressure might revert to hydration and hopefully bifunctionnal schemes might be contemplated.

One of us has investigated the effect of acidity changes on the activity and selectivity of Cu II Y zeolite. The acidity was modified by the usual procedure. While when no additional acidity was present besides that due to water heterolytic cleavage upon Cu II exchange (6-8 ions per unit cell) acrolein was the main product, increased acidity resulted in an overall decrease of both the activity and the selectivity towards acrolein production (30). Only further increase of copper II exchange level would restore the activity and selectivity values observed on Na Cu Y.

However the quantitative aspect of acidity is not the only parameter to affect both activity and selectivity. Comparison of copper II X and copper II Y zeolites showed drastic influence of the Si/Al ratio. Oxidation on Cu X needed significantly higher temperatures to proceed than on equally exchanged Cu Y. This extra rise in temperature to achieve comparable conversions resulted in significantly lower yields in acrolein and increased yields in CO_2. Hence the charge distribution among the lattice significantly influences the activity and selectivity as well (30). Another determining parameter is certainly the electronic structure of the TMI. While copper II produced acrolein, cobalt II seemed to yield essentially carbon dioxide and cobalt III amine allylic alcohol (34). Rh III Y and Pd Cu Y produced acetone selectively. Yet far too little reliable and detailed data are available in the litterature concerning mild oxidation so that no coherent straightforward correlation between the nature of the products and the electronic structure could be reasonably attempted.

2.1.3   Nature of the activated species.

The electronic structure of the transition metal ion is certainly of importance even though its effects could not be at present, correctly evaluated. In fact the activation mode of both reagents as well as the Redox potential which greatly affects the kinetic parameters are primarily determined by the TMI electronic structure.

. Activation of oxygen : In a number of studies oxygen was shown to be activated associatively or dissociatively on various TMI and TMIZ where the cations were usually at a low oxidation state. Kasai et al. (35) showed that the superoxide ion could be formed on a variety of exchanged zeolites. Species such as $\left|Cu-^{O}\smallsmile Cu\right|^{2+}$ (1,36) were shown to exist whose oxygen was found to be extremely reactive. Cobalt II ammine proved to undergo an oxidative addition of dioxygen and both $\left|Co\ III\ O_{2}^{-}\right|$ and $\left|Co\ III\ -O_2\text{-}Co\ III\right|^{4+}$ and $\left|Co\ III\ -O_2\text{-}Co\ III\right|^{5+}$ were characterized (37, 38). $(NiO_2)^{+}$ was also reported (39) rhodium I ions were shown to form μ peroxo complexes of Rhodium II (Rh II$-^{O}\diagdown_{O}-$Rh II) ; this species reacted readily with olefins including ethylene (18, 33). This brief survey simply shows that all possible oxygen species might be present in TMIZ, therefore it is not surprising that a coherent relation between selectivities and the nature of activated oxygen should be difficult to find out.

. Activation of propelyne : Propelyne also adsorbs in various ways on TMIZ thus equilibration of Pd II, Cu II, Ni II Y etc... with propelyne, invariably resulted in a one-electron reduction of the transition metal ion (33,32) and presumably in the formation of the propelyne positive radical. In fact the isotropic ESR line observed under these conditions lacked the necessary resolution to enable a clearcut identification of this radical. On the other hand in a number of other cases propelyne could adsorb in a significantly different manner. In particular hydride abstraction from the methyl group could well account for the transition metal ion one-electron reduction ; under these circumstances a π allyl cation would form which might undergo a variety of transformations, dimerisation, oxidation etc... Equilibration of copper II zeolites with $CD_2 = CH-CH_3$ showed that carbonium ions were formed (32). However, alternatively σ-allyl complexes of copper I are likely to form as an initial step, as suggested by EPR spectroscopy which provides evidence for copper II reduction and IR measurements which indicate the formation of hydroxyl groups. Schematically the various steps could be visualized as follows :

$$-O_L- Cu \; II + CH_2 = CH-CH_3 \longrightarrow -O_L- Cu(I) - CH_2 - CH^+ - CH_3$$

$$-O_L- Cu \; II - CH_2 - \overset{+}{CH} - CH_3 + -O_L- \rightleftharpoons -O_LH + O_L- Cu \; I - CH_2 -CH = CH_2$$

$$-O_L- Cu \underset{CH_2}{\overset{CH_2}{>}} CH \rightleftharpoons -O_L - Cu \; I \underset{CH_2}{\overset{CH_2}{<}} CH$$

Recent ^{13}C NMR studies involving Pt II Y, Ag I and Cu I Y zeolites (40,41) showed that propelyne also formed stable π complexes identical to those observed in solution (Zeiss salts). The π bond exhibited an assymetric character in that the CH_2 group was priviledged with respect to the methyne group. This assymetric bond may be interpreted as a trend to form a σ bond between the methylene group and the ion. A recent IR report confirmed the formation of strong π bond between Cu I and Ag I cations and propelyne in Y zeolites Thus again the complexity of propelyne activation reflected by the rather various species one might encounter does not provide a straightforward means to correlate product distribution to the structure of coordinated propelyne. However Mochida et al. (26) attempted such a challange and correlated the hardness of the transition metal ion with its overall activity in propene conversion. However the scattering of their experimental results on the graph, to our opinion, precludes reliable correlation. In fact too few studies were aimed at finding the possible relationship between the structure of activated propene and the catalytic activity and/or the products distribution.

2.1.4 Possible oxidation pathways.

In spite of the severe lack of experimental spectroscopic and kinetic data as far as selective oxidation is concerned one can tentatively postulate general mechanistic schemes relying particularly on widely accepted schemes in homogeneous catalysis for soluble homologues of the transition metal complexes hosted in zeolites.

Copper II zeolites proved to convert specifically propene to acrolein, although minor quantities of other carbonyls do form. Unfortunally no homogeneous oxidation of propene to acrolein was reported. The only catalysts which promote this reaction are heterogeneous oxides or mixed oxides such as CuO/Cu_2O, MoO_3/Bi_2O_3 and SnO_2/Sb_2O_5 etc...

Thorough mechanistic studies were performed which over-whelmingly pointed to a symmetric π allyl intermediate to acrolein formation. Adsorption of perdeutero propelyne reported by Leach et al. (32) indicated that $CD_2H-CH = CH_2$ was formed starting from $CD_2 = CH-CH_3$. This would be in agreement with a symmetric π allylic intermediate coordinated to copper ions, although other species should not be excluded. ^{13}C NMR studies suggested that π complexes are formed which involve a priveledged interaction with the methylene group. Such interaction could eventually labilize the allylic protons one of which would leave the molecule to form a hydroxyl group with a subsequent rearrangement of the copper olefin complex to a σ allyl in equilibrium with a symmetric π allyl according to the following scheme.

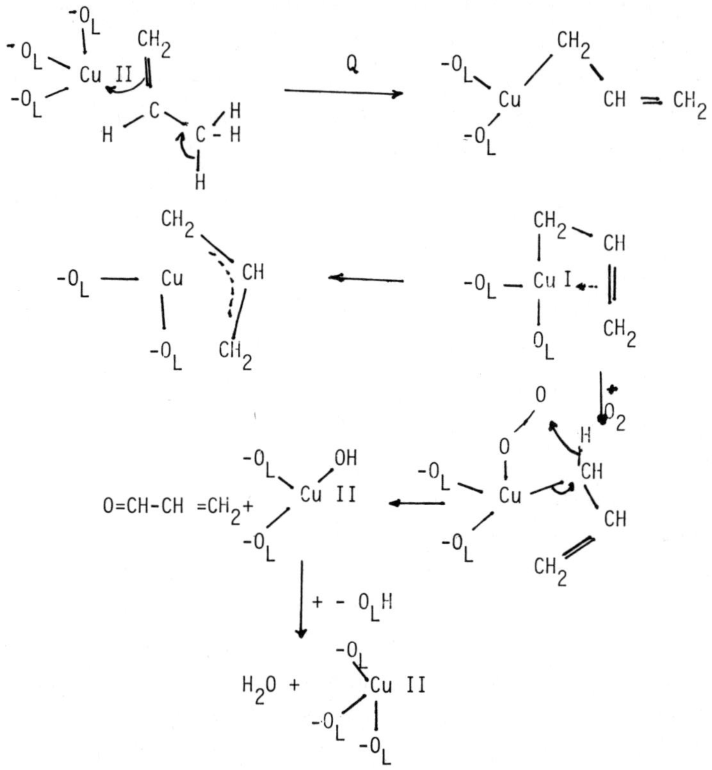

In the case of rhodium zeolite, acetone was selectively produced. This result is in line with consistent reports of acetone production in liquid media catalyzed by Rhodium chloride and copper chloride or Iron chloride and other additives. As it was shown in many instances that zeolite hosted rhodium behaved similarly to soluble rhodium salts in complex formation (carbonyls,μ peroxo species,olefin adducts etc...),here again we propose an overall scheme to account for propelyne conversion to acetone which bears close resemblance to homogenous oxidation of propelyne.

$(-O_L-)_n$ Rh II $\overset{O}{\underset{O}{\diagdown}}$ Rh (II) $- (O_L)-_n$ has been characterized by EPR technique and was shown to form readily upon activation of Rh III $(NH_3)_5$ Cl Y zeolite at 400-500°C under flowing oxygen.

This species readily reacts with propelyne resulting in the disappearance of the EPR signal. This could be visualized by the following reaction scheme

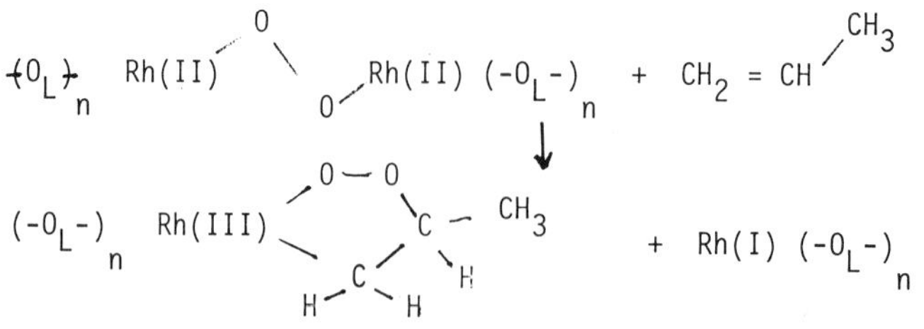

$$\{O_L\}_n \quad Rh(II) \overset{O}{\underset{O}{\diagdown}} Rh(II) \ (-O_L-)_n \quad + \quad CH_2 = CH \diagup^{CH_3}$$

$$\downarrow$$

$$(-O_L-)_n \quad Rh(III) \diagdown \overset{O-O}{\underset{\underset{H \diagdown C \diagup H}{C}}{\diagup}} \overset{CH_3}{\underset{H}{\diagdown C}} \quad + \quad Rh(I) \ (-O_L-)_n$$

all diamagnetic complexes

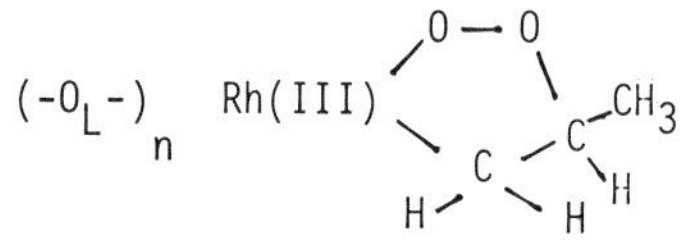

$$(-O_L-)_n \quad Rh(III) \diagdown \overset{O-O}{\underset{\underset{H \diagdown C \diagup H}{C}}{\diagup}} \overset{CH_3}{\underset{H}{C}}$$

The latter is analogous to various complexes of platinum which have been fully characterized by a number methods including X Rays diffraction (43-47). This complex would then rearrange leaving acetone and rhodium oxo species in the following way :

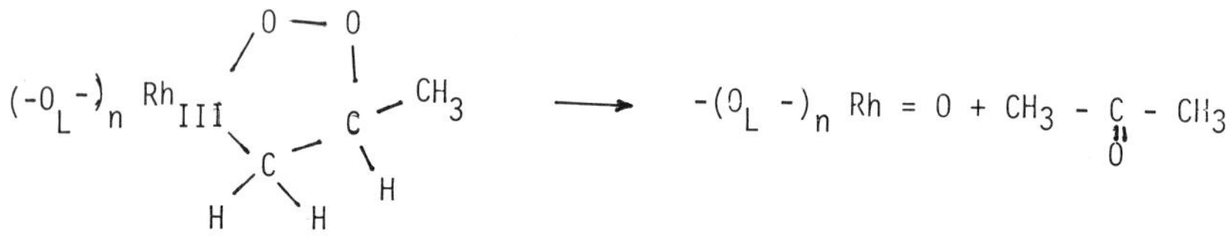

$$(-O_L-)_n \ Rh_{III} \diagdown \overset{O-O}{\underset{\underset{H \diagup \diagdown H}{C}}{\diagup}} \overset{CH_3}{\underset{H}{C}} \longrightarrow -(O_L-)_n \ Rh = O + CH_3 - \underset{O}{\overset{}{C}} - CH_3$$

The oxo species is easily polarised due to the zeolite acidity and the following scheme would account for further formation of acetone

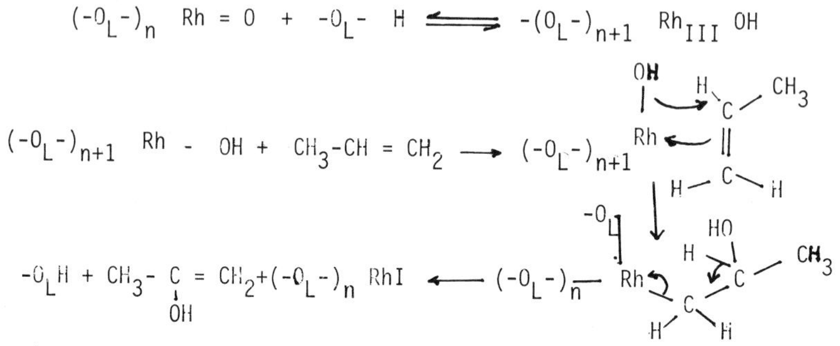

$$(-O_L-)_n \ Rh = O + -O_L- \ H \ \rightleftharpoons \ -(O_L-)_{n+1} \ Rh_{III} \ OH$$

$$(-O_L-)_{n+1} \ Rh - OH + CH_3-CH = CH_2 \longrightarrow (-O_L-)_{n+1} \ \overset{OH}{\underset{Rh \leftarrow}{\overset{|}{\diagdown}}} \overset{H \diagdown_C \diagup CH_3}{\underset{H-C \diagdown H}{\|}}$$

$$-O_L H + CH_3- \underset{OH}{\overset{}{C}} = CH_2 + (-O_L-)_n \ RhI \ \longleftarrow \ (-O_L-)_n \ Rh \overset{HO}{\underset{H \diagdown H}{\diagdown} \overset{H}{\underset{C}{\diagdown}} \overset{CH_3}{\diagup}}$$

the enolic form of acetone in acidic medium readily desorbs as acetone and Rh III is reduced to Rh I and oxidized again by molecular oxygen.

In this mechanism we postulated the involvement of acidic OH groups as a necessary step to further deplete activated oxygen. The experimental observation is in line with this assumption since while Na Rh Y did actually catalyse acetone formation the catalyst deactivated rapidly. By contrast Rh HY showed both higher activity and a remarkable stability (48). The most striking analogies and closest similarities between zeolite hosted transition metal complexes and their solution analogues would be perhaps even better illustrated in their behaviour in Wacker-like oxidation of ethylene to acetaldehyde.

2.2 Oxidation of ethylene.

In an early report Arai and Tominaga (29) reported the oxidation of ethylene to acetaldehyde on a series of Cu II Pd II exchanged Y zeolites. It was reported that neither Cu II Y nor Pd II Y alone could carry out this reaction under whatever conditions. It was shown that this reaction could be performed at as low temperatures as $110°C$, which is fairly close a temperature to that used in homogeneous processes. However, rapid pore filling apparently precludes further investigation of this catalyst. Recently this system was investigated (48) and it was observed that indeed the reaction rate decreased rapidly at $120°C$ but a substantial rise of the reaction temperature ($200-220°C$) resulted in a fairly stable catalyst over a very long period. The optimum Cu/Pd ratio appeared to be 4-5 : 1.

Recent EPR studies (48) showed that indeed upon adsorption of ethylene Pd II ions were readily reduced to Pd I ions even in the presence of excess O_2. interestingly enough, upon progressive coexchange of copper II into a Pd II Y zeolite at constant exchange level in Pd II, the reduction level was dramatically decreased in the presence of the reaction mixture untill a ratio of 4-5 copper ions per Pd ion was reached. In addition the acidity of the zeolite seemed to have a promoting effect on both the catalytic activity and also on the ease of reoxidation of Pd I species. This pattern of features fit in what is now generally accepted as to the role of each cationic component in this bicationic catalytic process. Indeed it is postulated, and several systematic studies confirmed, that Pd II ions were reduced to Pd(o) state essentially by ethylene while the cocatalyst is not affected by the addition of ethylene. The following scheme might illustrate the overall catalytic oxidation of ethylene.

$$-O_L- + (-O_L-)_n \; Pd \; II + H_2O \rightleftharpoons (-O_L-)_n \; Pd \; II - OH + -O_L-H$$

$$+ C_2H_4$$

$$(-O_L-)_n - Pd - OH$$

$$\left[(-O_L-)_n - \underset{CH_2}{\overset{HO}{PdI}} \overset{H}{\underset{H}{C}} \right] \longleftarrow \underset{H}{\overset{H}{C}} \rightleftharpoons \underset{H}{\overset{H}{C}}$$

$$\longrightarrow (-O_L-) \; Pd(0) + -O_L \; H + CH_2 = CHOH \longrightarrow CH_3 \; CHO$$

Subsequent reoxidation of Pd(o) to Pd I seems to be a too fast process so that no Pd(o) could be observed

$$Pd(o) + Cu^{2+} \xrightarrow{rapid} Pd^+ + Cu^+$$

$$Pd^+ + Cu^{2+} \longrightarrow Pd^{2+} + Cu^+$$

Finally oxidation of cupreous ions by molecular oxygen is promoted by Brönsted acidity

$$2 -O_L- H + 2 Cu^+ + 1/2 O_2 \longrightarrow H_2O + 2^{-O_L-} + 2 Cu^{2+}$$

It is noteworthy that in the case of soluble complexes there is no report of the presence of Pd I species. No mention of the rapid reoxidation of Pd(o) to Pd I in the case of Pd^{2+} doped vanadium pentoxide (49) where Pd I species could concievably be better isolated than in solution. This difference might indicate that higher reoxidation rate of Pd(o) is achieved in zeolite which would certainly have an enormous importance since it would definitely prevent aglomeration of metallic palladium whose reoxidation is more difficult and in a may prevent catalyst ageing.

As to the inertness of copper II to ethylene, it was evidenced by a remarkable steadiness of the Cu^{2+} EPR signal upon addition of ethylene (in sharp contrast to its high reactivity towards propelyne) and a recent report (42) confirmed that Cu II in Y zeolite interacted with ethylene by Van der Waals forces.

The involvement of protonic acidity in the reoxidation step emphasizes the very close similarity between the catalytic properties of transition metal ions in zeolites and theirs in solution clearly pointing out the potentialities of zeolite hosted transition metal complexes, while outlining the role of a convenient solid solvent of the zeolite which provide the opportunity to modify at one's will the acidity of the medium.

3 DEHYDROGENATION REACTIONS

These oxidation reactions without incorporation of the activated oxygen to the organic substrate fall into two categories. Those where the organic substrate is a plain hydrocarbon examplified by cyclohexane and those concerning fonctionalized substrates such as alcohols. The interaction between the catalyst and either types of substrates might be different and therefore result in two different reaction classes.

3.1 Dehydrogenation of alcohols.

Few studies were devoted to this reaction despite of its obvious practical importance. Again the competing dehydration reaction precludes thorough investigations of this reaction. One of the earliest studies was reported by Kazanskii and Coworkers (50) where ethanol dehydrogenation over Cu Y was performed at 200-300°C. Using EPR spectroscopy the formation of a copper II octahedral complex involving three lattice oxide ions and three oxygens from ethanol molecules was detected at S_{II} sites. The complex decomposed upon heating resulting in hydrogen evolution and acetaldehyde formation which desorbed from the catalyst. results set in table 2 show the relative evolution of dehydration and dehydrogenation reactions depending on the reaction temperature.

Table 2

Catalyst	H_2%			C_2H_4 %		
	200	250	300	200	250	300
Cu(-7-11 %) Y	traces	2	5	6	18	40
Cu(29-44 %) Y	1	5	7	12	45	43
Na Y	0	0	0	0	0	15
Ca Y	0	0	0	0	15	45

Relative percentages of H_2 and C_2H_4 deriving from catalytic conversion . of C_2H_5 OH.

The active species in dehydrogenation reaction was ascribed to copper II ions coordinatively insaturated presumably at S_{II} sites while dehydration activity was attributed to increased acidity upon exchange with divalent cations. The assumption that the strong elec-

electrostatic field might polarize the alcohol molecule was discarded since Ca Y does not initiate the dehydrogenation reaction. Recently (51) benzyl alcohol dehydrogenation to benzaldehyde was reported to occur on Cu II Y at temperatures in the range 300-350°C. Apart from benzaldehyde, CO_2 and H_2O were produced resulting from deep oxidation. The activity of a number of copper exchanged zeolites was compared to those of reduced copper Y, Na Y and HY. It appeared (table 3) that Cu^{2+} ions seemed to be highly active sites, although the selectivity to benzaldehyde was lower than that obtained with pure HY zeolite.

Table 3

Catalyst	Conversion mol. %	Yield of benzaldehyde mol. %	selectivity to benzaldehyde
Cu II Na Y-37	3.3	1.5	46
Cu II Na Y-45	4.7	1.6	34
Cu II Na Y-67	8.0	3.0	38
Cu I Na Y-37	1.7	0.8	47
Cu I Na Y-45	2.4	0.9	37
HY	1.0	1.0	100
Na Y	1.5	0.6	40

P_{O_2} = 140 Torr $P_{C_6H_5 CH_2 OH}$ = 20 Torr P_{N_2} = 600 Torr

from this table the authors deduced that Cu II was the most active species. However the fact that Cu I Y samples were found to be less active than fresh samples is rather puzzling since one would expect that under the working experimental conditions the same equilibrium between Cu II and Cu I should be reached at steady state conditions.

The effect of the exchange level was also investigated and it was apparently concluded that copper ions were inactive until an exchange level of 25% was reached. Then the activity increased linearly except for the totally exchanged zeolite where the activity dropped to quite a low value. However the selectivity to benzaldehyde did not follow the same trend although the yield in benzaldehyde increased slowly. Also the effect of the pretreatment temperature and atmosphere on the selectivity was investigated. Pretreatment with H_2 was shown to increase the selectivity to benzaldehyde although the overall conversion rate decreased. Thus apparently Brönsted acidity seemed to promote benzaldehyde formation which is consistent with the excellent selectivity obtained over HY. However no explanation for this phenomenon was attempted especially that acidity would be expected to promote side reactions such as dehydration etc... The authors (51) reported a dissimilar effect of pyridine and piperidine on the activity and selectivity:pyridine appeared to poison the activity and reduce the benzaldehyde yield. By contrast piperidine promoted drastically the activity while increasing also the benzaldehyde yield even though the selectivity was depressed. The authors related this contrasted behaviour to the higher covalent character of the Cu II - N bond in the copper piperidine complex (51). Finally the kinetic study showed that the conversion rate was first order in oxygen pressure and inverse first order with respect to benzyl alcohol suggesting that O_2 and benzaldehyde competed for the same adsorption site and that O_2 adsorbs associatively .

3.2   Dehydrogenation of cyclohexane

This reaction was investigated over a variety of TMIZ by two groups (29,52) Seiyama
et al.(52) found the conversion of cyclohexane to yield carbon dioxide and benzene. The
selectivity over Cu II Y using a wide range of cyclohexane : O_2 ratios remained essentially
constant at around 80 % for a reaction temperature of 320°C and a constant O_2 pressure
of 228 Torr. The kinetics showed a first order rate with respect to oxygen, irrespective
of cyclohexane pressure as far as benzene formation was concerned. The CO_2 rate by contrast
varied with the square root of oxygen partial pressure independently of the cyclohexane
pressure.

Pulse experiments showed that preliminarly adsorbed cyclohexane or cyclohexene produced,
when equilibrated with oxygen aliquots, mainly degradation products. On the other hand
preadsorbed oxygen led to essentially benzene when equilibrated with cyclohexane pulses and
to CO_2 when equilibrated with cyclohexene. Thus cyclohexene was discarded as a possible intermediate. This view was further confirmed by the different rate variations of cyclohexene
and benzene production as a function of the contact time thus denying the consecutive
formation of benzene from cyclohexene. The first order kinetic in oxygen pressure observed
for the reaction rate law together with the zero order in cyclohexane suggested that the
associative adsorption of oxygen is rate determining. The following scheme was deduced
from the kinetics.

$$O_2 \ + \ catalyst \longrightarrow O_2 - catalyst$$

$$C_6H_{12} \ + \ O_2 - catalyst \longrightarrow C_6H_8 \ + \ 2 \ H_2O \ + catalyst$$

$$C_6H_8 \xrightarrow{\ rapid\ } C_6H_6 \ + \ H_2$$

$$2 \ H_2 + O_2 - catalyst \xrightarrow{\ rapid\ } 2 \ H_2O \ + \ catalyst$$

Cyclohexadiene would dehydrogenate rapidly, evolved hydrogen would also consume adsorbed
O_2 in a fast step. This scheme was noticed by the authors to be at variance with Balandin's
theory which anticipates a sextet scheme. In fact this is not unexpected since Balandin's
theory was concerned essentially with dehydrogenation on metal surfaces and certainly not
on isolated ions in solid matrices.

An alternative mechanism has also been proposed involving a Redox process of Cu II/Cu I
and Fe II/Fe III and assuming a dissociative activation of oxygen and the formation of cyclohexene as an intermediate (29). In the absence of more detailed experimental data it
is not clear whether the apparent discrapancies could be accounted for by different experimental conditions. In particular whether or not cyclohexene could be a possible intermediate
seems difficult to settle. On the contrary depending on the pretreatment conditions, on
whether the steady state has been reached or not, and on the reaction temperature and on
the hydrocarbon to oxygen ratio, associative and dissociative oxygen activation are conceivable as already outlined in a former paragraph.

As activated O_2 seemed to play a major role in the kinetics of benzene production a correlation between O_2 sorptive capacities and the activity and selectivity of a series of
TMIZ was attempted (52). Among the ions investigated i.e. Cu II, Pd II, Cr III, Zn II, Ni II
and Ag I, only Cu II Y zeolite adsorbed a significant amount of oxygen. Its sorptive capacity was found to be two orders of magnitude that of any other catalyst in the series. Significantly, the Cu II Y zeolite was by far the most active. However the larger sorptive

capacity of Cu II Y is hardly concievable without the assumption that at least partial reduction of copper II to copper I ions has occured prior to O_2 chemisorption measurements. Similar measurements were conducted by Arai and Tominaga (29) who showed that Cu II Y and Fe III Y hardly adsorbed oxygen while Cu I Y and Fe II Y adsorbed half an oxygen atom per cation aproximatly at 400°C under a 100 Torr of oxygen. Thus it seems reasonable to expect that reduced forms of TIMZ adsorb oxygen and that the nature of adsorbed O_2 might vary with the experimental conditions so that associative as well as dissociative activation of oxygen are not contradictory.

Whatever the actual nature of the adsorption site of oxygen, all experimental data suggest that it is the adsorbed oxygen which reacts with the hydrocarbon substrate. This is in sharp contrast with the general view that <u>on metal oxides, it is the lattice oxide ion that is incorporated</u> in the organic substrate or removed as water. Thus the role of oxygen should rather be compared to that of oxygen activated by homogeneous complexes such as cobalt salen, Rhodium complexes, metallo-porphyrine etc... This feature again relates zeolite hosted transition metal ions to their solution analogues and illustrates the specific solvent character of the zeolite.

3.3   On the copper exchange Level effect

The effect of the exchange Level on the activity of various TMIZ in propelyne oxidation (28), cyclohexane dehydrogenation (52), benzyl alcohol dehydrogenation (51) and on a number of other reactions has been quite often mentioned. The general idea refered to, in order to account for the usual disturbing parabola shaped relation between the activity and the exchange level, is the sacrosanct cation inaccessibility principle. This very principle has been refered to even in the case of decationated zeolites where protons should not be considered in any way as a fictitious hidden cation. In fact it seems that this principle was the legitimate descendant of the well known steric hindrance so relied upon in organic chemistry before electronic considerations were put forward.

In the field of zeolite, although still in honour, this dogma seems inacceptable at least in two cases : (i) protons which are probably the most mobile species in chemistry (besides smaller particles) and which mobility in the case of zeolite has been unambiguously proved and publicised by innumerable NMR studies (ii) transition metal ions in zeolites and particularly copper II ions which high mobility at rather low temperature was demonstrated both by direct and indirect measurements (1-4). Yet the experimental data even though seeming to hint to the cation inaccessibility scheme, some other interpretation must be attempted which take in to account the well established mobility of transition metal ions within the zeolite lattice. At this stage of knowledge of zeolite behaviour it would seem closer to the truth to hypothise intrinsic cation properties changes with the exchange level, together with the concomittant influence of other parameters which inevitably vary with the exchange degree. For example as water is present in most of oxidation processes, water heterolytic splitting occurs inevitably. The formation of hydroxyl groups, quantitatively and qualitatively affects the charge distribution within the whole lattice and certainly modifies the charge density at the external sites as well. This in turn is bound to modify the nature of the cation to the zeolite bond and, consequently, affect the Redox potential and other physico-chemical characteristics of the cation. Grossly with the exchange level varying, the electronic character of the macroscopic ligand to the

cation changes thus affecting the properties of the coordinated cation. Such changes in
electronic properties of the cation with exchange level have already been witnessed.
Cu II ions, for instance, form N-copper bonds with pyridine at the low exchange level to
form the square planar tetrapyridine copper II complex (2). At higher exchange levels
π-bonds involving the pyridine aromatic ring are formed (4) ; and EPR investigation showed
that charge transfer complexes were formed involving one electron transfer from the π cloud
of the pyridine ring to cupric ions to form pyridine cupreous complexes (53). Thus, where
inaccessibility of copper ions to the incoming pyridine molecule, cannot be refered to,
it was obvious that copper II ions behaved in rather different manner due likely to important changes in their electronic properties, Redox potential, acidic properties etc...
Therefore we think there are better grounds than the dubious steric excuse to account for
particular variations of the TMIZ activities as a fonction of the exchange level.

5. MICELLANEOUS OXIDATION REACTIONS

5.1 Olefin ammoxidation

Actually few studies were devoted to the oxidative amination of olefins on transition metal ion exchanged zeolites. The early study by Margolis and coworkers (54) was
performed on a variety of Fe III exchanged zeolites namely X, A and Y zeolites. The starting Na and Ca-zeolites were also tested. The major products from propelyne conversion at
480°C were acrylonitrile, CO and CO_2. Increased exchange level resulted in increased activity and selectivity. The activity pattern among the different types of zeolites was
Fe Na X > Fe Na Y > Fe Na A. Apparently replacement of sodium ions by calcium does not affect
significantly the selectivity and activity. Thus the zeolite acidity does not seem to play
a major role,which is not surprising since the presence of excess ammonia is expected to
neutralize the framework acidity and therefore to hinder any possible effect. The observed
activity pattern was simply explained in terms of spatial availability for the reaction
since the supercages of zeolite X were slightly larger than in the Y type and of course
than those of the A type.

In a more recent investigation (55) Zn Y zeolite was reported to yield acetonitrile
at temperatures ranging 350-480°C from propelyne conversion. It was suggested that acetaldehyde was the intermediate product in propelyne conversion. It was shown using IR spectroscopy that acetaldehyde was converted to acetonitrile in presence of O_2 and NH_3 on Zn Y.
In the absence of ammonia, acetaldehyde was converted to degradation products (which again
is in agreement with Rudham and coworkers (28) observation that deep oxidation was promoted
by increased acidity of the zeolite though it is at variance with the proposition that both
deep and selective oxidation might proceed via closely related intermediates presumably
induced by Brönsted sites.It is in line however with the view expressed in a former paragraph
that acidity might most probably result in depressed selectivities towards mild oxidation
products because of acid induced subsequent transformations of the carbonyl compounds.

5.2 Ammonia oxidation by nitric oxide.

The reduction of nitric oxide to yield nitrogen and water has been extensively studied
(56-59) due to its significance in the removal of nitrogen oxides from industrial plants
effluents. This reaction appeared to be conviently carried out on transition metal ion
exchanged zeolites and particularly copper II amine complexes, which not only exhibited high

turnover numbers but higher stability and enhanced activity at lower temperatures than usual metal or metal oxide catalysts. Co II and Cu II Y zeolites were shown to be good catalysts with higher performance of the latter. The nature of the active species and the mechanistic aspects were investigated using kinetic results and joint EPR and IR spectroscopy. In the case of Cu II Y zeolite which optimum operation temperature was around 110°C, EPR studies showed that the $|Cu(NH_3)_4|^{2+}$ complex was the active precursor at moderate temperatures (57). IR studies conclusively demonstrated that no mixed nitrosyl amine complex was formed under the reaction mixture. On the other hand isotopic labeling showed that while N_2 was derived from a NH_3 molecule and a NO molecule, when N_2O was formed, it was derived from the coupling of two NO molecules. At lower temperatures the rate determining step was thought to be the reaction of gas phase NO with coordinated ammonia while at higher temperature i.e. T >110°C a significant reduction of copper II to copper I occurs which slows down the reaction rate. Moreover substitution of ND_3 to NH_3 resulted in a neat kinetic isotopic effect of $\sim$ 1.5 indicating that N-H (or N-D) bond rupture is rate determing. On the other hand it was observed that the rate constant increased linearly with increased copper content. Cobalt Y zeolite was active at significantly higher temperatures around 200°C and the decomposition of NO to N_2O increased significantly since the N_2O/N_2 ratio was always more than unity (58). Again N_2 was found to be formed through the combination of a NH_3 nitrogen atom and a NO nitrogen atom. Both mixed mono nitrosyl amine and dinitrosyl amine complexes were detected. The latter complex well accounts for increased N_2O yield.

6 CARBONYLATION REACTIONS

6.1 Carbonylation of alcohols

Among alcohols, methanol carbonylation is probably the most interesting due to the large scale industrial use of acetic acid as a solvent and as a starting material for the synthesis of vinyl acetate. Acetic anhydride also finds its use as an acetylating agent and in the manufacture of cellulose acetate, aspirin and other organic chemicals.

The operating homogeneous processes are using cobalt based catalyst developed by BASF while Monsanto has developed a rhodium based catalyst. The latter operates under significantly milder conditions i.e. at 180°C under 30-40 atmospheres providing a high selectivity to acetic acid.

The actual starting materials are rhodium salt and methyl iodide as a promoter. The catalyst activation presumably involved a reduction of rhodium III salt to Rh(I) carbonyl $|Rh(CO)_2I_2|^-$ under CO pressure. Subsequent oxidative addition of methyl iodide to the carbonyl produces methyl rhodium III compounds which could insert CO into the C-Rh bond and coordinate an additional CO ligand, thus producing the acethyl-rhodium carbonyl complex. This latter ultimatly reductively eliminates acetyliodide. The overall scheme is pictured in figure 1.

6.1.1 The structure of rhodium and iridium catalysts precursors in zeolites.

Carbonylation of methanol at atmospheric pressure has been successfully attempted on rhodium zeolite based catalysts (60-65) and further extended to higher alcohols (60,61). Several investigation means were employed in order to identify the active species and to elucidate the reaction mechanism. In particular IR and ESCA studies, isotopic labeling and

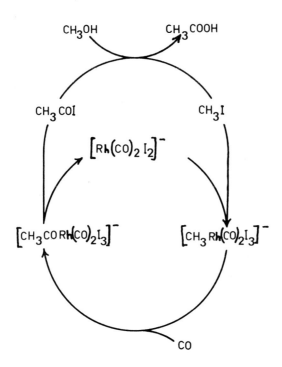

Figure 1 : Schematic representation of rhodium catalyst cycles in acetic acid synthesis

kinetic investigations were performed. More recently iridium complexes solvated in zeolites were also successfully used to carry out methanol carbonylation (66). Rhodium exchanged zeolites (X and Y) were prepared by the usual exchange procedure via the chlororhodium pen-tammine chloride (66,67) or directly by impregnation with a rhodium salt solution (60,62) which might not provide an exchanged cation but yet an effective catalyst.

Treatment with CO of activated rhodium exchanged or impregnated zeolite was shown, using ESCA, IR and volumetric measurements, to produce rhodium dicarbonyl attached to the zeolite framework or simply occluded within the cavities (67).

Iridium exchanged zeolite was prepared by the usual ion exchange procedure using an aqueous solution of the chloroiridium III pentammine   chloride. The complex $|Ir(NH_3)_5Cl|^{2+}$ is then substituted to sodium ions as evidenced by IR measurements (ν N-H and δ NH_3) and ESCA results for iridium, chlorine and nitrogen (68). Upon heating in flowing oxygen at 250°C and subsequent evacuation ammonia ligands were removed while chlorine atoms were unaffected according to ESCA results. Equilibration with CO, at the same temperature as that used during methanol carbonylation i.e. 170°C, resulted in the development of two twin strong bands at 2102-2086 and 2030-2001 cm^{-1} along with characteristic absorptions dues to CO_2 and H_2O.

The appearance of this IR pattern is interpreted as arising from the concomitant reduction of Ir III species and the formation of Ir I(CO)$_x$ carbonyl according to the following scheme.

$$Cl\ Ir\ (OH)_2\ +(x + 1)\ CO\ \rightarrow\ Cl\ Ir(CO)_x + CO_2 + H_2O$$

Volumetric measurements of the CO uptake and the CO_2 yield showed that 4 CO molecules were consumed per initial iridium and one CO_2 molecule formed. Hence the stoechiometric scheme was shown to be the following.

$$Cl\ Ir(OH)_2\ +\ 4\ CO\ \rightarrow\ Cl\ Ir(CO)_3 + CO_2 + H_2O$$

or alternatively

$$-O_L-Ir\ (OH)_2\ +\ 4\ CO\ \rightarrow\ -\ O_L\ -\ Ir(CO)_3 + CO_2 + H_2O$$

The IR spectrum in the carbonyl region is in fact consistent with a dicarbonyl complex in a C_{2v} symmetry or a tricarbonyl complex in a C_{3v} symmetry.

Thus the number of carbonyl ligands i.e. the nature and the symmetry of the carbonyl complex was elucidated using two independent procedures.:

(i) the number of CO ligands per iridium was determined by equilibrating a known amount of the complex by a measured quantity of ^{13}CO molecules. After heating replacement of the ^{12}CO of the complex by ^{13}CO molecules present in the gas phase occured and was monitored by IR spectroscopy. When exchange equilibrium was reached, as deduced from repeated IR measurements, the gas phase composition was analyzed by mass spectrography thus allowing the equilibrium isotopic ratio to be determined $^{12}CO/^{13}CO$. Assuming there is no isotopic effect the number of CO ligands per iridium was found to be equal to three.

(ii) when an equimolecular mixture $^{12}CO/^{13}CO$ was used to generate the complex, substitution of a number of ^{12}CO by ^{13}CO lowered the local symmetry of the complex and absorptions at 2086, 2075, 2060, 2001, 1978 and 1956 cm^{-1} were now observed figure 2.

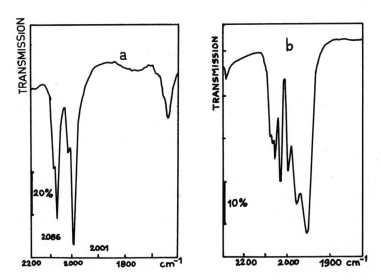

Figure 2 : IR Spectrum of Iridium Carbonyl a: ^{12}CO complex,
b : $^{12}CO/^{13}CO$ complex

Use of 93 % ^{13}C enriched CO resulted in twin strong bands at 2036 and 1956 cm^{-1}.
The interpretation of the ^{12}CO/^{13}CO was carried out on the basis of local symmetry of
the carbonyl group where the number of infrared active stretching vibrations, assuming a
C_{2v} or a C_{3v} symmetry, could be predicted and fitted to the experimental values. Comparison
of the experimental values and the predicted values for C_{2v} and C_{3v} symmetry respectively
table 4 and 5 ruled out the dicarbonyl structure and confirmed the tricarbonyl nature of the
iridium complex.

Table 4

Comparison of the experimental CO frequencies with those calculated for $L_x M(CO)_2$ species.

Isomers	$L_x M(^{12}CO)_2$		$L_x M(^{12}CO)(^{13}CO)$		$L_x M(^{13}CO)_2$	
Symmetry	C_{2v}		C_1		C_{2v}	
CO stretching modes	A_1		A'		A_1	
	B_1		A'		B_1	
Calculated frequencies cm^{-1}			a) 2072	b) 2068	a) 2046	b) 2036
			1971	1972	1952	1956
Experimental frequencies cm^{-1}	2086		2075		2036	
	2001		1978		1956	
			2060			
			1966			

a) force constant : 17.38 mdyne. $\mathring{A}^{-1}$ for ^{12}CO and ^{13}CO as calculated from frequencies
of unsubstituted $L_x M(^{12}CO)_2$.
b) Force constant : 17.38 mdyne. $\mathring{A}^{-1}$ for ^{12}CO and 17.31 mdyne. $\mathring{A}^{-1}$ for ^{13}CO as calculated
from totally ^{13}CO exchanged compound.
θ) 105°5 as calculated from I_{A_1}/I_{B_1}.

Hence both rhodium and iridium exchanged zeolites do form metal carbonyls structurally
well defined when treated with CO under conditions close to those experienced in the
methanol (or alcohols) carbonylation reaction. These carbonyl complexes, interestingly
enough have similar structures to those known in liquid media. The possible difference probably resides in that monomers were entrapped within the zeolite while usually dimers and
higher polymers are most frequently encountered for rhodium and iridium carbonyls.

6.1.2 Kinetics of methanol carbonylation

Carbonylation of methanol proceeded readily at atmospheric pressure at around 150-180°C
on rhodium or iridium loaded zeolites. With the exception of trace dimethyl-ether, methyl
acetate was by far the major product ∿ 90 % of converted methanol, as long as the conversion
level was set below 50 %. At higher conversion levels acetic acid could also be detected. In
the absence of methyl iodide, however, there was no production of methyl acetate. Thus zeolite hosted rhodium and iridium complexes showed not only the same selectivities as their

Table 5

Comparison of the experimental CO frequencies with those calculated for $L_x M(CO)_3$ species.

Isomers	$L_x M(^{12}CO)_3$	$L_x M(^{12}CO)_2\,^{13}CO$		$L_x M(^{12}CO)(^{13}CO)_2$		$L_x M(^{13}CO)_3$	
Symmetry	C_{3v}	C_s		C_s		C_{3v}	
CO stretching modes	A_1 E	A' A" A'		A' A" A'		A_1 E	
Calculated frequencies cm^{-1}		a) 2079 2001 1964	b) 2075 2001 1966	a) 2065 1957 1977	b) 2060 1956 1979	a) 2038 1957	b) 2036 1957
Experimental frequencies cm^{-1}	2086 2001	2075 2001 (1966)		2060 1956 1978		2036 1956	

a) force constant = 16.64 mdyne. $\overset{\circ}{A}^{-1}$ for ^{12}CO and ^{13}CO as calculated from frequencies of $L_x M(^{12}CO)_3$.

b) force constant = 16.64 mdyne. $\overset{\circ}{A}^{-1}$ for ^{12}CO and 16.61 mdyne. $\overset{\circ}{A}^{-1}$ for ^{13}CO as calculated from frequencies of totally ^{13}CO exchanged compound.

θ) 87°2 as calculated from I_{A_1}/I_E.

solution analogues but also, unfortunally, their very drawbacks suggesting most probably identical reaction pathways.

It was also reported that while iridium zeolites with high iridium loads deactivated rapidly, the activity loss in the case of rhodium was almost unperceptible. At low loadings both types of catalysts exhibited higher specific activities than for high loadings with a significantly higher effeciency in the case of iridium which exhibited a stable activity as long as the iridium content was kept below 1.7 % (66). The kinetics of methanol carbonylation on rhodium and iridium zeolites closely paralleled those reported for soluble rhodium and iridium complexes respectively. Interestingly,while the rate law was first order in methyl iodide and zero order in methanol and carbon monoxide in the case of Rh zeolites, it was first order in methanol and zero order in methyl iodide and carbon monoxide in the case of iridium zeolites. Hence apparently, while methyl iodide interaction with rhodium seemed to be rate determining, methanolysis of the resulting intermediate to form methyl acetate was a fast process. On the contrary, interaction of methyl iodide with the iridium complex seemed to be a fast process, while at least a reaction step involving methanol appeared to be rate determining.

The kinetic data were interpreted in terms of fast oxidative addition of methyl iodide to the iridium carbonyl, while in the case of rhodium it would be a rather slow process. This step was experimentally witnessed for both metal complexes using IR and ^{13}C NMR spectroscopy.

Also, it was shown that, in the case of rhodium zeolites, carbonylation of CD_3OD in the presence of CH_3I as promotor resulted in fact in the carbonylation of CH_3I and subsequent esterification to give CH_3COOCD_3 even though exchange between CH_3I and CD_3OD to give CD_3I and CH_3OD gave rise to increased amounts of CD_3COOD_3, CD_3COOCH_3 and also CH_3COOCH_3, as conversion proceeded (64). On the other hand carbonylation of ethyl alcohol promoted by methyl iodide resulted exclusively in ethyl acetate at a high rate while carbonylation of methanol promoted by ethyl iodide proceeded at a lower rate mainly producing methylacetate with little ethyl acetate and significant amounts of propionic acid and methyl propionate ; ethyl acohol carbonylation promoted by ethyl iodide proceeded with more difficutly (64).

This set of data clearly indicates that in fact, carbonylation involves the iodide rather then the acohol and that high exchange rates seem to favour the carbonylation of lower alkyl iodides which therefore exhibit an increasing reactivity towards the carbonyl as the alkyl carbon number decreased. The observation that isopropanol was not carbonylated in the presence of isopropyl iodide while isopropyl propionate was formed with a selectivity of 13% when ethyl iodide was the promoter indicated that isoalkyl iodides were more inert than their linear analogues.

6.1.3   Possible reaction pathways - Structure of the reaction intermediates.

As suggested from IR evidence upon addition of methyl iodide, rhodium dicarbonyl characteristic νCO bands at 2110 and 2043 cm^{-1} vanished while new bands at 2090 and 1710-1685 cm^{-1} develop (66,69). A similar situation is produced upon equilibrating rhodium III loaded zeolite (exchanged or impregnated) by the CO + CH_3I mixture (65). UV measurements showed that rhodium I initially present as rhodium dicarbonyl has been oxidized to rhodium III (69). Also ^{13}C NMR studies of CH_3I addition to $Rh(CO)_2Y$ showed the appearance of a cationic carbonyl species due to the binding of a carbon monoxide molecule to a rhodium III cation (70). The IR bands at 2090 and 1710-1685 cm^{-1} are consistent with a single linear CO ligand and an acetyl group both bound to a rhodium III central ion. Thus the proposed scheme involves an oxidative addition of methyl iodide to the initial dicarbonyl rhodium I complex as described below :

The alkyl carbonyl rearranges rapidly to form the acetyl rhodium III carbonyl. Reductive elimination of $CH_3-\overset{O}{\underset{\|}{C}}-I$ would regenerate the rhodium I species in the presence of excess CO.

Methanolysis or hydrolysis of CH_3COI yields either methyl acetate or acetic acid. Even though kinetic studies hinted to CH_3I addition as a rate determining step, no spectroscopic evidence showed the assisted character of this addition. However kinetic and spectroscopic time scales are so different that it is hardly conceivable to observe identical effects. In the case of iridium zeolite a similar pattern of results seems to suggest a quite similar mechanism.

Addition of methyl iodide to $XIr(CO)_3$ resulted in the shift of the characteristic doublet at 2086 and 2001 cm^{-1} to higher frequency (2148 and 2100 cm^{-1}) again suggesting an oxidative addition of methyl iodide. On the other hand, ^{13}C NMR showed a carbon resonance due to iridium alkyl δ CH_3 = - 22 ppm together with the appearance of a cationic of iridium III δ CO = 132 ppm (71). Hence again methyl iodide oxidative addition to the $XIr(CO)_3$ precursor is evidenced as a primary step accounting for the role of the promoter.

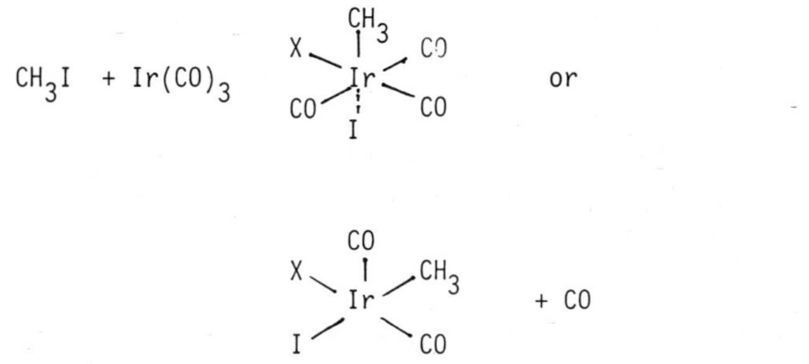

$$CH_3I \;+\; Ir(CO)_3$$

No subsequent transformation was observed. In particular no band appeared in the vicinity of 1700cm^{-1} indicating that methyl migration to form iridium acetyl should be a slow activated process, which is in line with earlier findings concerning mixed alkyl carbonyls of iridium in solution. Addition of methanol at room temperature resulted in the appearance of ^{13}C resonance at δ CO = 178 ppm (71) which could be ascribed to an iridium III alkoxy carbonyl derivative formed as follows :

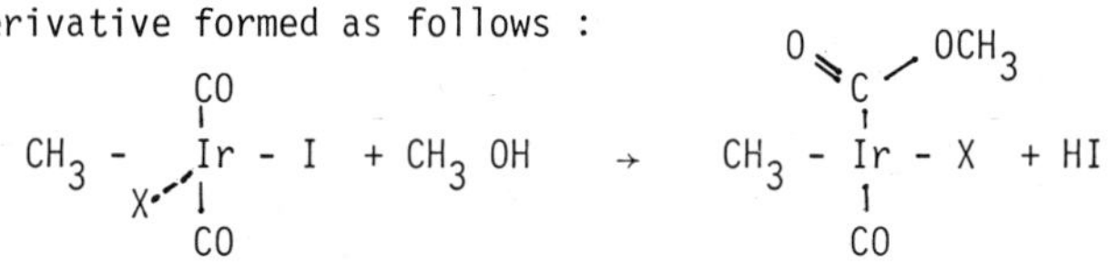

Heating at 170°C resulted in the appearance of reaction products. Thus in the case of iridium the electrophilic cationic carbonyl carbon seemed to undergo a nucleophilic attack by a methoxy group generated from methanol and that the nucleophilic character of the carbon, in the case of iridium, is not sufficient to induce the methyl migration as it occurs readily in the case of rhodium III alkyl carbonyl. This latter step may well account for the first order rate with respect to methanol observed in the case of iridium.

Reductive elimination of CH_3-COOCH$_3$ would follow and subsequent carbonylation is achieved in presence of excess CO to regenerate the active species. The differences found between rhodium and iridium zeolite catalysts are presumably accounted for by the propensity of iridium compounds to undergo oxidation to iridium III complexes readily, while rhodium I species are the most stable rhodium complexes. On the other hand, the other major difference between those two catalysts is the higher intrinsic activity exhibited by iridium catalysts under identical conditions. Yet while iridium catalysts usually deactivate when the iridium content was forced beyond 2 %, high rhodium loads do not exhibit this unfortunate trend

6.2   Hydroformylation of olefins

While carbonylation catalysts have been thouroughly studied, only a single report was devoted to liquid phase hydroformylation of hexene-1 using a rhodium zeolite catalyst. The reaction was carried out under rather moderate conditions 50-100 atom. of CO : H$_2$ mixture at temperatures ranging 50-130°C. (72)

The catalyst was prepared by subjecting chloro rhodium pentammine   exchanged Y zeolite to heating at 130°C under 80 atm of CO : H_2 mixture. IR analysis of the resulting compound showed νCO sharp absorptions at 2095, 2080 (sh) and 2060 cm^{-1} (w) on the one hand typical of linear carbonyls bound to zerovalent metal atoms in polynuclear complexes and a band at 1765 cm^{-1} characteristic of bridging carbonyl.

These bands were interpreted as an evidence for the formation of a polynuclear rhodium (0) carbonyl formed in situ. Such a compound seemed to be tightly held within the cavities as only faint losses of rhodium were observed not exceeding a few percent of the initial content over 8 cycles while rhodium on carbon catalyst experienced heavy losses over two cycles of course no mechanistic studies were attempted in this recent report (72), presumably due to the use of high pressures. And yet high selectivities towards aldehydes, and the iso-to-normal aldehyde ratio quite similar to those observed for rhodium soluble complexes strongly suggest that similar reaction mechanisms might be operating in the case of zeolite entrapped rhodium carbonyls. Further studies might be needed in order to ascertain the similarities or possible differences between soluble and zeolite hosted-hydroformylation catalysts. In particular, the porous structure of the zeolite might well induce certain specific reactivity features.

7   HYDRIDE SHIFT REACTIONS

7.1   Amination of acetelynes.

As for ammoxidation of propelyne to form acetonitrile, ZnY appeared essentially as a good amination catalyst rather than an oxidation catalyst since while oxidation proceeded with difficulty, amination of acetaldehyde proceeded readily. A typical amination reaction was reported earlier (73) that is amination of acetylenes according to the reaction scheme.

$$RNH_2 + R'C = CR'' \rightarrow R'CH_2 - CR'' = NR$$

Various amines as well as various acetylenes were employed with various degrees of success. The reaction was thought to proceed via enamines which later tautomerize by a hydrogen shift to the ketimine. Such a mechanism, if it were proved, might account for the inertness of secondary and tertiary amines where an alkyl shift would necessarily be more demanding. On the other hand further isomerisation may give aldimines such as :

$$CH_2 = N - CH \underset{CH_3}{\overset{CH_3}{<}}$$

giving rise to high boiling condensation products.

Besides ZnY, a series of transition metal ion exchanged zeolites were tested. Only cadmium exhibited satisfactory activity while the other d^{10} ions Cu(I) and Ag(I) were reduced to metals as well as Hg II.

Among the various treatments adopted, flush vacuum proved to be the most effective presumably because Zn ions were stripped of a number of their ligands before they could migrate to internal sites. Also the zeolite type proved to significantly influence the nature of the final products. X zeolite favoured further condensation.

An intriguing feature is the superiority exhibited by "Td" ions over "Oh" ions which usually showed more interesting catalytic properties in most reactions.

7.2 Alcohol - Ketone disproportionation

The typical reaction reported as catalysed by TMIZ is the methyl ethyl ketone and isopro-panol disproportionation :

$$CH_3 - \underset{\underset{O}{\|}}{C} - C_2H_5 \quad + \quad CH_3 - CHOH - CH_3 \quad CH_3 - \underset{\underset{O}{\|}}{C} - CH_3 \quad + \quad CH_3 - CHOH - C_2H_5$$

which also involves a hydrogen transfer between two molecules in stead of an intramolecular rearrangement in the case of acetelyne amination.

This disproportionation reaction was claimed to be carried out at 150-210°C on nickel (II) X and A type zeolites (74). The active centres were identified to nickel ions which could be poisoned by trace amounts of water molecules originating from the dehydration side reaction. Although the authors did not engage in mechanistic scheme this reaction may well proceed via a concerted mechanism which seems quite likely in organic reactions

7.3 Water splitting.

Among the novel reactions which attracted a considerable interest in the last few years is water splitting into O_2 and H_2. Because of the energy gap, seeking hydrogen from water upon illumination of appropriate catalyst by solar light seems a very attractive way.

In fact two types of systems were investigated. On one hand, there are systems, in par-ticular silver-exchanged zeolites which can release oxygen upon reductive addition of water. Silver ions were thereupon reduced to silver metal. Hopefully the stored hydrogen could be transferred to an organic substrate. In fact the catalyst could only be regenera-ted at quite high temperature (75) which is rather energy consuming.

More interesting seems to be the more recent report which claimed that under illumina-tion by sunlight, Titanium III exchanged A zeolite saturated with water gave rise to gas evo-lution which composition was 40 % H_2 and O_2 and N_2 (76). Apparently water splitting could be achieved repeatedly upon illumination by solar rays. Interestingly, it was reported that contrary to zeolite A, type X and Y zeolites also exchanged with T^{3+} ions did not catalyse water spilitting into its elements nor did they give, under the conditions quoted for TiA, to any gas evolution.

REFERENCES

1   C.M. Naccache and Y. Ben Taarit, J. Catal., 1971, 22, 171
2   C. Naccache and Y. Ben Taarit, Chem. Phys. Letters, 1971, 11, 11
3   P. Gallezot, Y. Ben Taarit and B. Imelik, C.R. Acad. Sci., 1971, 272C, 261
4   P. Gallezot, Y. Ben Taarit and B. Imelik, J. Catal., 1972, 26, 295
5   P. Gallezot, Y. Ben Taarit and B. Imelik, J. Catal., 1972, 26, 481
6   P. Gallezot, Y. Ben Taarit and B. Imelik, J. Chem. Phys., 1973, 77, 2364
7   C.M. Naccache and Y. Ben Taarit, J.C.S. Faraday I, 1973, 69, 1475

8   C. Naccache, M. Che and Y. Ben Taarit, Chem. Phys. Letters, 1972, 13, 109
9   K.R. Laing, R.L. Leubner and J.H. Lunsford, Inorg. Chem., 1975, 14, 1400
10 Y. Ben Taarit, H. Praliaud and G.F. Coudurier, J.C.S. Faraday I, 1978, 74, 3000
11 J.C. Vedrine, E.G. Derouane and Y. Ben Taarit, J. Chem. Phys., 1974, 78, 531
12 J.H. Lunsford and E.F. Vansant, J.C.S., Chem. Commun., 1972, 830
13 E. Garbowski and J.C. Vedrine, Chem. Phys. Letters, 1977, 48, 550
14 M. Che, J.F. Dutel, P. Gallezot and M. Primet, J. Phys. Chem., 1976, 80, 2371
15 P. Gelin et al. to be published
16 G. Coudurier, P. Gallezot, H. Praliaud, M. Primet and B. Imelik, C.R. Acad. Sci., 1976, 282C, 311
17 P.H. Kasai and R.J. Bishop, J. Am. Chem. Soc., 1972, 94, 5560
18 C. Naccache, Y. Ben Taarit and M. Boudart,"Molecular Sieve II" (J.R. Katzer, ed). ACS Symposium series No 40, 1977, 156
19 C. Naccache, J.F. Dutel and M. Che, J. Catal., 1973, 29, 179
20 Y. Ben Taarit, J.C. Vedrine, J.F. Dutel and C. Naccache, J. Magn. Reson., 1978, 31, 251
21 C.R. Adams and T.J. Jennings, J. Catal., 1964, 3, 549
22 G.W. Keulks, L.D. Krenzke and T.M. Notermann, Adv. Catal., 1978, 27, 183
23 G.K. Boreskov, Kinet. i. Katal., 1973, 14, 7
24 I. Mochida, S. Hayata, A. Kato and T. Sciyama, J. Catal., 1969, 15, 314
25 I. Mochida, S. Hayata, A. Kato and T. Seiyama, J. Catal., 1970, 19, 405
26 I. Mochida, S. Hayata, A. Kato and T. Seiyama, J. Catal., 1971, 23, 31
27 R. Rudham and M.K. Sanders, J. Catal., 1972, 27, 287
28 S.J. Gentry, R. Rudham and M.K. Sanders , J. Catal., 1974, 35, 376
29 H. Arai and H. Tominaga, Asahi Garasu, Ind. Techno. Prom. Res. Reports 1975, 27, 93.
30 Y. Ben Taarit et al., to be published
31 H. Praliaud et al. unpublished data
32 C.S. John and H.F. Leach, J.C.S. Faraday Trans., 1977, 73, 1595
33 C. Naccache and Y. Ben Taarit, unpublished data.
34 H. Praliaud et al. in journées sur les zéolithes Villeurbanne, December 14-15 1978
35 J.A. Rabo, C.L. Angell, P.H. Kasai and V. Shomaker, discussions Faraday Soc., 1966, 41, 328
36 C.C. Chao and J.H. Lunsford, J. Chem. Phys., 1972, 57, 2890
37 R.F. Howe and J.H. Lunsford, J. Am. Chem. Soc., 1975, 97, 5156
38 H. Praliaud, Chem. Phys . Letters, 1979, 66, 407
39 D. Olivier, M. Richard and M. Che, Chem. Phys. Letters, 1978, 60, 77
40 D. Michel, W. Meiler and H. Pfeifer, J. Mol. Cat., 1975/76, 1, 85
41 Y. Ben Taarit unpublished data
42 Y. Y. Huang, J. Catal. 1980, 61, 461
43 R. Ugo, F. Conti, S. Cenini, R. Mason and G. Robertson, J.C.S., Chem. Commun., 1968,1498
44 G.M. Zanderighi, R. Ugo, A. Fusi and Y. Ben Taarit, Inorg. Nucl. Chem. Lett. 1976, 12, 729
45 R. Ugo, Engel. Ind. Techn. Bull. XI, 1971, 2, 45
46 G. Wilke, H. Scholt, P. Heimbach, Angew. Chem. Intern. Edition, 1967, 6, 92
47 F. Igersheim and H. Mimoun, Nouveau journal de Chimie, 1980, 4, 161
48 Y. Ben Taarit et al. to be published
49 A.B. Evnin, J.A. Rabo and P.H. Kasai, J. Catal., 1973, 30, 100
50 I.D. Mikheikin, Yu. I. Pecherskaya and V.B. Kazanskii, Kinet. i. Katal. 1971, 12, 191
51 S. Tsuruya, Y. Okamoto and T. Kuwada, J. Catal., 1979, 56, 52
52 I. Mochida, T. Jitsumatsu, A. Kato and T. Seiyama, J. Catal., 1975, 36, 361
53 C. Naccache et al. unpublished data.
54 L.V. Skalkina, I.E. Kolchin, L. Ya Margolis, N.F. Ermolenko, S.A. Levina and L. N. Malashevich, Kinet. i. Katal., 1971, 12, 242
55 R. Grabowski, J. Haber, J. Komorek, J. Ptaszynski, T. Romotowski and J. Słoczynski, Proceedings of the symposium on zeolites, Szeged, Hungary, 1978, 141.
56 a. T. Seiyama, T. Arakawa, T. Matsuda and N. Yamazoe, Preprint 9-1 Japan-USA seminar on catalytic Nox reactions, Susono, Japan, nov. 1975.
    b. T. Seiyama, T. Arakawa, T. Matsuda, Y. Takita and N. Yamazoe, J. Catal. 1977, 48, 1
57 W.B. Williamson and J.H. Lunsford, J. Phys. Chem., 1976, 80, 2664
58 K.A. Windhorst and J.H. Lunsford, J. Am. Chem. Soc., 1975, 97, 1407
59 M. Mizumota, N. Yamazoe and T. Seiyama, J. Catal., 1978, 55, 119
60 B. Christensen and M.S. Scurrell, J.C.S. Faraday I, 1977, 73, 2036
61 B. Christensen and M.S. Scurrell, J.C.S. Faraday I, 1978, 74, 2313
62 T. Yashima, Y. Orikasa, N. Takahashi and N. Hara, J. Catal., 1979, 59, 53
63 B.K. Nefedov, N.S. Sergeeva, T.V. Zueva, E.M. Schutkina and Ya T. Eidus, Izvest. Akad. Nauk. SSSR, Ser. Khim., 1976, 582
64 N. Takakashi, Y. Orikasa and T. Yashima, J. Catal. 1979, 59, 61
65 S.L. Anderson and M.S. Scurrell, J. Catal., 1979, 59, 340

66 P. Gelin, Y. Ben Taarit and C. Naccache, 7th Intern. Congress on Catalysis, Tokyo, Japan July 1980
67 M. Primet, J.C. Vedrine and C. Naccache, J. Mol. Cat., 1978, 4, 411
68 P. Gelin, G. Coudurier, Y. Ben Taarit and C. Naccache, J. Catal., in press
69 M. Primet and E. Garbowski, Chem. Phys. Letters, 1980, in press
70 Y. Ben Taarit, Chem. Phys. Letters, 1979, 62, 211
71 P. Gelin et al., to be published
72 E. Mantovani, N. Palladino and A. Zanobi, J. Mol. Cat., 1977/78, 3, 285
73 R.S. Neale, L. Elek and R.E. Malz Jr, J. Catal., 1972, 27, 432
74 Z.V. Gryaznova, K.A. Baskunyan and I.A. Mel'Nichenko, Kinet. i. Katal., 1971, 12, 1471
75 P.A. Jacobs, J.B. Uytterhoeven and H.K. Beyer, J.C.S. Chem. Commun., 1977, 128
76 S.M. Kuznicki and E.M. Eyring, J. Am. Chem. Soc., 1978, 100, 6790

B. Imelik *et al.* (Editors), *Catalysis by Zeolites*
© 1980 Elsevier Scientific Publishing Company, Amsterdam — Printed in The Netherlands

ACETYLENE CONDENSATION CATALYZED BY CrY ZEOLITES : FORMATION OF ALKYLAROMATICS

J. LEGLISE, Th. CHEVREAU and D. CORNET

"Structure et réactivité d'espèces adsorbées" (ERA 824)

I.S.M.R.A., Université de Caen - 14032 CAEN CEDEX (FRANCE)

Transition metal ions fixed on solid supports catalyze the cyclotrimerization of acetylene into benzene at room temperature. However, both the rate and selectivity of the reaction decrease at higher temperature [1]. Among the by-products, alkylaromatics are of special interest, and it was thought that an acidic catalyst such as the chromium-exchanged Y zeolite would be suitable for their formation. Accordingly as shown by Pichat [2] benzene is not the only reaction product of acetylene over ion-exchanged faujasite.

Acetylene condensation catalyzed by CrY zeolites undergoes a gradual selectivity change at ca. 150°C [3]. The main features of the reaction above this temperature are reported here : the composition of the alkylaromatic fraction will be examined for different catalysts and reaction conditions, and the influence of the zeolite structure will be emphasized.

EXPERIMENTAL

The starting material for catalyst preparation was a Y zeolite with unit cell composition $Na_{56}(AlO_2)_{56}(SiO_2)_{136}$, nH_2O. Ion-exchange with a chromic nitrate solution yielded exchanged forms containing 4, 8 or 12.9 Cr, per unit cell (x in Cr_xY).

The reaction was performed in a closed pyrex vessel (400 ml). The catalyst was first activated (37.5 mg after dehydration). The amount of acetylene introduced into the reactor is expressed as n_o C_2H_2 molecules per unit cell and the initial pressure in the absence of any adsorption is P_o. Usually, n_o = 133 molecules/cell and P_o = 4.5 kPa for a reaction performed at 350°C.

A Riber mass spectrometer connected to the reaction vessel through a permanent leak allowed a continuous analysis of the gaseous products above the catalyst : the measured amounts of gaseous acetylene $n_g(A)$, benzene $n_g(B)$, toluene $n_g(T)$ etc... are expressed as eq. C_2H_2 reacted per zeolite unit cell. The adsorbed phase as a whole is noted $n_a(C_2H_2/cell)$. After two hours, the reaction is stopped and the hydrocarbons are condensed in a cold trap for 1 hour, while the zeolite is kept at the reaction temperature. The fraction of the adsorbed phase that cannot be desorbed in this way is $n_i(C_2H_2/cell)$. GLC analysis of the recovered fraction gives a value $(n_g + n_d)$ for every product, i.e. the sum of the gaseous and desorbed amounts. From these, a "recovered" conversion n may be calculated.

RESULTS

a) Nature of the products

Aromatic hydrocarbons only were detected in the gas phase when acetylene reacted over $Cr_{12.9}Y$. Fig. 1 shows the evolution of the gas phase in an experiment at 350°C with n_0 = 505 C_2H_2/cell : benzene is the major reaction product, but toluene, ethylbenzene, xylenes and higher aromatics Ar-9 and Ar-10 altogether exceed benzene. Traces of naphtalene also appear. After two hours, the observed conversion into gasphase products, is rather limited although it is 15 times higher than for the nonexchanged NaY zeolite. The quantity of adsorbed hydrocarbons, deduced from the gasphase mass-spectrum, amounts to n_a = 91 eq. C_2H_2/cell for the $Cr_{12.9}Y$, compared to 16 for the NaY zeolite.

The contribution of the adsorbed phase to the final "recovered" conversion n may be judged from the results of Table I. It is important at all temperatures. In the aromatic Ar-8 fraction, styrene is abundant over NaY, but almost absent over CrY. In addition to aromatic hydrocarbons, some aliphatics are formed. Vinylacetylene is the main aliphatic in the reaction over NaY, but the distribution is more diversified in the case of CrY : by decreasing importance, the aliphatics are C_6 (from C_6H_{14} to an hexadienyne) C_4 (butane to vinylacetylene), and then C_5 (C_5H_{12} to C_5H_8) and C_3 (propane to propyne). The H/C ratio in this fraction exceeds 1.

TABLE I

Analysis (eq. C_2H_2/cell) of the hydrocarbons recovered by trapping after 2 hours reaction over $Cr_{12.9}Y$.

Temp. °C	n_0	Products (gaseous + desorbed)						Conversion	
		A	B	T	Ar-8	Ar-9	Aliph.	gas n_g	recovered $(n_g + n_d)$
160	142	88.5	38.6	0.16	< 10^{-3}	-	0.2	16.2	38.8
250	147	102	10.8	1.5	0.3	-	0.1	7.5	12.7
350	123	77	6.6	3.4	7.4	4	0.2	12.7	21.6
350*	144	143.5	0.16	-	0.01	-	0.36	0.5	0.5

* Reaction over NaY zeolite.

b) Effect of activation, pressure and temperature

The effect of the activation procedure has been examined. First, CrY zeolite samples were dehydrated under vacuum at several temperatures between 200 and 550°C, and then acetylene condensation was run for 2 hrs at 350°C. Fig. 2 shows the evolution of the gaseous products for two dehydration temperatures : the 550°C activated catalyst is less active, but has a higher selectivity for benzene than the 350°C-activated one. It is likely that benzene and alkylaromatics are produced by two different mechanisms : the production of alkylaromatics is favoured at moderate dehydration temperature, i.e. when the zeolite contains many Bronsted acidic sites. The adsorbed amount n_a and the irreversible residue n_i follow a different trend, since they are maximum after activation at 450°C and decrease at higher temperature (table II).

The catalyst has also been reduced by hydrogen, or oxidized by oxygen at 350°C,

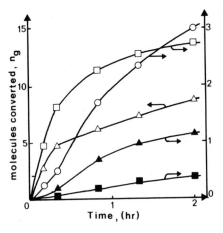 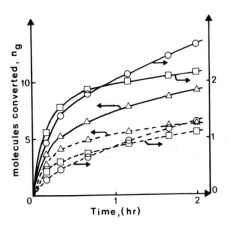

Number of C_2H_2 converted into gaseous products per unit cell $Cr_{12.9}Y$ at 350°C.
Left scale : $-\triangle-$ benzene. Right scale : $-\square-$ toluene ; $-\bigcirc-$ Ar-8 ; $-\blacktriangle-$ Ar-9 ; $-\blacksquare-$ Ar-10.
Fig. 1. Activation 350°C, n_0 = 505   Fig. 2. Activation 350°C, n_0 = 123 (solid lines) ;
activation 550°C, n_0 = 120 (dotted lines).

before evacuation, and acetylene reacted at the same temperature : in both cases activity as well as selectivity into benzene are lowered (Table II). A similar behavior was noted for ethylene polymerization [4]. Pre-reduction of the catalyst leads to a higher residue n_i ; pre-oxidation tends to favour aliphatics production, and some ethanal is detected. Thus, the standard activation adopted was vacuum treatment at 350°C.

TABLE II
Analysis (eq. C_2H_2/cell) of the reaction products of C_2H_2 over $Cr_{12.9}Y$ at 350°C.
Effect of activation process and acetylene pressure.

Activation process. °C	P_0 (kPa)	n_0	n_a (2 hrs)	Products (gaseous + desorbed)				n_i
				A	B	Ar-7$^+$	Aliph.	
Vac. 350	18.6	505	91	401	9.2	15.8	4.3	75
Vac. 350	4.5	123	37	76	6.6	14.8	0.2	25
Vac. 450	4.7	129	51	71	6	5.8	4.3	42
Vac. 550	4.4	120	40	74	9.7	5.9	0.3	30
H_2 then Vac. 350	19.5	512	88	433	6	9.2	4.1	60
	4.6	131	61	78	2.3	3.8	0.5	47
O_2 then Vac. 350	4.5	131	45.3	102.8	1.8	2.6	5.7	18.1

Due to experimental reasons, it was not feasible to change the initial acetylene pressure within wide limits. Table II shows the effect of a fourfold increase in initial pressure for a reaction at 350°C using $Cr_{12.9}Y$. With the vacuum-activated zeolite, the conversion after 2 hrs increases by only 20 %, and there is almost no change in alkylaromatics production, while the adsorbed amount n_a and the irreversible residue

n_i shows considerable variation. For pre-reduced zeolites, the effect of pressure upon conversion is more pronounced.

The effect of reaction temperature upon activity is shown on Fig. 3. The cumulative curves represent the percent conversion into benzene, alkylaromatics, aliphatics and irreversible fraction. The recovered conversion first increases up to 150°C, decreases from 150 to 300 and then again increases. It is clear that the amount of alkylaromatics is important above 250°C and this is confirmed by plotting the conversion n into benzene, toluene and Ar-8 (Fig. 4). Toluene appears at temperatures lower than those required for ethylbenzene and xylenes, but it is rapidly offset by the Ar-8 above 300°C.

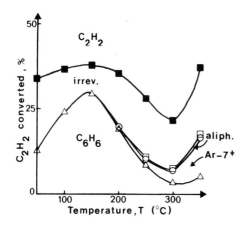

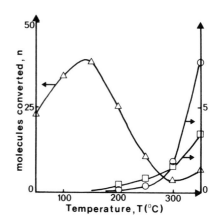

Effect of temperature on the products (gas + desorbed) and residue formed over $Cr_{12.9}Y$. Reaction time : 2 hrs.

Fig. 3 Percent C_2H_2 converted (cumulative curves for products).

Fig. 4. $n(C_2H_2)$ converted : individual curves. Left scale : —△— benzene. Right scale : —□— toluene ; —○— Ar-8.

c) Influence of the chromium content.

In the case of NiY zeolites, the rate of acetylene cyclotrimerization sharply increased at high nickel loading [2]. Similarly the yield of benzene n(B) at low temperatures over the CrY increases as the Cr content x goes from 4 to 12.9 [3]. But above 200°C, n(B) is independent of x. A different situation arises with alkylaromatics. The conversions per chromium ion $\frac{n}{x}$ into toluene and Ar-8 are plotted in Fig. 5 and 6 for several temperatures. Up to 300°C, the horizontal lines show that the yields of toluene and Ar-8 vary almost linearly with Cr content, but at 350°C the variation of n(Ar-8) is steeper, while n(T) decreases slowly with x. This again suggests different mechanisms for benzene and alkylaromatics formation : obviously, the acidity of the zeolite has to be invoked, since it is enhanced by Cr-exchange.

In order to ascertain the influence of zeolite acidity, the relative amounts of ethylbenzene and the three xylenes in the Ar-8 fraction have been examined (table III).

In each experiment, the amount of ethylbenzene is higher than required by thermodynamic equilibrium, and the sum of the xylenes lower. At 350°C, the Ar-8 fraction sharply rises with the Cr content x, and the xylenes percentage steadily increases with conversion whereas ethylbenzene falls down. It may be concluded that the most

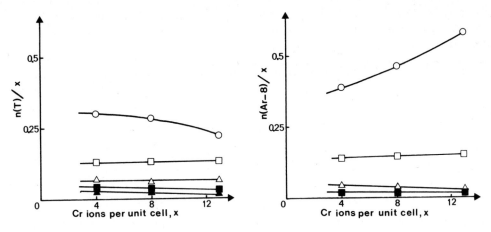

Fig. 5. Number of C_2H_2 molecules converted into toluene, per Cr ion in Cr_xY Reaction temperature (°C) ▲ 150 ; ■ 200 ;

Fig. 6. Number of C_2H_2 molecules converted into Ar-8, per Cr ion in Cr_xY △ 250 ; □ 300 ; ○ 350.

TABLE III

Distribution of alkylaromatics in the Ar-8 fraction[*]

T°C	250		300		350				
Catalyst Convers. n(Ar-8)	Cr_8 0.16	$Cr_{12.9}$ 0.32	Cr_8 0.8	$Cr_{12.9}$ 1.2	NaY[**] 0.17	Cr_4 1.7	Cr_8 3.8	$Cr_{12.9}$ 7.4	Equ. -
% Ethyl-B	30.5	40.5	29.8	20.2	47.8	36.7	27	15.5	6.4
o-xyl	26	19.7	26.5	24.7	18.7	22.4	22	21.7	21.8
m-xyl	27.4	20.2	30	40.4	13.7	26.5	38.2	45.8	49.5
p-xyl	16.4	19.6	13.7	14.6	19.8	14.3	12.9	16.9	22.2

[*] excluding styrene ; [**] zeolite mass 150 mg.

part of the xylene originates from ethylbenzene, which then undergoes isomerization. In this process, o-xylene reaches its equilibrium value faster than the other isomers ; at low conversion levels, the amount of m-xylene is particularly small.

d) Irreversibly adsorbed fraction.

The products that cannot be recovered by condensing the gas phase for 1 hr are shown on Fig. 3 (part comprised between the two upper curves). This amount is almost independent of temperature.

Performing the reaction in a vacuum microbalance allowed a separate study of this residue, whose composition was found temperature dependent, in accordance with Venuto [5]. The residue at 70°C completely comes off as benzene upon heating at 360°C. A similar treatment performed on the residue at 200°C leads to only 80 % desorption, under the form of C_3-C_6 aliphatics (36 %) toluene (16 %), Ar-8 (15 %), Ar-9 (14 %), benzene (10 %) and acetylene (8 %). For higher reaction temperatures (300 and 360°C) no change in mass in noted after prolonged heating, but some hydrogen is evolved.

Measuring this amount of hydrogen, and taking into account the hydrogen content of the previously desorbed products leads to the following average composition of the residue per unit cell after 2 hrs reaction and desorption :

Reaction temperature	15 mn desorption	15 hrs desorption
303°C	$C_{30.4} H_{26.9}$	$C_{30.4} H_{14.6}$
368°C	$C_{30.9} H_{25.3}$	$C_{30.4} H_{12.8}$

This hydrogen loss probably plays some role in the overall transformation of acetylene.

INTERPRETATION
Mechanism of alkylaromatics formation

The complex reaction network observed as acetylene interacts with CrY zeolite may be summarized as follows :

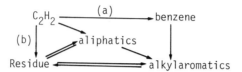

Up to 200°C, acetylene undergoes mainly a cyclization into benzene, catalyzed by Cr ions (path a) ; the decrease in conversion between 150 and 300°C was attributed to a partial reduction of the Cr^{3+} ions [3]. Meanwhile a high molecular weight polymer (path b) builds up in the channels. Free access to the cavities is thus prevented and the overall conversion nearly stops after two hours.

The precursor of all the alkylaromatics is thought to be styrene, arising from condensation of benzene and acetylene, catalyzed by Cr ions. Dimerization of the thermally produced vinylacetylene may also occur to some extent. Styrene indeed was detected in our system, and its concentration was found even greater in an experiment starting with benzene + acetylene. Further transformation of styrene into alkylaromatics, which are not oligomers of C_2H_2 involves a different type of catalytic activity, presumably connected with acidic properties. The aquo ligands of the Cr^{3+} ions generate protons upon heating at 350°C.

$$O{\Large\diagdown}\!\!\!-\!\!Cr(H_2O)_n^{3+} \longrightarrow O{\Large\diagdown}\!\!\!-\!\!Cr(OH)_n^{(3-n)+} + nH^+ \qquad \text{with } n = 1 \text{ or } 2$$

The above results suggest that Bronsted acid sites, located in the large cavities, play an important role in alkylaromatic formation. Apparently Lewis sites formed at higher temperatures do not show the same activity.

Toluene. As C-C bond breaking is a necessary step in the formation of odd carbon number molecules, we propose that toluene arises from cracking of 1,2-diphenylethane (D.P.E.) which is itself a C_2H_2 oligomer :

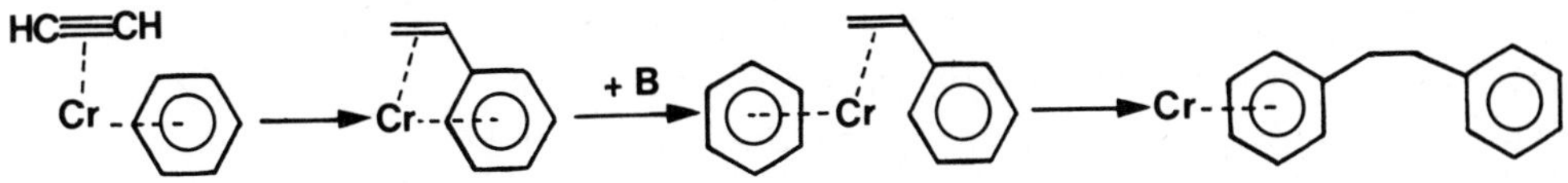

Such an intermediate has been isolated in homogeneous systems [6] but not in our case, probably because it is easily cleaved. This rupture may be acid-catalyzed :

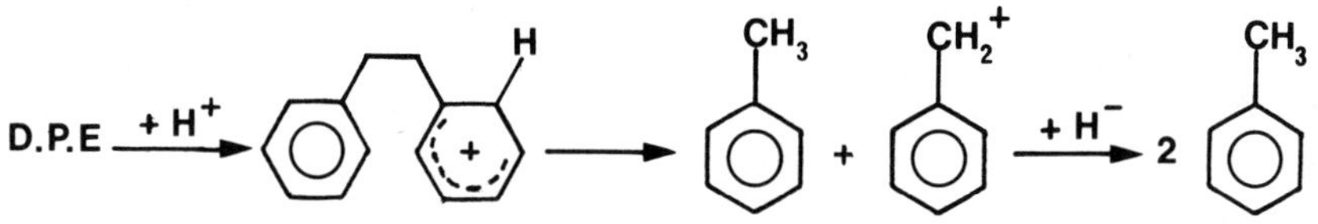

The hydride donor is another hydrocarbon molecule, or the non-desorbable residue. Another possibility, which is considered less likely, involves molecular hydrogen [7] which then reacts with the C-C bond :

$$\equiv Si - OH + HR \longrightarrow \equiv SiO^- + H_2 + R^+$$

Ethylbenzene and xylenes. Appearance of Ar-8 alkylaromatics at about 300°C suggests that, at this temperature, styrene is easily hydrogenated into ethylbenzene. Hydrogen used up in this process may be provided as H^+ and H^- added in two steps, or as molecular hydrogen. Then ethylbenzene is successively transformed into ortho-xylene, then meta and para, presumably via a classical carbenium mechanism [8].

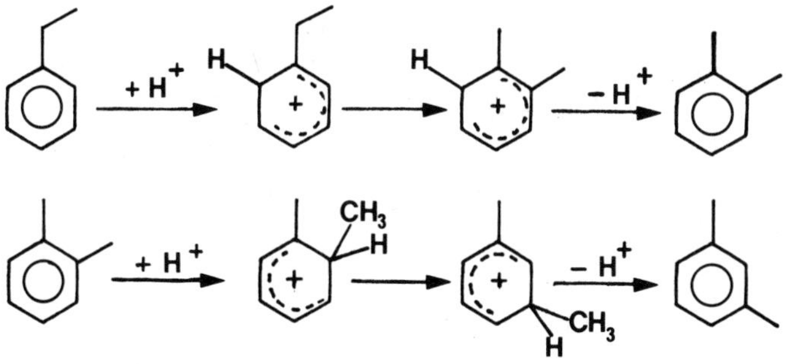

The rapid increase in xylene production with Cr content in the zeolite suggest that, at high chromium loading, strongly acidic sites are available in the cavities.

The occurence of some bimolecular processes over the $Cr_{12.9}Y$ is evidenced by reacting toluene at 350°C : 1.6 % disproportionation (into benzene and xylenes) and 1.8 % cracking (into benzene) are observed in 2 hrs. The ability of the CrY to catalyze the disproportionation of toluene was already noted by Merrill [9] : some xylene, mostly the para isomer, stems from this reaction.

Other reactions. Some other processes are characterized during the reaction of C_2H_2 at 350°C : i) xylene cracking which was found faster than (xylene + benzene) joint disproportionation - ii) heavier alkylaromatics production (Ar-10$^+$) followed by some chain cracking giving aliphatics - iii) hydrogen transfers, whereby chain cyclization into condensed aromatics is an hydrogen source.

The simultaneous formation of a high polymer is a general feature of reactions with unsaturated hydrocarbons [10]. For this reason, the active catalytic sites are progressively blocked. Moreover, protonic sites involved in hydrogenation processes are not renewed. As a result, acetylene conversion progressively slows down and cannot be made complete.

REFERENCES

1 G. Belluomini , A. Delfino, L. Manfra and V. Petrone, Int. J. of Appl. Rad. Isot., 29 (1978) 453.
2 P. Pichat, J.C. Vedrine, P. Gallezot and B. Imelik, J. Catal., 32 (1974) 190.
3 J. Léglise, Th. Chevreau and D. Cornet, Results to be published.
4 T. Yashima, J.I. Nagata, Y. Shimazaki and N. Hara, A.C.S. Symposium Series 40 (1977) 626.
5 P.B. Venuto, L.A. Hamilton, P.S. Landis and J.J. Wise, J. Catal., 4 (1966) 81.
6 G.A. Chukhadzhyan, Zh. I Abramyan, V.G. Grigoryan, Zh. Org. Khim., 9 (1973) 632.
7 A.P. Bolton, Zeolite Chemistry and Catalysis, A.C.S. Monograph, 171, J.A. Rabo, Editor (1976) 746.
8 M.L. Poutsma, Ibid., p. 496.
9 H.E. Merrill and W.F. Arey, Preprints, Amer. Chem. Soc., Petr. Div. (1968) 193.
10 E. Garbowski and H. Praliaud, J. Chim. Phys., 76 (1979) 687.

B. Imelik *et al.* (Editors), *Catalysis by Zeolites*
© 1980 Elsevier Scientific Publishing Company, Amsterdam — Printed in The Netherlands

A NEW METHOD FOR THE DEALUMINATION OF FAUJASITE-TYPE ZEOLITES

HERMANN K. BEYER and ITA BELENYKAJA

Central Research Institute for Chemistry of the
Hungarian Academy of Sciences, Budapest (Hungary)

INTRODUCTION

In the last few years highly siliceous zeolites with SiO_2/Al_2O_3 ratios over 30 have been synthesized, e.g. ZSM-5 (ref. 1) and Nu-1 (ref. 2). These materials possess unique catalytic properties and offer the possibility of new catalytic processes, especially the conversion of methanol to gasoline (ref. 3).

Knowing the relation between the SiO_2/Al_2O_3 ratio of a zeolite and the acid strength of its hydrogen from (ref. 4) it is obvious that highly siliceous H-zeolites are extremely strong Brönsted acids. In order to decide whether the unique catalytic behaviour is connected to the acid strength or rather to shape-selectivity effects it would be of interest to study other zeolite structures with high SiO_2/Al_2O_3 ratios, especially the well known and from catalytic viewpoints also intensively studied faujasite-type zeolites. However, no methods are known for the synthesis of faujasite with extremely high silicon contents. Consequently, the preparation of highly siliceous faujasite by dealumination of synthetic faujasite-type zeolites is of great importance.

It has been reported that during hydrothermal treatment of ammonium-exchanged Y zeolites part of the framework aluminium migrates into lattice cation positions (ref. 5), nevertheless framework vacancies are not or only scarcely formed because of the migration of silicon and oxygen atoms under these experimental conditions (ref. 6, 7). In the last years methods have been developed for the preparation of faujasite-type zeolites with SiO_2/Al_2O_3 ratios over 100 basing on repeated ammonium-exchange and hydrothermal treatment combined with acid extraction of the nonframework aluminium (ref. 8, 9). However, these procedures are time- and labour-consuming and result in the formation of a mesopore-system probably due to the elimination of part of the cubooctaeders (ref. 10).

This paper deals with the dealumination of faujasite-type zeolites basing on a completely different chemical reaction. Starting from any cation form the dealumination process is performed in only one step and in absence of water steam.

MATERIAL AND METHODS

Na-Y zeolite used in most of the experiments was obtained from VEB Elektrochemisches Kombinat Bitterfeld, GDR ($Na[Al_{1.0}Si_{2.52}]-Y,FAU$). From this sample a ($NH_4$,Na)-Y zeolite was prepared by ion exchange with NH_4Cl solution (degree of exchange: 65 %). Further experiments were carried out with Na-Y, Na-X and L-type zeolite from Union Carbide (Linde Division) and Na-mordenite from Norton. Calcium-exchanged X zeolite and ammonium-exchanged L zeolite were also studied.

X-ray diffraction pattern of the original and dealuminated zeolites were taken on a Phillips diffractometer PW 1130/00. The mid-infrared transmission spectra were recorded with a Nicolet 7199 FT-IR spectrophotometer using the KBr pellet technique.

DEALUMINATION PROCEDURE

Already more than 100 years ago it has been reported that alumina reacts at "red heat" with silicon tetrachloride forming volatile $AlCl_3$ and SiO_2 (ref. 11, 12). Inspired by this early observation we treated Y zeolites at high temperatures and we obtained the expected but nevertheless surprising result that under certain experimental conditions aluminium is substituted in the framework by silicon. This reaction may be formally described by the equation

$$M_{1/n}[AlO_2 \cdot (SiO_2)_x] + SiCl_4 \longrightarrow 1/n\ MCl_n + AlCl_3 + [(SiO_2)_{x+1}] \qquad (1)$$

in which M is some lattice cation.

The zeolites are placed in powder form in a vertical quartz tube reactor (bed height about 3 cm) and dehydrated for two hours at about 650 K in dry nitrogen streaming through the zeolite bed. After dehydration the nitrogen stream is saturated at room temperature with silicon tetrachloride and the temperature of the reactor is rised at a constant rate of 4 $K \cdot min^{-1}$ till a choosen final level between 730 and 830 K. Using Y zeolites as starting material a white smoke of aluminium trichloride appears in the effluent gas at temperatures around 730 K which indicates the beginning of the substitution reaction. The silicon tetrachloride treatment is carried on for two hours at the final temperature. Then the product is purged with dry nitrogen, washed until complete disappearence of any chlorides in the washing water and dried at about 400 K. High-crystallinity faujasite-type structures are obtained.

However, if dehydrated Y zeolite is first heated to temperatures over 750 K and then brought into contact with silicon tetrachloride vapour a violent exothermal reaction proceeds resulting in the formation of an amorphous product. This indicates that the incorporation of silicon into framework vacancies left by dealumination is a slower process than the formation of $AlCl_3$. Consequently, at high dealumination rates the concentration of vacancy sites reaches a level where the structure becomes unstable.

The degree of dealumination depends mainly on the reaction time and less on the final temperature. Treatment of Y zeolites in the temperature range from 780 till 830 K for 2 hours leads to products with SiO_2/Al_2O_3 ratios from about 40 till 100. however, starting from Na-Y zeolite only part of the aluminium leaving framework positions is escaping in form of volatile aluminium chloride. That is easy to explain. Equation (1) shows that sodium lattice cations are transformed to NaCl. On the other hand it has been found long ago that $Na[AlCl_4]$ is formed by melting NaCl with $AlCl_3$ (ref. 13) or by passing gaseous $AlCl_3$ over heated NaCl (ref. 14). It is obvious that part of the $AlCl_3$ reacts with the formed sodium chloride to Na $AlCl_4$ which is not volatile under the reaction conditions but can be removed by washing. Indeed, the washing water contains always aluminium and it is acidic due to the hydrolytic decomposition of the complex.

If the aluminium would escape quantitatively as $AlCl_3$ the progression of the dealumination reaction could be followed by gravimetric measurements using a suitable thermobalance or a special McBain balance. Unfortunately, the secondary reaction of part of the $AlCl_3$ with the formed NaCl excludes the application of this simple method. At present we don't see any possibility to study the kinetics of the dealumination of metal cation containing zeolites. However, it should be possible to study the dealumination of completely ammonium-exchanged Y zeolite by gravimetric methods.

It should be noted that Y zeolites can be also dealuminated using trichlorosilane as dealuminating agent. This compound is, however, less thermostable than silicon tetrachloride. It is necessary to carry out the process in a hydrogen atmosphere but even observing all precautions it was not possible to avoid completely the decomposition of trichlorosilane. The dealuminated products were always contaminated with traces of metallic silicon.

(NH_4,Na)-Y zeolite can be dealuminated under the same conditions as the sodium Y. However, attemps to dealuminate L-type zeolite and its NH_4-form as well as Na-mordenite and Na-X failed. (Ca,Na)-X zeolite could be partially dealuminated under the conditions given above for Na-Y. It is obvious that the described method is not applicable for the dealumination of zeolites with narrow pore systems in which silicon tetrachloride cannot enter. We suppose that in the case of X zeolites the framework is shielded from the attack of the silicon tetrachloride by lattice cations present in higher concentration as in Y zeolites.

CHARACTERIZATION AND PROPERTIES OF DEALUMINATED Y ZEOLITES

The dealuminated samples have very good crystallinity as reflected by the X-ray diffractograms (fig. 1). As expected the diffraction peaks are shifted towards higher 2θ values indicating the contraction of the unit cell. Unit cell size and residual aluminium content of the dealuminated samples fit exactly the relation given by BRECK and FLANIGEN (ref. 15).

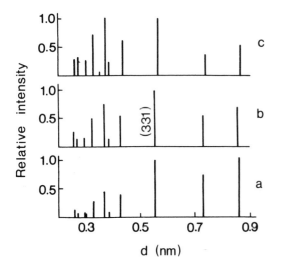

Fig. 1. Relative intensites of selected X-ray diffraction peaks (related to the intensity of the 111 reflection).

(a) Y zeolite dealuminated by silicon tetrachloride; SiO_2/Al_2O_3 ratio = 46.

(b) Y zeolite dealuminated by hydrothermal treatment combined with acid extraction (from a diffractogram given in ref. 8). SiO_2/al_2O_3 ratio = = 192.

(c) Na-Y zeolite; SiO_2/Al_2O_3 ratio = 5.04.

All diffraction peaks appearing in the X-ray diffractogram of the original Na-Y zeolite are also present in the diffractograms of the dealuminated samples but remarkable intensity differences are observed. The peaks belonging to crystal planes with higher indices are more intense in the diffractogram of Na-Y. On the contrary, the (111), (220), (311) and (331) reflections are much more intense in the case of dealuminated samples. The structure factors calculated from the intensity of these peaks are in good agreement with the contribution of the framework to the corresponding structure factors of hydrated Na-Y zeolite given by VARGA et al. (ref. 16).

The obtained X-ray results are consistent with the substitution of aluminium by silicon in the zeolite framework. As for the intensity of the individual diffraction peaks Y zeolite dealuminated by hydrothermal treatment combined with acid extraction (fig. 1b) is situated between the original Na-Y zeolite (fig. 1c) and the samples dealuminated by silicon tetrachloride (fig. 1a) though the acid extracted sample has the highest dealumination degree. That indicates structural differences between the products obtained by the two fundamentally different dealumination procedures. A detailed structure factor analysis may give more information.

It has been reported that aluminium removal from the framework of (NH_4,Na)-Y zeolites and elimination of lattice vacancies created by dealumination shift the framework vibration bands in the mid-infrared spectrum toward higher frequencies and result in an increase of the band-sharpness (ref. 6, 10). The frequency shift can be explained by the shorter average T-O bond distances in the aluminium-free structure. The increasing band sharpness points to a higher "degree of ordering".

Itcan be seen from the spectra given in fig. 2 that the dealumination of Y zeolite by reaction with silicon tetrachloride results also in sharper bands and shifts the absorption maxima to higher frequencies. The

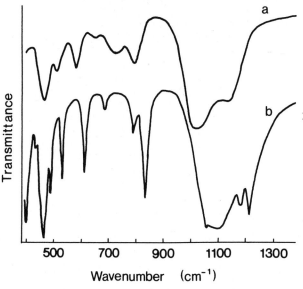

Fig. 2. Mid-infrared spectra of
(a)   Na-Y zeolite,
      SiO_2/Al_2O_3 ratio = 5.o4;
(b)   Y zeolite dealuminated by
      silicon tetrachloride,
      SiO_2/Al_2O_3 ratio = 46.

spectrum (b) of the dealuminated product is quite similar to the spectra of Y zeolites after hydrothermal treatment without (ref. 6) and with (ref. 10) acid extraction of the formed nonframework aluminium. However, the spectrum of the sample dealuminated with silicon tetrachloride is much better resolved, it reveals more details and the bands are considerably sharper. Consequently, the infrared spectrophotometric results indicate a higher "degree of ordering" in this sample.

Zeolites dealuminated by silicon tetrachloride show an extremely high thermal stability and resistance to mineral acids. As reflected by X-ray diffractograms and adsorption data the crystallinity of a dealuminated sample with a SiO_2/Al_2O_3 ratio of about 50 remains unchanged even after heating at 1370 K. The structure collapses only at 1450 K as indicated by an exothermal DTA peak appearing at this temperature. The crystallinity is also fully maintained after boiling in 6 N hydrochloric acid for two hours. In this respect the samples dealuminated by silicon tetrachloride or under hydrothermal conditions behave in the same way (ref. 6, 10).

The dealuminated zeolites do not contain sodium and aluminium in equimolar amounts. It is assumed that part of the sodium is exchanged during the washing process.

Acid treatment of samples dealuminated by silicon tetrachloride results in a further removal of framework aluminium remaining after the substitution reaction. For example the SiO_2/Al_2O_3 ratio of a dealuminated Y sample could be increased from 27 to 550 by repeated treatment with 1 N hydrochloric acid at 353 K.

It is well known that the adsorption capacity of Na-Y zeolite for water and ammonia amounts to about 0.33 cm^3 liquid adsorbate per g adsorbent and that it is reached already at very low relative pressures. In contrast to the preference of zeolite surfaces for polar molecules, highly dealuminated Y zeolites have an extremely low selectivity for the adsorption of water and ammonia (fig. 3) and behave in this respect like silicalite, the aluminium-free homologue of ZSM-5 zeolite (ref. 17).

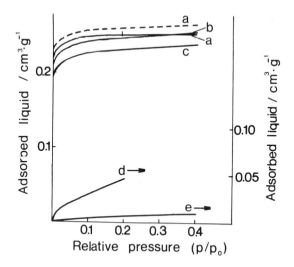

Fig. 3. Adsorption isotherms on Na-Y (broken line) and dealuminated Y zeolite (full lines); Si/2Al ratio = 44. (a) = n-hexane; (b) = n-butane; (c) = benzene; (d) = ammonia; (e) = water.

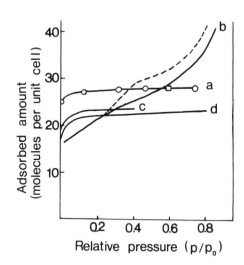

Fig. 4. Adsorption isotherms of n-hexane at 300 K on (a) Na-Y (points from data given in ref. 10); (b) Y zeolite dealuminated under hydrothermal conditions (from ref. 10); (b), (c) Y zeolite dealuminated with silicon tetrachloride, Si/2Al ratios 44 and 60, respectively.

The isotherms for the adsorption of organic compounds (n-butane, benzene, n-hexane) on dealuminated Y zeolite (fig. 3) show the near rectilinear shape typical for the volume filling of micropores and known from the adsorption on zeolites. The adsorption capacity of dealuminated zeolites is only slightly lower than that of the original zeolite (see "a" curves in fig. 3). A more realistic comparison of the adsorption capacities is possible if the adsorbed amount is given in molecules per unit cell (fig. 4). As can be seen from the adsorption isotherms in fig. 4 the dealuminated Y samples adsorb about 15 % less molecules in the unit cell than Na-Y zeolite. About half of this difference is due to the diminuation of the unit cell volume caused by the lattice contraction during the substitution of framework aluminium by silicon. It is not yet clear whether the remaining difference of 7-8 % in the adsorption capacity reflects a crystallinity loss or a less dense packing of the molecules in the supercages. This difference, however, is not enough pronounced to attribute to it a fundamental significance.

As for the adsorption of n-hexane (and other hydrocarbons) zeolite Y dealuminated under hydrothermal conditions shows quite another behaviour (see fig. 4. curve b) which points to a bidisperse pore system (ref. 10). The volume of the faujasite-type micropore system amounts to 75 % compared with Na-Y. Secondary pores formed during the dealumination process have radii in the range from 1.5-1.9 nm. It is assumed that silicon atoms migrating into the framework vacancies left by dealumination come from positions inside the crystal in such a manner that structural units, e.g. whole sodalite units, disappear (ref. 10).

CONCLUSION

Y zeolites can be dealuminated by reaction with silicon tetrachloride without collapse of the crystal structure. X-ray diffraction pattern and mid-infrared spectra point to a high degree of ordering in the framework of the dealuminated products and adsorption of hydrocarbons reveals no significant differences in the pore system compared with Na-Y. In contrast to the described new dealumination method, dealumination under hydrothermal conditions changes the pore system and the structure of the framework to a certain degree.

Y zeolites dealuminated by silicon tetrachloride may offer new application possibilities as hydrophobic adsorbents. It can be expected that the hydrogen form of partly dealuminated zeolite Y is an active catalyst for hydrocarbon reactions with excellent thermal and hydrothermal stability.

ACKNOWLEDGEMENT

The authors are indebted to the X-Ray Diffraction and Optical Spectroscopy Groups of the Central Research Institute for Chemistry of the Hungarian Academy of Sciences for X-ray diffraction and infrared measurements, respectively. Technical assistance of Mrs. I. Szaniszló and Mr. I. Csorba is gratefully acknowledged.

REFERENCES

1   G.T. Kokotailo, S.L. Lawton, D.H. Olson and W.M. Meier, Nature, 272 (1978) 437-438.
2   M.S. Spencer and T.V. Whittam, Acta Phys. Chem. (Szeged), 24 (1978) 307-311.
3   S.L. Meisel, J.P. McCullough, C.H. Lechthaler and P.B. Weisz, Chemtech, 6 (1976) 86-89.
4   D. Barthomeuf, Acta Phys. Chem. (Szeged), 24 (1978) 71-75.
5   G.T. Kerr, J. Phys. Chem., 71 (1967) 4155-4156.
6   J. Scherzer and J.L. Bass, J. Catalysis, 28 (1973) 101-115.
7   P. Gallezot, R. Beaumont and D. Barthomeuf, J. Phys. Chem., 78 (1974) 1550-1553.
8   J. Scherzer, J. Catalysis, 54 (1978) 285-288.
9   U. Lohse, E. Alsdorf and H. Stach, Z. anorg. allg. Chem., 447 (1978) 64-74.
10 U. Lohse, H. Stach, H. Thamm, W. Schirmer, A.A. Isirikjan, N.I. Regent and M.M. Dubinin, Z. anorg. allg. Chem., 460 (1980) 179-190.
11 M. Daubrée, Comt. Rend., 34 (1854) 135-140.
12 L. Troost and P. Hautefeuille, Compt. Rend., 75 (1872) 1819-1821.
13 F. Wöhler, Pogg. Ann., 11 (1827) 155.
14 H. Rose, Pogg. Ann., 96 (1855) 157.

15 D.W. Breck and E.M. Flanigen, Molecular Sieves, The Chemical
    Society, London, 1968, pp. 47-60.
16 K. Varga, I. Kiricsi and Gy. Argay, Acta Phys. Chem. (Szeged),
    25 (1979) 69-77.
17 E.M. Flanigen, J.M. Bennett, R.W. Grose, J.P. Cohen, R.L. Patton
    and R.M. Kirchner, Nature, 271 (1978) 512-516.

B. Imelik *et al.* (Editors), *Catalysis by Zeolites*
© 1980 Elsevier Scientific Publishing Company, Amsterdam — Printed in The Netherlands

SPECTROSCOPIC INVESTIGATIONS ON COBALT(II) EXCHANGED A-TYPE ZEOLITES.
COORDINATION SPHERE OF THE CATIONS; SORPTION AND ISOMERIZATION OF
n-BUTENES

H. FÖRSTER, M. SCHUMANN and R. SEELEMANN
Institut für Physikalische Chemie der Universität Hamburg,
Laufgraben 24, D-2000 Hamburg 13, Bundesrepublik Deutschland

INTRODUCTION

Apart from cracking, also isomerization plays an important role in petroleum refining processes. Since zeolites have been introduced as catalysts due to their higher activity, greater stability and reduced coke
formation, a great number of investigations has been performed in order
to elucidate the active centers, responsible for catalysis. For isomerization frequently Brønsted centers are favoured, although there is much
evidence that also cations are of catalytic importance.

As more basic research seemed to be necessary, we studied the behaviour of some alkali and alkaline earth ion exchanged forms of the simply-
structured zeolite A on low temperature n-butene isomerization, cocata-
lyzed by sulfur dioxide. It is known that on other zeolites often a linear relationship between the electrostatic properties of the cations and
the catalytic activity can be established (ref.1), which we could also
prove for zeolites type A (ref.2 and 3). As transition metal ion exchanged zeolites show a deviation from this relationship, and as it should be
possible that Brønsted acidity is introduced by reaction of these cations
with adsorbed water, we extended our study to these ion exchanged forms.

In this paper we present the results of our investigations of sorption
and isomerization of n-butenes on A-type zeolites, exchanged to different
degrees with Co^{2+} ions.

MATERIALS

The samples were prepared, starting from zeolite NaA and replacing the
sodium by cobalt(II) ions by stirring, overnight, the zeolite powder in a
10^{-4} - 10^{-2}m aqueous solution of cobalt nitrate. Although ion exchange was
complete, the composition was checked once more by neutron activation analysis. In this way specimens with the formulae $CoNa_{10}A$ (I), Co_2Na_8A (II),
$Co_{2.5}Na_7A$ (III), and Co_4Na_4A (IV) were obtained. The preservation of crystallinity after ion exchange was assured by X-ray diffraction and nitrogen adsorption at liquid nitrogen temperature.

cis-But-2-ene and trans-but-2-ene (Linde), stated purity 99%, but-1-
ene (Linde) 99.7% and sulphur dioxide (Deutsche L'Air Liquide) 99.9% were
used without further purification.

METHODS

n–Butene adsorption could be monitored by spectral changes either in the visible (CARY 17) or in the infrared range (PERKIN ELMER Model 225, DIGILAB FTS 14), using the intrazeolitic cobalt complex or the butene molecule as a probe. Both kinds of spectroscopy were applied for

- characterizing the sites of interaction between adsorbate and adsorbent,
- following the translational and rotational motion of the adsorbate inside the zeolitic cavities, and
- observing the catalytic reaction actually occurring on the zeolite surface.

The latter was achieved using the dynamic Fourier transform infrared spectroscopic technique.

Prior to the adsorption and kinetic studies, the zeolites were compressed into wafers of about 5 mg/cm^2 and activated overnight at 625 K and 10^{-6} Pa in UHV cells equipped with quartz and KBr windows, respectively.

RESULTS AND DISCUSSION

As it is assumed that acidic hydroxyl groups are formed by dissociation of water on Co^{2+} ions (ref.4), also the dehydration of the samples was studied by UV-VIS and IR spectroscopy.

From the IR spectra two kinds of water can be distinguished: an unspecifically adsorbed one with stretching frequencies at 3450 and 3290 cm^{-1} and a bending frequency at 1645 cm^{-1}, which can be easily removed by room temperature evacuation, and a second one directly interacting with the Co^{2+} ion, giving rise to a more pronounced blue shift of its fundamentals ($\nu_{stretch}$ = 3670 cm^{-1}, ν_{bend} = 1685 cm^{-1}). Upon complete removal of the water no indication of hydroxyl groups was found in accord with Wichterlová et al. (ref.5), so that other centers must be responsible for the formation of an activated complex with the reactant.

Whereas for the fully hydrated zeolite Co_4Na_4A from X-ray analysis two different cobalt-water complexes are assumed (ref.4), from electronic spectra there is only evidence for one octahedral complex, a so-called "floating" complex.

During dehydration the coordination changes over a tetrahedral one to a bare cobalt(II) ion on S1 site, surrounded by the three nearest oxygen atoms of the six-ring in D_{3h} symmetry, which is proved by the occurrence of the 25000 cm^{-1} band (see Fig.1). Upon butene uptake this band disappears, indicating the formation of a near-tetrahedral wall complex of C_{3v} symmetry with butene as the fourth ligand. On cobalt-rich samples a total disappearance of this band, i.e. the complete coordination of each cobalt ion by butene molecules cannot be achieved due to their size and geometry. This is only possible with smaller adsorbate molecules, e.g. CO,

SO_2, and H_2O. The slight displacement of the bands, when SO_2 is coadsorbed, exhibits a distortion of the ligand field by the cocatalyst.

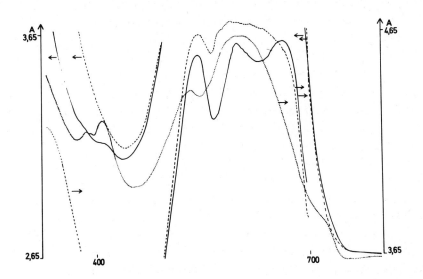

Fig. 1. Change of coordination sphere as revealed from the electronic spectra. Disappearance of the 25000 cm^{-1} band indicating the change of a bare cobalt(II) ion (D_{3h}) to a wall complex of near-tetrahedral symmetry (C_{3v}). ...... activated at 625 K and 1 μPa,——admission of 665 Pa trans-butene, —————— admission of 665 Pa trans-butene + 65 Pa SO_2. Abscissa in nanometers.

As the diameter of the CO molecule exceeds that of the six-ring, which forms the entrance from the supercage to the sodalite unit, penetration of the carbon monoxide into the β cages is impossible. Therefore monitoring the intensity of the band of CO adsorbed on the cation with respect to thermal pretreatment gives information about the cation content of the supercage. Decreasing intensity would indicate migration into inaccessible positions. In a preliminary experiment on zeolite Co_4Na_4A, gradually dehydrated between room temperature and 625 K, upon successive CO admission the strong band of the $CO-Co^{2+}$ complex at 2192 cm^{-1} grew slightly in intensity with increasing dehydration temperature. This experiment shows only the removal of the competing adsorbate water, but gives no evidence for a migration of the Co^{2+} into the sodalite cages.

In general, the penetration rate decreases from trans-butene over but-1-ene to cis-butene. On zeolites with high cobalt content n-butenes were adsorbed without any hindrance, indicating vacant S2 positions. As these get more and more occupied with diminishing degree of exchange, diffusion especially of cis-butene and but-1-ene gets rendered difficult.

The general outline of IR spectral changes upon adsorption on cobalt ion exchanged zeolites is the same as on alkali or alkaline earth ion containing samples, dealt with elsewhere (ref.2). All fundamentals get infrared active due to symmetry lowering by adsorption. Their band positions are shifted mainly to lower frequencies compared to the gas phase. The

drastic loss of rotational structure indicates complete quenching of free
rotational motion. Especially informative is the study of the C=C stretching band, the red shift of which is particularly pronounced and depends on
the cation so that e.g. butene molecules adsorbed on sodium or cobalt
ions can be distinguished IR-spectroscopically (see Tab.1). Because of
less obscuration this band is well suited for the quantitative evaluation
of the kinetics of isomerization by dynamic Fourier transform infrared
spectroscopy. In preceding investigations (ref.2 and 3) the sorption bond
was interpreted in terms of an electron donation from the butenic π-bond
to the cation, which explains the linear correlation between the C=C band
shift and the charge/radius ratio of the cation. In case of cobalt this
shift is far too large as to be explained only by π-bonding; an additional

TABLE 1. Band shift of the butenic C=C stretching fundamental on zeolites
containing different cations.

Cation	Charge/radius V	But-1-ene cm^{-1}	Cis-but-2-ene cm^{-1}	Trans-but-2-ene cm^{-1}
Na^+	15.16	− 10	− 15	− 17
Li^+	24.00	− 16	− 18	−117
Ca^{2+}	29.09	− 25	− 25	− 32
Mg^{2+}	44.36	− 28	− 33	− 47
Co^{2+}	40.05	− 38	− 50	− 57

back-donation of d-electron density into antibonding π^*orbitals has to be
considered.

Isomerization of the n-butenes was studied mainly at room temperature.
The course of the catalyzed reaction on the zeolite surface was directly
observed by IR spectroscopy. Under these mild conditions on alkali and alkaline earth ion exchanged zeolites A a different behaviour has been observed (ref.3): Only rarely spontaneous isomerization occurs, i.e. isomerization without any additional assistance. Frequently isomerization must
be induced by a cocatalyst, e.g. sulfur dioxide, and at least there are
zeolites, completely inactive in isomerization, e.g. zeolite NaA.

In case of the cobalt ion exchanged zeolites A spontaneous isomerization is only observed on sample IV with the highest cobalt content, as
evidenced from the changing spectra (see Fig.2), while on samples I − III
all n-butenes are stable. The failing isomerization of cis- and trans-butene on zeolite $CoNa_{10}A$ has been already confirmed by Klier et al.(ref.6).
As reaction products only the isomeric n-butenes but neither the thermodynamically most stable isobutene nor other by-products such as fragments
or polymers could be detected by infrared spectroscopy.

Such a spontaneous release of isomerization has already been observed
on zeolite Caα with all n-butenes (ref.3), whereas on zeolite $Na_{3.2}Zn_{4.4}A$
only trans-butene and on zeolite $Na_{1.1}Zn_{11.4}A$, which has a non-stoichio-
metric zinc excess, trans-butene and but-1-ene do spontaneously react.

In the first case only the change of the Si/Al ratio from zeolite CaA to Caα causes the reaction proceeding spontaneously. In all cases the spontaneous character of isomerization can be correlated with a high content of the catalytically active cation.

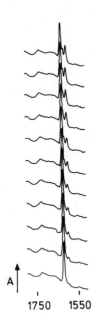

1750   1550

Fig. 2. Dynamic infrared spectra of adsorbed trans-but-ene on Co_4Na_4A in the region of the C=C stretching bands. Time increases from bottom to top, 2 hours between each spectrum.

These investigations reveal that spontaneous isomerization depends on both the zeolite and the butene, of which trans-butene is preferentially activated followed by but-1-ene. This can be understood taking into account a closer approach of the cation by the trans-butene molecule due to its lacking dipole moment, which is reflected in the infrared spectra by a more distinct frequency shift. Therefore it can only be hardly realized that on zeolite $CoNa_{10}A$ the trans-isomer shows a less ability to split the 17300 cm^{-1} band compared to cis-butene. This has been explained by an increased sterical hindrance in forming the trans-butene-Co^{2+} bond (ref.6). Sterical hindrance does not matter the adsorption of but-1-ene, as recent CNDO/2-SCF and PCILO calculations make an "end on" interaction wich the cation more likely (ref.7).

Is there any evidence in the spectra for the different behaviour of the inactive and the spontaneously isomerizing samples? The electronic spectra of the trans-butene complexes of zeolites $CoNa_{10}A$ and Co_4Na_4A (see Fig.1) are very similar and resemble those of tetrahedral cobaltous complexes. This means that in both samples the bonding of the butenes to the cobalt ion is essentially the same. But on zeolite Co_4Na_4A the 25000 cm^{-1} band does not completely disappear, indicating that not all cobalt ions are coordinated by the hydrocarbons. Furthermore, the 17300 cm^{-1} band, which has been assigned to transitions from the $^4A_2(F)$ ground state to nondegenerate

$^4T_1(P)$ states (ref.6), shows a more distinct splitting in case of zeolite CoNa$_{10}$A compared to Co$_4$Na$_4$A, which may be interpreted by a more tight bonding between the butene and the cobalt ion. Whatever the reason for this band splitting will be, from Klier's (ref.6) and our results the tendency for spontaneous isomerization must be correlated with a decreasing splitting of the $^4T_1(P)$ states.

However, in the infrared region no drastic differences are observed concerning the overall spectrum and the C=C band shift, which is identical for both cobalt samples within the margin of error. Comparing the IR spectra of the n-butene complexes on zeolites CaA and Caα, slightly differing shifts of the bands in the CH stretching and bending regions were interpreted in terms of a stronger interaction with the zeolitic framework in case of Caα, caused by a small change in cation position. On zeolites CoNaA only with but-1-ene a band at 1415 cm^{-1}, which can be assigned to the CH$_2$ in plane deformation vibration, increases in intensity parallel to the cobalt content. Therefrom a stronger interaction can be derived. In order to elucidate the change in catalytic properties of zeolites with respect to the degree of cobalt exchange the complex application of several experimental techniques is required. Electronic and infrared spectroscopy separately can only make their limited contribution hereto.

Coadsorption of sulfur dioxide induces or enhances the isomerization of the three n-butenes on all cobalt-containing zeolites at room temperature. In analogy to the spontaneous reaction only the isomeric n-butenes are found by infrared spectroscopy. Compared to former results (ref.3), the kinetic data reveal that at low levels of ion exchange the reaction rate is limited by diffusion, as especially in case of formed cis-butene fast penetration through the eight-membered rings is hindered by sodium ions on S2 positions. As soon as these sites become unoccupied, the isomerization rate gets determined by the surface reaction.

While on other zeolite samples a reaction order of one with respect to both, n-butene and sulfur dioxide, has been observed (refs.3,8), Co$_4$Na$_4$A behaves completely different. Instead of the expected rate constant, linearly increasing with the amount of sulfur dioxide, a deviation to higher rates gets obvious (see Fig.3). In a sequence of runs on the same zeolite wafer, after reactivation at 525 K in UHV the isomerization reaction was studied under identical conditions. With growing number of runs the isomerization rate increases continuously, indicating an irreversible change of the catalyst's properties. This is further confirmed by a darkening of the zeolite and by electronic spectra, in which the intensity of the 25000 cm^{-1} band gradually decreases after each cycle, i.e. after reactivation the trigonal coordination spheres of the Co^{2+} ions on S1 sites are not originally restored.

After drastic heating of the sample in a sulfur dioxide atmosphere the 25000 cm^{-1} band is totally absent, the X-ray diffraction pattern shows a

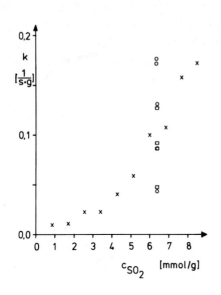

Fig. 3. Plot of the isomerization rate constant, obtained from a reversible first order law, vs. the amount of sulfur dioxide (x). The □ and ○ indicate the results of successive runs under identical conditions (butene : SO_2 = 5 : 4).

very high background, and the surface area has decreased to about one third of its original value. These facts can be interpreted in terms of a breakdown of the zeolite lattice, increasing upon successive sulfur dioxide treatment.

From this point of view the growing activity of zeolite Co_4Na_4A may be explained by a partial breakdown of the zeolite structure, leading in the beginning to a better accessibility of the cations to adsorbed butene molecules. But a more detailed investigation is required for further information.

Concerning the active centers responsible for catalysis, a comparison of the initial rates of the isomerization reaction of but-1-ene on zeolites containing lithium, magnesium, calcium and zinc ions shows an increase of the rate parallel to the polarizing power of the cations (refs. 3,8). This clearly points out that the interaction of reactant and cations in conjunction with the directly intervening sulfur dioxide plays an important role for the catalytic activity. If in a first approach the increase of activity is neglected, also isomerization on cobalt-containing zeolites A fits this scheme.

As already mentioned, after activation neither water nor hydroxyl groups could be detected. Because also during the course of isomerization no OH groups emerged, it must be concluded that the cations are the active sites responsible for catalysis.

The cocatalytic effect of the sulfur dioxide seems to depend on a direct intervention into the catalytic reaction rather than on an overall change of the zeolite properties upon SO_2 adsorption, as can be infered from the reaction rate, which depends on the SO_2 concentration. Furthermore, electronic spectra confirm that isomerization proceeds via a near-

tetrahedral cobalt complex, in which the three basal ligands are the pro-
ximal oxygens of the six-ring and the fourth one is represented by butene
interacting with SO_2. After admission of sulfur dioxide to the butene–CoA
complex the tetrahedral symmetry is retained, while the entrance of the
SO_2 into the coordination sphere is indicated by a slight decrease of the
17300 cm^{-1} band splitting (see Fig.1).

Summarizing these findings, on cobalt-containing zeolites A upon n–butene adsorption π complexes of near-tetrahedral symmetry are formed, in
which at high cobalt contents trans-butene and but–1-ene spontaneously
react to the other n–isomers. At lower degrees of exchange the attack of
sulfur dioxide is required for inducing isomerization of all n–butene ligands via a concerted reaction mechanism. In disagreement with others
(ref.9), our adsorption data and infrared results confirm that isomerization essentially proceeds inside the zeolitic pore system.

REFERENCES

1   W.F. Kladnig, Acta Cient. Venezolana, 26 (1975) 40–69
2   H. Förster and R. Seelemann, J.C.S., Faraday I, 74 (1978) 1435–43
3   H. Förster and R. Seelemann, J.C.S., Faraday I, 75 (1979) 2744–52
4   P.E. Riley and K. Seff, J. Phys. Chem., 79 (1975) 1594–1601
5   B. Wichterlová, L. Kubelková and P. Jirů, Coll. Czech. Chem. Commun.
    45 (1980) 9–16
6   K. Klier, R. Kellerman and P.J. Hutta, J. Chem. Phys., 61 (1974)
    4224–34
7   R. Lochmann ,W. Meiler and K. Müller, Z. phys. Chemie (Leipzig), 261
    (1980) 165–170
8   H. Förster and R. Seelemann, in preparation
9   E. Detreköy and D. Kallo, in J.R. Katzer (Ed.), ACS Symp. Ser. 40,
    ACS, Washington, D.C., 1977, pp. 549–558

B. Imelik *et al.* (Editors), *Catalysis by Zeolites*
© 1980 Elsevier Scientific Publishing Company, Amsterdam — Printed in The Netherlands

CATALYTIC HYDRODESULFURIZATION OF DIBENZOTHIOPHENE OVER Y TYPE ZEOLITES

M.L. VRINAT, C.G. GACHET and L. de MOURGUES
Institut de Recherches sur la Catalyse, 2, avenue Einstein 69626 Villeurbanne Cédex France

1   INTRODUCTION

Few communications on zeolite catalysts used in hydrodesulfurization processes (HDS) are reported in the litterature and most of these studies were undertaken in order to evaluate new catalysts. For example, a high activity was pointed out by Mays "et al" /1/ on a cobalt exchanged and impregnated by MoO_3 zeolite, and ion exchanged forms of Co and Ni zeolites, impregnated with solution of molybdenum salts, were tested for thiophene conversion /2/. More recently, on mordenite and faujasite Y metal loaded zeolites, Brooks /3/ indicated that these zeolites have a dual functionnality in which both the transition metal and the acid zeolite substrate contribute to the observed thiophene hydrodesulfurization. But, compared with commercial catalysts, all these zeolite catalysts show lower activity and generally fast deactivation.

However, zeolite as a support may be interesting for a better understanding of the support influence on HDS reactivity and the specific interaction between Co and Mo species, interaction generally called "synergy".

For that purpose, sodium Y zeolite and decationized Y zeolite with or without metal loadings Co and (or) Mo were prepared in our laboratory and their reactivities were studied in the HDS of dibenzothiophene (DBT) in flow experiment in a hydrogen carrier at 613 K.

2   EXPERIMENTAL PART

2.1   Materials

The original zeolite was a synthetic NaY zeolite SK 40 with unit cell formula $(NaAlO_2)_{56}$ $(SiO_2)_{136}$, nH_2O from Linde.

The H-Y zeolite was prepared by exhaustive exchange of the Na-Y zeolite with ammonium chloride followed by heating for 3 hours under O_2 flow at 623 K.

The CoNa-Y and CoH-Y zeolites were prepared by conventional ion exchange of the parent zeolites (NaY and NH_4-Y) with a cobalt sulphate dilute solution. The samples were then thoroughly washed with deionised water and air dried at 373 K.

The Mo-NaY catalyst was obtained from Na-Y zeolite previously subjected to vacuum at 553 K and by sublimation of a known weight of commercial $Mo(CO)_6$ followed by heating in a closed reactor at 600 K to ensure thorough decomposition. The MoH-Y and CoMo-HY zeolites were prepared in the same way from H-Y and Co-H-Y precursors.

The extent of exchange was determined by analysis of the cations in the solution of dissolved zeolite by atomic absorption spectrometry. Composition and unit cell formula are given in table I.

For comparison with alumina supported catalysts, $Co/\gamma Al_2O_3$, $Mo/\gamma Al_2O_3$ and $CoMo/\gamma Al_2O_3$ (HR 306) were also tested.

TABLE 1

Sample	Catalyst	% Co	% Mo	Unit cell formula
1	Na-Y	0	0	Na_{56}-Y
2	H-Y	0	0	$Na_4 H_{52}$-Y
3	Co-NaY	4.40	0	$Co_{12.7} Na_{30.6}$ Y
4	Co-HY	1.47	0	$Co_4 H_{44} Na_4$ Y
5	Mo-NaY	0	4.74	$Mo_{8.8} [Na_{56}$ Y$]$
6	Mo-HY	0	3.41	$Mo_{5.8} H_{17.2} Na_4$ Y
7	CoMo-HY	1.47	3.22	$Mo_{5.6} Co_{4.1} H_{10.2} Na_4$ Y
8	$Co/\gamma Al_2O_3$	2.75	0	-
9	$Mo/\gamma Al_2O_3$	0	9.6	-
10	$CoMo/\gamma Al_2O_3$	2.36	9.33	$CoMo/\gamma Al_2O_3$ HR 306 PROCATALYSE

2.2 Activity measurements

The test reaction is the hydrodesulfurization, in vapor phase, of the dibenzothiophene (DBT), chosen as a model molecule.

Experiments were carried out in a conventional flow microreactor, at 613 K, under a hydrogen flow of 25 cm^3.mn^{-1}, a DBT partial pressure of 116 Pa and under atmospheric pressure. The activity is defined as being the specific rate (by g of catalyst) of DBT desulfurization into biphenyl and H_2S (production of phenylcyclohexane is negligible). On line chromatographic analysis are made every 15 minutes.

2.3 Pretreatments of the catalyst

In this work, catalytic activity measurement was carried out not only on different catalysts, but also after different pretreatments of the same catalyst.

Four types of (pre) treatment (a, b, c, and d) were employed, namely :

a) Heating of the catalyst load under pure H_2 until the desired temperature 613 K was reached,

b) Reduction under pure H_2 at 673 K for 2 hours,

c) Sulfidation in a flow of H_2S and H_2, with a H_2S/H_2 volume ratio of 15/85 and a flow rate of 25 cm^3.mn^{-1} at 673 K for 2 hours,

d) successive applications of (b) and (c)

After the chosen pretreatment(s), the temperature was lowered to 613 K and DBT was introduced in the reactor.

3 RESULTS AND DISCUSSION

3.1 Hydrodesulfurization over NaY, CoNaY and MoNaY catalysts

Figure 1 shows catalytic activity of NaY, CoNaY and MoNaY zeolites during the first 15 hours after the (a) pretreatment.

The striking points to be noticed are : the high activity of NaY zeolite, the low activity of CoNaY and MoNaY catalysts and the activation of all these catalysts during the first hour of the reaction.

All these points can be explained by the specific influence of Na^+ cation in the zeolite.

Effectively, Imai and Habgood /4/ observed after adsorption at room temperature of H_2S on silica rich faujasite type zeolite (NaY) actived and treated by H_2, an ESR spectrum which leads to consider a dissociative chemisorption of H_2S according $H_2S_{ads} \rightarrow HS \cdot + H \cdot$.

Moreover, Mars "et al" /5/ studied H_2S adsorption at 423 K under N_2 atmosphere on a NaX zeolite and correlated the ESR sulfur signal obtained to sulfur chain radicals. These sulfur radicals ($\cdot Sx \cdot$, $HSx \cdot$) have been identified and described by Dudzik /6/ in H_2S oxidation reaction over NaX zeolite.

The activation of the catalyst during the first hour of the run can be explained if Na^+ may be considered as a precursor of an active site by H_2S interaction ; the low activity of the Co-NaY zeolite may be correlated to a more difficult access of hydrocarbon molecules to Na^+ cations.

For the Mo-NaY catalyst, the black color of the catalyst is probably related to the formation of Mo aggregates which may close the pores of the catalyst.

It can be noticed that no influence of preliminary reduction or sulfidation was observed for Co-NaY and Mo-NaY catalysts.

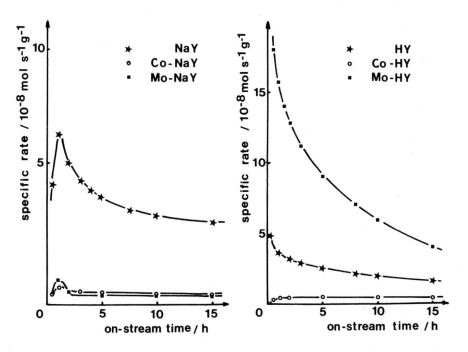

Fig. 1 Specific rate of HDS reaction v on-stream time for NaY zeolites

Fig. 2 Specific rate of HDS reaction v on-stream time for HY zeolites

(Reactions conditions : 613 K, 116 Pa pressure of DBT,(a) pretreatment)

3.2 Hydrodesulfurization over HY, CoHY and MoHY catalysts

Compared to Na-Y type zeolites, the Mo-HY catalyst present a high activity ; moreover a high conversion of DBT is obtained from the beginning of the reaction on HY and MoHY catalysts (Figure 2).

Differences between HY and NaY zeolites can be explained by their different properties : HY is a strong protonic acid and NaY has a relatively weak, if not negligible protonic acidity. So, initial activity of HY zeolites must be due to a protonic interaction. That was reported by Deo, Dalla Lana and Habgood [7] who studied the interaction of H_2S with sodium and decationized Y zeolites and found some hydrogen sulfide hydrogen-bonded to the surface hydroxyl groups of the HY zeolite.

In CoHY, exchange Co^{++} would probably stop that protonic interaction.

On MoHY catalyst, the high initial activity and its rapid decrease must be explained by the presence of two types of site : a protonic interaction like in HY zeolite and a specific activity of Mo.

UV and IR investigations of the Mo-HY zeolite show Mo^{6+} species, some of them can be reduced to Mo^{5+}, but no stronger reduction can be found by H_2 treatment (CO adsorption fails after H_2 reduction).

X-Ray diffraction shows that Mo species are well dispersed in the zeolite, which can explain the high initial activity of Mo-HY compared to that of the Mo-NaY catalyst. Perhaps some oligo-sulfur is induced on Mo species.

Moreover, hydrogen pretreatment (b) or sulfidation (c) have not influence on CoHY and MoHY catalytic activity.

3.3 Hydrodesulfurization over CoMo-HY zeolite

The CoMo-HY catalyst subjected to the (a) pretreatment shows no initial activity like the Co-HY catalyst, but reaches the level of the Mo-HY catalyst activity after a 10 hours on stream time (figure 3). That fact seems to confirm the negative influence of Co at the beginning of the reaction. However, compared to that of the Mo-HY catalyst, the CoMo-HY catalytic activity appears to be more stable after 10 hours of work and Co seems to stabilize the catalyst.

Moreover, it is interesting to notice the influence of sulfidation on this catalyst (Figure 3, pretreatment (c) and (d)) which was not observed with the Mo-HY catalyst.

So, we think that on the MoHY catalyst, some active sites are created in the first few hours of the run and that on CoMo-HY catalyst, Co delays the formation of active and more stable sites for the HDS reaction. To corroborate this hypothesis, more experiments with different Co contents CoMo-HY catalysts are to be carried out.

3.3 Hydrodesulfurization over Alumina-supported catalysts

In order to compare the zeolite catalysts with alumina-supported catalysts which are more classical for these studies, tests were carried out on $Co/\gamma Al_2O_3$, $Mo/\gamma Al_2O_3$ and $CoMo/\gamma Al_2O_3$.

Results of figure 4 show some parallelism between HY and γAl_2O_3 catalysts : low activity of Co catalysts, high initial activity of Mo catalysts, and a more stable activity of CoMo catalysts.

The influence of H_2 pretreatment and the deactivation of these alumina supported oxide catalysts have been previously studied in our laboratory /8/ and the results suggest the presence of two types of site for HDS : strongly reduced molybdenum ions, active but fragile and got only by reduction, and sulfided molybdenum less active but more resistant obtained by concerted reduction and sulfidation.

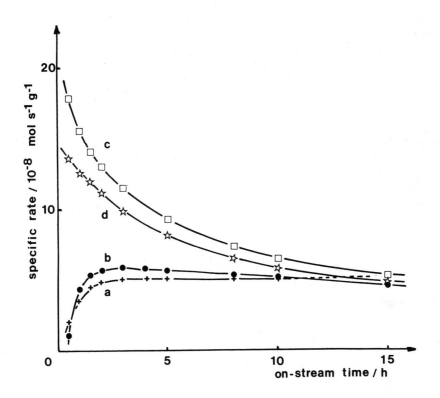

Fig. 3 Specific rate of HDS reaction v on-stream time for CoMo-HY zeolite : influence of the pretreatment. (pretreatment conditions are described in the text)

224

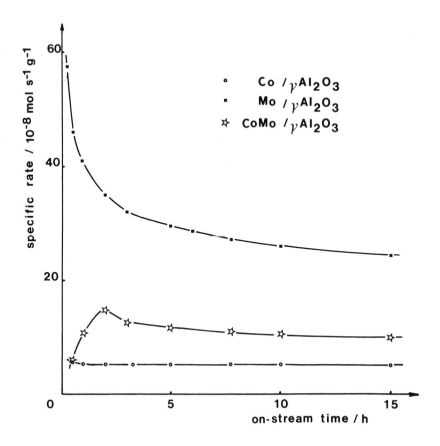

Fig. 4 Specific rate of HDS reaction v on-stream
time for alumina-supported catalyst

4.   CONCLUSION

The catalytic HDS of DBT over Y zeolites and decationized Y zeolites, with or without
metal loadings, shows that these two catalysts have quite different behaviours. Attention
must be especially paid to the initial activity of these catalysts and to the activity
after 15 hours of work. In particular , though that the three zeolites with Co have very
low initial activity, after 15 hours of work the CoMo-HY zeolite is the most active and
the most stable catalyst. So, neither the Co, nor the association Co-Mo are active by
themselves, but the Co has a main influence on the activation of the catalyst during the
sulfidation. Moreover, some parallelism can be done between the HY and the alumina-
supported catalysts and that work seems a good approach for a better understanding of the
interaction of Co and Mo species.

ACKNOWLEDGEMENTS

The authors thank Drs G. COUDURIER and C. NACCACHE for some helpful suggestions.

REFERENCES

1 R.L. Mays, P.E. Pickert, A.P. Bolton and M.A. Lanewala, Preprints 30th API Mtg Refining (Montreal) 1965.
2 A.V. Vysotskii, N.A. Chuikova and V.G. Lipovich, Kinet. Katal, 18, (1977), 1106-1107.
3 C.S. Brooks, VIth North-American Meeting of the Catalysis Society, Chicago, March 18-22, (1979).
4 T. Imai and H.W. Habgood, J. Phys. Chem. 77, (1973), 925-931.
5 M. Steijns, F. Derks, A. Verloop and P. Mars, J. Catal. 42, (1976), 87-95
6 a) Z. Dudzik, M. Bilska and J. Czeremuzinska, Bull. Acad. Polon. Sci. Ser. Sci. Chim. 22, (1974), 307.
   b) Z. Dudzik and M. Bilska-Ziolek, Ibid 23, (1975), 699.
7 A.V. Deo, J.G. Dalla-Lana and H.W. Habgood, J. Catal. 21, (1971), 270-281.
8 C.G. Gachet, E. Dhainaut, L. de Mourgues, M. Vrinat, International Symposium on Catalyst Deactivation, Antwerp, Belgium, 13-15 october 1980.

B. Imelik *et al.* (Editors), *Catalysis by Zeolites*
© 1980 Elsevier Scientific Publishing Company, Amsterdam — Printed in The Netherlands

CATALYTIC ACTIVITY OF VERY SMALL PARTICLES OF PLATINUM AND PALLADIUM SUPPORTED IN Y-TYPE ZEOLITES.

P. GALLEZOT

Institut de Recherches sur la Catalyse - C.N.R.S. - 2, avenue Albert Einstein, 69626 VILLEURBANNE CEDEX

1 INTRODUCTION

Apart from their present and potential applications as industrial catalysts reviewed by Bolton (ref. 1) platinum and palladium faujasites are very useful materials for fundamental studies on highly divided metals. Different homogeneous states of dispersion with particle sizes ranging down to the atomic scale can be otained by proper preparative techniques, even with metal concentration exceeding 10 wt % (refs. 2-4). High metal loadings present little interest as for as economy and catalysis are concerned but they make possible solid state studies involving physical methods which would not be sensitive enough for characterizing metals at low concentration on supports. Thus the accurate particle size measurements by electron microscopy and small samples X-ray scattering (refs. 2, 3), the determination of cluster structure by radial electron distribution (refs. 5, 6) and by EXAFS (ref. 7), the evaluation of the electronic state of these particles by XPS (ref. 8) and by X-ray absorption edge spectroscopy (ref. 9) have been performed successfully on PtY zeolites. Correlations between the structure and the catalytic properties of small particles can therefore be established.

One may expect that the geometric arrangement and the electronic structure of palladium and platinum undergo large change when the particle sizes decrease down to a few Angstroms. However the intrinsic effect of particle size on the catalytic activities is difficult to evaluate because the particles are embedded in the zeolite matrix which adds its own effects. Firstly, metal-reagent accessibility must be taken into account since encaged metal particles can be hidden or partially hidden from the incoming molecules. Secondly, the reactivity of the sorbed molecules and of the metal can be modified because of the strong electrostatic fields generated by the zeolite framework itself, by the lattice defects (Al^{3+} or O^{2-} vacancies) by the hydroxyls groups and by the multivalent cations. These effects can be studied systematically by changing these species, so that metal zeolites are well suited to investigate the effect of the acidic and redox properties on the catalytic activities.

This article will not review in details the large body of litterature devoted to PtY and PdY zeolites since this work has been done recently (ref. 10). New contributions on the relations between the dispersion, the geometric and electronic structure of the metal on the one hand and the catalytic properties on the other hand will be discussed.

2 DISPERSION, METAL ACCESSIBILITY AND CATALYTIS

2.1 Platinum and palladium dispersion in Y zeolites

The different states of dispersion of platinum (refs. 2, 3) and of palladium (ref. 11) in Y zeolite are given in tables 1 and 2 respectively. In both cases the final dispersion and location of the metal depend upon the pretreatment under oxygen which governs the cation positions. Palladium ions or atoms enter the sodalite cage, are reduced and migrate at much lower temperature than platinum ions or atoms do, so that the conditions of formation of the isolated atoms in sodalite cages, of the 20 Å particles in zeolite bulk and of the 10 Å particles in supercages are somewhat different. These states of dispersion are very similar except that the 10 Å Pd particles are clustered in contiguous supercages whereas the 10 Å Pt particles are homogeneously scattered throughout the zeolite crystals. It is noteworthy that the encaged Pd particles can be reoxidized into cations by oxygen at comparatively low temperature whereas larger particles are converted into PdO oxide.

2.2   Metal-reagent accessibility and catalysis

The interpretation of kinetic data on metal zeolite catalyst must take into account the possible inaccessibility of metal particles to reagent molecules. The encaged metal might be unavailable because reagent molecules are too bulky to enter the main pore system or diffuse too slowly. Thus benzene has been selectively hydrogenated in presence of triethyl-benzene over Pt faujasites (ref. 12). However shape selectivity plays a minor role in faujasites because of the comparatively large aperture of the main pore system. Nevertheless pores can be blocked and reactions poisoned by strongly adsorbed molecules such as water or by carbonaceous residues. Even if the pores are not blocked poor activities may result from very low diffusion coefficients of incoming molecules, in that respect Turkewich (ref. 13) suggested that ethylene hydrogenation in liquid phase could be used to check whether the metal is on the external surface or in the zeolite crystals because of the poor accessibility in the latter case.

Reagent molecules entering and diffusing in the main pore system have to reach the metal particles trapped in the cages. There is not enough room between the 8-12 Å particle and the supercage walls to accomodate reagent molecules, therefore adsorption and reaction on metal surface have to take place through the four 7.5 Å apertures of the cages giving access to about half of the particle surface. It is more difficult to envision the accessibility of metal particles trapped in the sodalite cages since metal-reagent interaction through the 2.2 Å cage window is questionable. However this possibility cannot be ruled out at present, further work is needed to check the catalytic activity of Pt and Pd atoms isolated in the sodalite cages. Finally one may wonder if some catalytic reactions on metal zeolite system proceed necessarily via a direct metal-reagent bonding. Thus Bandiera (ref. 14) suggested that toluene hydrogenation on PtY zeolites is bifunctional in the sense that toluene could be bonded on the acidic sites of the zeolite and hydrogenated by dissociated hydrogen spilled over from the metal in neighbouring cages.

3   STRUCTURE OF METAL PARTICLES AND CHEMISORPTION

Except the obvious presence on the metal surface of a high fraction of edge and corner atoms little is known on the structure of particles smaller than 15 Å. Cluster calculations indicate that instead of the normal structure of bulk metal they may exhibit a non-crystal structure (polytetrahedral or icosahedral arrangement) (ref. 15), at any rate one should expect a contraction of the interatomic distances. These theoretical forecasts are based on calculations made on isolated clusters whereas metal particles in supported catalysts are in close contact with a support and are covered with chemisorbed reagents. Experimentally

TABLE 1

States of platinum dispersion in PtNaHY zeolite (refs. 2, 3).

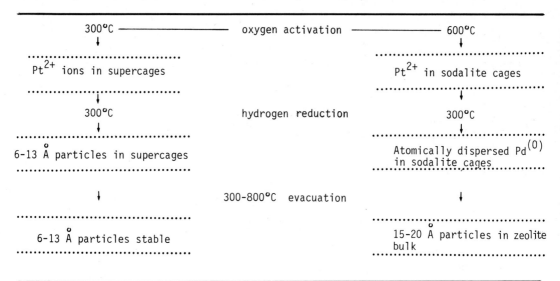

TABLE 2

States of palladium dispersion in PdNaHY zeolite (ref. 11).

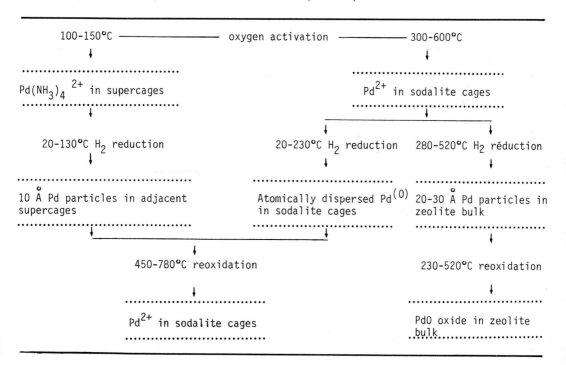

230

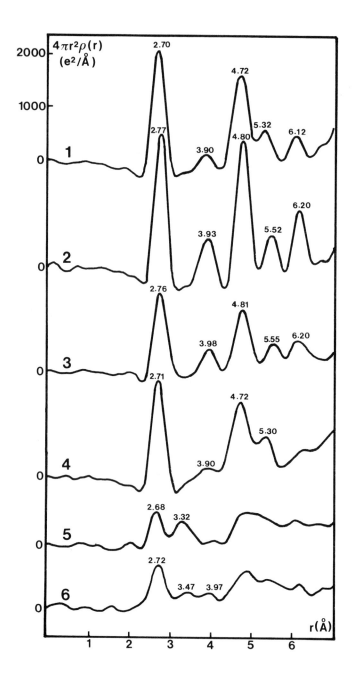

Fig. 1. Radial distribution function for PtY zeolites (r.e.d. from the support has been substracted)-1.1., outgassed zeolite kept under Ar-1.2., covered with H_2-1.3., covered with CO-1.4. covered with C_6H_6-1.5., covered with O_2-1.6. covered with S.

the structures of encaged 10 Å Pt particles covered with various adsorbates have been determined by the radial electron distribution (RED) method (refs. 5, 6). Some of these results and more recent ones are summarized on figure 1. The distribution function corresponding to the Pt particles covered with H_2 (Fig. 1.2) shows the normal f.c.c. structure and the interatomic distances of bulk metal. This is also the case for the particles covered with CO (Fig. 1.3) but unexpectedly the magnitudes of all the Pt-Pt peaks are reduced. This could be due to the fact that CO bonding to the metal surface atoms causes both elongations and contractions of bond lengths. Peak magnitudes are lower because of the electron density spreading but the mean value of the interatomic distances is almost normal. This is exactly the situation encountered in $(Pt_{19} CO_{22})^{4-}$ cluster where the mean nearest neighbours distances is 2.77 like in bulk Pt but individual Pt-Pt distances are within 2.52-2.95 Å (ref. 16). The particles outgassed and kept under Ar (Fig. 1.1) have a distorted structure with respect to the normal f.c.c. structure and the distances are contracted in agreement with the results obtained by EXAFS on similar catalysts (ref. 7). The structure appears intermediate between the f.c.c. and the icosahedral arrangement however it is still speculative to conclude that the distorsion is an intrinsic size effect or that it is due to a perturbation induced by the particle environment. The adsorption of H_2 (Fig. 1.2) produces a relaxation of the structure toward a regular f.c.c. configuration whereas benzene adsorption (Fig. 1.4) produces an additional perturbation. However the largest perturbations are caused by O_2 adsorption (Fig. 1.5) and H_2S adsorption (Fig. 1.6). In both cases, there is an important structural disorder since former Pt-Pt peaks are very weaks and new ordered structures appear characterized by the Pt-Pt distances at 3.32 Å (Pt-O-Pt) and at 3.47 Å (Pt-S-Pt) respectively.

The important conclusion of these studies is that the structure of metal particles are modified by the chemisorption of atoms and molecules, therefore metal surfaces cannot be regarded as inert substrates but they rearrange upon reagent chemisorption. Further studies should be carried out to check that they rearrange in the course of catalytic reactions. These conclusions obtained on Pt particles in zeolites can probably be extended to other types of metal dispersion and to other supports. At any rate, correlating the structure and the catalytic properties of less than 20 Å particles will not be an easy task especially if rearrangements of surface atoms in the course of reactions would complicate the problem of characterizing the structure of these particles. However methods like RED and EXAFS which can be applied under reaction condition will be useful.

4   RELATIONS BETWEEN ELECTRONIC AND CATALYTIC PROPERTIES

The electronic structure of small metal particles have received much attention in recent years both on theoretical and on experimental grounds. Studies of metal clusters by the SCF-Xα method (ref. 17) predict a high ionization potential and a net effective positive charge on surface atoms so that one can expect a high electron affinity for these clusters. The trend can be further increased by charge transfer from metal to support (ref. 18). These theoretical forecasts have been checked experimentally with 10 Å Pt particles in Y zeolite. Thus, three independant methods -I.R. spectroscopy of adsorbed probe molecules (ref. 19), XPS (ref. 8) and X-ray absorption edge spectroscopy (ref. 9)- have shown that the platinum is electrophilic. However the relative importance of the intrinsic size effect and of the support effect in inducing this electrophilic character is difficult to estimate since acidity is always present in zeolites but none of the support effect would appear without the Pt particles being small enough. However a recent investigation (ref. 20) led to the conclusion that the support effect is predominant as far as the 10 Å Pt particles are considered.

The effect of the electrophilic character (e.c.) of the metal on "structure insensitive" reactions such as the hydrogenation of aromatics hydrocarbons or olefins is generally weak. Thus the turnover frequency (N) for ethylene hydrogenation is the same on Pt/SiO_2 and on PtNaY (ref. 21) and for benzene hydrogenation N is comparable on Pt/Al_2O_3 and PtNaY (ref. 19). A possible interpretation is that the hydrocarbons with aromatic nucleus or double bond are strongly bonded to the platinum whatever its electronic state so that increased metal-reagent affinity has little effect on the activity (zero order reaction). However the rates of hydrogenation reactions are increased when multivalent cations are exchanged in the zeolite (refs. 21, 22). One can wonder if in this case, the rate enhancement is due to the increase of the metal e.c. or to a direct effect of the electrostatic fields (zeolite crystal field and cation field) which could polarize and destabilize the reagent molecules thereby increasing their reactivity. This last interpretation holds certainly in the case of the cyclopropane hydrogenation (ref. 23) which is 100 times faster on 10 Å or 20 Å particles in Y zeolite than on Pt/Al_2O_3.

On the other hand the rates of "structure sensitive" reactions such as the hydrogenolysis and isomerisation reactions are expected to depend much on the electronic structure of the metal. Indeed the hydrogenolysis rates of neopentane (ref. 21) and ethane (ref. 23) are at least one order of magnitude faster on electron deficient Pt in Y zeolite than on other supported catalysts. A recent study dealing with n-butane conversion on 10 Å Pt particles in PtY zeolite (ref. 20) has shown that the hydrogenolysis activities can be correlated with the electrophilic character of platinum. Non-electrophilic platinum (20 Å Pt/SiO_2, 10 Å PtNaHY + NaOH) exhibits a weak activity at low butane pressure and a positive reaction order whereas electrophilic platinum (10 Å PtNaHY, 10 Å PtCeY) exhibits an inverse behaviour. These results were interpreted in terms of the competitive adsorption equilibrium of H_2 and C_4H_{10} on the metal surface. Hydrocarbon intermediates are strongly bonded on the surface of electrophilic platinum even at low C_4H_{10} pressure and the activity is high. The increase of pressure results in a saturation of the surface by the adsorbed hydrocarbons resulting in a self-poisoning and a negative reaction order. On the other hand, the sticking coefficient of C_4H_{10} on non-electrophilic platinum is small and the hydrogen adsorption predominant so that slow rates at low C_4H_{10} pressure and positive reaction order are observed. The selectivity is also modified because the increased strength of the hydrocarbon bonding to the electrophilic platinum can change the various reaction pathways. Thus in the case of n-butane conversion the ratio isomerization/hydrogenolysis is 0.08 on PtCeY compared to 0.45 on Pt/SiO_2. Similar results have been obtained by Foger and Anderson (ref. 24) on PtLaY zeolite with the neopentane conversion, high hydrogenolysis activities were also interpreted in terms of the platinum electron deficiency.

The catalytic activities of metals depend to a large extent upon deactivation processes and therefore to the poison sensitivity of the metal. Early, Rabo et al (ref. 25) have shown that Pt particles encaged in Y zeolite are sulfur resistant. According to these authors the atomically dispersed platinum combine less strongly with sulfur than larger platinum particles do, it was later suggested (ref. 21) that weaker bonding was due to the electron deficient character of the platinum. The study of the H_2S poisoning of n-butane conversion reported in this meeting (ref. 26) shows that the higher the electrophilic character of platinum the higher is the resistance to sulfur poisoning because of the weaker the Pt to electronegative S bonding. The resistance to sulfur can therefore be increased by ion exchanging the PtY zeolite with multivalent cations which increases the electrophilic character. Consistently, it was shown that the electrophilic platinum is more readily poisoned by electron donor molecule, like ammonia (ref. 19).

5 CONCLUSION

When interpreting kinetic data on metal zeolite one must first take into account all the possible causes of metal inaccessibility as discussed in section 2. This requires a good knowledge of the metal dispersion and location not only before but after reaction. An other point in that respect is the possible limitation of reaction rates by diffusion in zeolite channels, the question whether or no the effectiveness factors are the same for a particle in the core of the zeolite crystal or near the outer surface is mostly unsettled. Therefore the real turnover frequencies on metal zeolite could be higher than those measured both because of a partial inaccessibility of the metal surface and because of possible diffusion limitations.

At any rate, it turns out that the catalytic behaviour of these catalysts are often different from that of other supported catalysts. It is still premature to establish correlations between catalysis and particle structure because the information on surface structure are still very scarce but structure determination methods like RED and EXAFS could prompt rapid progress. So far the catalytic activities of PtY and PdY zeolites have been interpreted in terms of the electron deficient or electrophilic character of the metal. Indeed, this property has been forecast by theory and checked by several methods, furthermore electrophilic character account for the enhanced hydrogenolysing activities and for the sulfur resistance which are the most salient feature reported. The electrophilic character is due both to the particle size and to the particle environment, the latter being more easily tuned to obtained a given result.

Finally the direct effect on reagent molecules of the zeolite crystal field and of the acidity of the zeolite have not been considered in this survey although it can play an important role in all the so-called bifunctional reactions carried out on zeolites.

REFERENCES

1 A.P. Bolton, Zeolite Chemistry and Catalysis, American Chemical Society, Washington, D.C., 1976, 714 pp.
2 P. Gallezot, A. Alarcon-Diaz, J.A. Dalmon, A.J. Renouprez and B. Imelik, J. Catal., 39 (1975) 334.
3 P. Gallezot, I. Mutin, G. Dalmai-Imelik and B. Imelik, J. Microsc. Spectrosc. Electron, 1. (1976) 1.
4 P. Gallezot and B. Imelik, Adv. Chem. Ser., 121 (1973) 66.
5 P. Gallezot, A. Bienenstock and M. Boudart, Nouv. J. Chim., 2 (1978) 263.
6 P. Gallezot, 5th Int. Conf. on Zeolites, Naples, 1980.
7 B. Moraweck, G. Clugnet and A.J. Renouprez, Surf. Sci., 81 (1979) L 631.
8 J. Vedrine, M. Dufaux, C. Naccache and B. Imelik, J. Chem. Soc., Far. Trans I, 74 (1978) 440.
9 P. Gallezot, R. Weber, R.A. Dalla Betta and M. Boudart, Z. Naturforsch., 34 A (1979) 40.
10 P. Gallezot, Catal. Rev. Sci. Eng., 20 (1979) 121.

234

11 G. Bergeret, P. Gallezot and B. Imelik, J. Catal. (submitted).
12 V.J. Frillette and P.B. Weisz, U.S. Patent 3,140,322 (1964).
13 R.S. Miner, K.G. Ione, S. Namba and J. Turkewich, J. Phys. Chem., 82 (1978) 214.
14 J. Bandiera, J. Chim. Phys. (in press).
15 J.J. Burton, Catal. Rev.-Sci. Eng., 9 (1974) 209.
16 D.M. Washecheck, E.J. Wucherer, L.F. Dahl, A. Ceriotti, G. Longoni, M. Manasserao, M. Sansoni, P. Chini, J. Am. Chem. Soc., 101 (1979) 6110.
17 R.P. Messmer, S.K. Knudson, K.H. Johnson, J.B. Diamond and C.Y. Yang, Phys. Rev. B, 13 (1977) 1396.
18 R.C. Baetzold, J. Chem. Phys., 80 (1976) 1504.
19 P. Gallezot, J. Datka, J. Massardier, M. Primet and B. Imelik, Proc. 6th Int. Cong. Catalysis, Chemical Society, London, 1976, 696 pp.
20 Tran Manh Tri, J. Massardier, P. Gallezot and B. Imelik, Proc. 7th. Int. Cong. Catalysis, Tokyo 1980, paper A 16.
21 R.A. Dalla Betta and M. Boudart, Proc. 5th Int. Cong. Catalysis, Vol. 2, North Holland, Amsterdam, 1973, 1329 pp.
22 F. Figueras, R. Gomez and M. Primet, Adv. Chem. Ser., 121 (1973) 480.
23 C. Naccache, N. Kaufherr, M. Dufaux, J. Bandiera and B. Imelik, Molecular Sieves II, American Chemical Society, Washington D.C., 1977, 538 pp.
24 K.F. Foger and J.R. Anderson, J. Catal., 54 (1978) 318.
25 J.A. Rabo, V. Schomaker and P.E. Pickert, Proc. 3rd Int. Cong. Catalysis, Vol. 2, North Holland, Amsterdam, 1965, 1264 pp.
26 Tran Manh Tri, J. Massardier, P. Gallezot and B. Imelik, This meeting.

B. Imelik *et al.* (Editors), *Catalysis by Zeolites*
© 1980 Elsevier Scientific Publishing Company, Amsterdam — Printed in The Netherlands

FORMATION AND CATALYTIC PROPERTIES OF NICKEL METAL PARTICLES SUPPORTED ON ZEOLITE.

D. DELAFOSSE

Laboratoire de Chimie des Solides - ER 133 CNRS, Université P. et M. Curie - T. 55-54,
4, place Jussieu 75230 - Paris Cedex 05

1 INTRODUCTION

Dispersed nickel supported on various oxides was used as catalysts specially in the hydrogenation, dehydrogenation, hydrogenolysis reactions (1-5). The importance of support is now
well known. It has multiple effects -on the formation of metal particles, it is available
to stabilize the nickel in a dispersed state and inhibit its sintering- on the catalytic
reaction, it is available to provide acidic sites and to play a role in dual functional
catalysis. (6,7)

The peculiar properties of zeolites are so interesting that they are used like support of
a lot of metal in reforming catalysis.

Indeed zeolites contain cavities or tubes with defined size within metal particles can be
entrapped and their strong acidic sites are available for dual functional catalysis (8,9).
Taking in account the interaction between metal and support, zeolites could be able to provide a synergetic effect for some reactions catalysed by metals. Few years ago, several studies concerning the formation (15) and catalytic properties of nickel metal in type A,X,Y
and Z zeolites were reported (10-14). In these works, reduced nickel has been found to sinter easily. As a result, large metal particles are formed at the external surface of zeolite crystals. (14). It is unlikely that for this dispersion the zeolite support may effect
drastically the catalytic properties of the metal.

Attempts have been reported to prevent sintering of nickel towards the external zeolite
surface and to precise (10, 16, 17, 18, 19) the various factors which interact on the formation and stability of nickel particles inside the framework.

Indeed two main problems influence the formation of small particles of supported nickel :
the first is the poor reducibility of nickel cation ; the second is the mobility of nickel
particles with heating and their ability to form large cristallites. In order to study the
catalytic action of metal catalysts, it is necessary to obtain a high dispersed state.(20,
21, 22).

The extensive body of literature of catalytic properties of metal zeolites has been reviewed in detail by Minachev and Isako (23) ; the mechanism of metallic cluster formation
in zeolites has been reviewed recently by J.B. Uytterhoeven (24). Our aim is to present now
a report of recent results on the formation, stability and catalytic properties of nickel
metal in zeolite framework, and to show how it is possible to obtain entrapped nickel clusters within zeolite cavities and obtain the way to a comparison (25) of properties of these
aggregates with the large and heterodispersed particles ones.

2   PREPARATION OF NICKEL ZEOLITES

2.1   Introduction of Nickel compounds

Impregnation, Ion exchange adsorption from the gaseous phase, adsorption of metal vapour can be used to introduce active compounds into zeolite cavities or on their crystal surface.

Derouane and al (26) have reported to obtain Mo, Re, Ru, Ni in Y zeolites from metal carbonyl adsorption and thermal decomposition. In the case of Mo the oxide was obtained.

A second method is the interaction of bis π allyl nickel with OH groups of the support (27). It was available to obtain small highly and homogeneous dispersed Ni° particles on silica after hydrogen reduction at 400°C. However this method has not been yet used on zeolite support.

It is also possible to obtain nickel particles from decomposition and reduction of nickel hexamine complex. This complex is introduced by exchange from SCN^- compound in liquid ammonia into NaX zeolite free of water (28).

In the general case, Ni^{2+} is introduced by ion exchange from nitrate or chloride (29). The size and dispersion of metal particles after reduction depends on the reducibility of Ni^{2+} ions, the reducing agent and the nature of cations other than Ni^{2+}.

2.2   Reducibility

Many factors are involved in the nickel cations reducibility. First of all is the initial location of Ni^{2+}. In faujasite zeolite Ni^{2+} ions are distributed in the SI, SII, SI', SII' sites. It is obvious that the more reducible are located in sites II. This location determined by Xray method depends on Si/Al ratio, exchange degree, pretreatment conditions, and nature of other cation present with Ni^{2+} (30) and Table 1.

TABLE 1

Initial location of cations (30, 31, 32) after pretreatment under vacuum at 773°K.

Samples	I	I'	II
$Ni_{13}Na_{30}Y$	11 Ni^{2+}	2 Ni^{2+}	25 Na^+
$Ni_{20}Na_{16}Y$	12 Ni^{2+}	4 Ni^{2+}	3 Ni^{2+} + 16 Na^+
$Ni_8Na_{64}X$	5 Ni^{2+}	20 Na^+	30 Na^+
$Ni_{31}Na_{24}X$	12 Ni^{2+}	10 Ni^{2+}	6 Ni^{2+} + 24 Na^+
$Ni_{10}Ca_{20}Na_{24}H_{12}X$	12 Ca^{2+}	8 Ni^{2+}	2 Ni^{2+} + 24 Na^+
$Ni_{14}La_{15}Na_X$	11 Ni^{2+}	9 La^{3+}	3 Ni^{2+} + 14 Na^+
$Ni_{24}Ce_6Na_X$	12 Ni^{2+}	3 Ce^{3+}	12 Ni^{2+} + 18 Na^+
$Ni_8Ce_5Na_{16}H_{40}X$	6 Ni^{2+} + 4 Ce^{3+}	1 Ce^{3+}	18 Na^+ + 2 Ni^{2+}
$Ni_{21}Ce_{6,5}Na_{20}H_{4,5}X$	4,5 Ce^{3+} + 6 Ni^{2+}	8 Ni^{2+} + 2 Ce^{3+}	18 Na^+ + 5 Ni^{2+}

Initial location of cations after pretreatment under vacuum at 623°K.

$Ni_8Ce_5Na_{16}H_{40}X$	3 Ce^{3+} + 5 Ni^{2+}	3 Ce^{2+} + 1 Ni^{2+}	2 Ni^{2+} + 18 Na^+
$Ni_{21}Ce_{6,5}Na_{20}H_{4,5}X$	1 Ce^{3+} + 10 Ni^{2+}	5,5 Ce^{3+}	18 Na^+ + 5 Ni^{2+}
			(II') 4 Ni^{2+} + 13 H_{20}

Initial location of cations after pretreatment with CO at 373°K.

$Ni_{10}Ca_{20}Na_{24}H_{12}X$	12 Ca^{2+}	Ca^{2+} + Na^+	10 Ni^{2+} + Na^+

The reducibility of Ni^{2+} ions is too depending on the composition and redox properties of zeolite framework. It is known now, the Ni^{2+} is more reducible in framework A than X and Y zeolite. (17). In NiNaX, the sites are attacked simultaneously by molecular hydrogen with an activation energy of 28 Kcal/mole. On the contrary, in Y zeolite, the Ni^{2+} in sites I are very difficult to reduce (31) with an activation energy of 40 Kcal/mole. As the concentration of proton of zeolite framework increases, the reducibility of Ni^{2+} decreases (30, 32). According to Rickert (10) in zeolite media, the redox equilibrium interacts always

$$Z - O \diagdown Ni^{2+} + H_2 \rightleftharpoons 2\ ZOH^+ + Ni° \quad (1)$$
$$Z - O \diagup$$

and is displaced towards the left by a high proton concentration.

- the presence of other cations modify drastically the behaviour of Ni^{2+} towards hydrogen. From a $Ca_{20}Ni_{14}H_{12}Na_{24}X$ sample reduced in mild conditions, it was obtained Ni^+ species according to

$$Ni°_{II} + Ni^{2+}_{I'} \xrightleftharpoons{H_2} 2\ Ni^+_{I'} \quad (2) \quad (33)$$

In presence of La^{3+} which hinders the dehydration, it is assumed the formation of $La^{3+} - O - Ni^{2+}$ entities into supercages with extraframework oxygen (34), the presence of Ce^{3+} modifies the redox properties of zeolite. By studying the adsorption of TCNE and Perylen on samples NiCeNaX, it was shown that the reducing properties are increased with the

Table II Ni^{2+} Reducibility

Samples	Dehydrated 773 K Reduced 573 K P=50 Torrs, time 25 h α*	Dehydrated 773 K Reduced 573 K P=50 Torrs, time 25 h D Å	Dehydrated 773 L Reduced 623 K, time 18 h H₂flow α	Dehydrated 773 L Reduced 623 K, time 18 h H₂flow D Å	Dehydrated 623 K Reduced 623 K, time 18 h H₂flow α	Dehydrated 623 K Reduced 623 K, time 18 h H₂flow D Å	Dehydrated 773 K O₂773 K.Reduced 623 K, time 18 h H₂ flow α	Dehydrated 773 K O₂773 K.Reduced 623 K, time 18 h H₂ flow D Å	Dehydrated 773 K O₂773 K + CO 373 K Reduced 273 K, time 6 h H° flow α	Dehydrated 773 K O₂773 K + CO 373 K Reduced 273 K, time 6 h D Å
$Ni_8Na_{64}X$	0,3	90 % 60 / 10 % 15							1	10
$Ni_{31}Na_{24}X$	0,19	80 % 150 / 20 % 15							0,6	10
$Ni_{13}Na_{27}Y$	0,06	idem								
$Ni_{20}Na_{16}Y$	0,1	-								
$Ni_{10}Ca_{20}Na_{14}H_{12}X$	0,07	Ni⁺							1	10
$Ni_{14}La_{15}Na_{11}X$	0,05	bidispersed								
$Ni_{10}H_{62}Na_4X$	0,036	-								
$Ni_{17}Pt_{0,4}Na_{10}H_{11}X$	0,2	25	0,95	25						
$Ni_{14}Pd_6Na_{12}H_{34}X$	0,26	30								
$Ni_{24}Ce_6Na_{18}X$	0,26	85 % 7 / 15 % 30	0,9	85 % 7 / 15 % 30	0,95	81 % 7 / 19 % 30	0,4	45 % 25 / 55 % 7		
$Ni_{28}Ce_{0,5}Na_{15}H_{13}X$			0,4 **	92 % 30 / 8 % 8	0,5	80 % 20 / 20 % 15	0,6	40 % 150 / 7		
$Ni_8Ce_5H_{40}Na_{16}X$			0,93	24	0,2	7	0,9	7		
$Ni_{21}Ce_{6,5}Na_{20}H_{4,5}X$			0,45	53 % 25 / 40 % 7	0,95	13 % 30 / 87 % 7	0,36	38 % 25 / 55 % 7		

* $\alpha = \dfrac{Ni°}{Ni_{tot}}$

** Reduced during five days

ratio Ce^{3+}/Ni^{2+} and the oxidating ones are increased with the Lewis sites number (35). Thus with a ratio $Ce^{3+}/Ni^{2+} \sim 1$ and a peculiar location of Ce^{3+} (in sites I'), it was possible to obtain a complete reduction of Ni^{2+} ions into $Ni°$.

Small quantities of $Pt°$ or $Pd°$ in the lattice increases the Ni^{2+} reducibility ; the higher is the $Pd°$, $Pt°$ concentration, the greater will be the reducibility (Table II).

Ni^{2+} reducibility depends on the nature of reducing agent. The influence of K, H_2, NH_3 on the degree of reduction and on the dispersion of $Ni°$ has been studied in NiY zeolite (36) The results are summarized in Table III.

TABLE III

| Sample | Condition of Dehydration | Reduction | | | | Size |
		Agent	Temp. °K	Time /Hours	degree of red.	DÅ
$Ni_{15,4}Na_{15,2}Y$	673 under vacuum	K	403	24	0,1	99% < 20
	flow He 1h 823K	H_2	723	-	0,3	20% $\sim$ 60
	flow He 1hh823K	NH_3	823	-	0,8	62% $\sim$ 42
						20% $\sim$ 10

A new and very interesting method to increase the reducibility of Ni^{2+} encaged into zeolite is to use atomic hydrogen (38, 39) at very low temperature (273°K).

This method unables a better accessibility of cation towards hydrogen, as atomic hydrogen is able to enter into sodalite cavities. Besides, since the reduction temperature is low, the mobility of cations and metallic atoms is inhibited. Thus it was obtained from $Ni_{10}Ca_{20}Na_{24}H_{12}X$ sample a complete reduction of Ni^{2+} into $Ni°$ (Table II) (38).

3 NICKEL METAL DISPERSION

In general case nickel particles obtained by reduction with molecular hydrogen of Ni^{2+} ions located in faujasite or mordenite zeolite are inhomogeneous (11, 13, 14, 15). Romanovski (16) was the first to provide evidence for the existence of a bidispersed metal size particles distribution in Y zeolite. Quantitative method has been developed to determine the amount of $Ni°$ outside and inside the Y zeolite (39). Recently Jacobs and al (40) have evidenced using FMR, TPR and TPO, a bidispersion.

For NiNaX samples, it was found also an heterodispersion with small particles into the framework and the larger ones in the outside (Table II).
However Guilleux and al (19) had reported that it is possible to obtain homogeneous and small nickel particles inside the framework in two cases. In the presence of very small amounts of $Pt°$ or $Pd°$, it was obtained cristallites of 25 Å inside the framework by breakdown of a window between two supercages (41). In the presence of Ce^{3+}, it was obtained homogeneous particles size the value of which depends on the Ce^{3+}/Ni^{2+} ratio and on the acidic sites concentration (Table II) (35). In all cases, the best results are obtained when reduction with molecular hydrogen was performed in dynamical conditions. Che and al (38) have obtained an homodispersion of $Ni°$ with a size of 10 Å by reducing at 273K $Ni_{10}Ca_{20}H_{12}Na_{24}X$ with atomic hydrogen. This sample was pretreated by CO at 373 K in order to obtain all the Ni^{2+} in supercages.

Thus it was possible under severeand peculiar conditions to obtain homogeneous and small nickel particles.

Their size depends also on the nature of other elements introduced with nickel : Ce^{3+} located in site I' is available to stabilize very small aggregates (7 Å) and inhibits their migration with atomic hydrogen reduction, the size was restricted to the dimension of one supercage (10 Å) and in the presence of small amount of Platinum to that of two super-cages (25 Å).

4   STABILITY OF NICKEL PARTICLES

FMR study (40) shows that Ni° formed outside the framework corresponds to large particles They do not strongly interact with the support and exhibit some shape anisotropy. These particles are stable under vacuum and they are reoxidized by molecular oxygen above 573°K into Nickel oxide.

Ni° encaged in the supercages are assumed to interact strongly with the support and a charge transfer between the Nickel particles and the zeolite framework may occur. In agree-ment with other searchers (10, 39, 40) we found these particles are reoxidized by molecular or activated zxygen or NO to Ni^{2+} cations which go back into their cristallographic sites.
$$2 \; Z\text{-}OH + Ni° \xrightarrow{1/2 \; O_2} Ni^{2+} + H_2O \quad (3)$$

The stability of these small particles under vacuum depends on the reduction conditions. So for NiCeX reduced by molecular hydrogen, the nickel particles of 7 Å (Table II) are very sensitive to the presence of protons in the lattice and they are reoxidized at low tempera-ture ; the equilibrium (1) is displaced towards the left. Some various experiments were performed to determine the state of aggregates < 10 Å. Thus from the L_{III} emission spectrum both for small nickel particles (10 Å) in zeolite prepared by atomic hydrogen reduction and for bulk nickel, the 3d filled distribution has been obtained. But while for the bulk nickel a structureless band is observed, structures are clearly apparent for small particles. The comparison between the L_{III} spectra of NiO and bulk Ni evidences that the aggregates are pure Ni°. These results can be explained in terms of a decrease of interatomic interactions between 3d electrons and suggests an infinite to finite system behaviour transition (42).

5   CATALYTIC PROPERTIES

Most of the studies on the nickel loaded zeolite were relative to very large metallic particles outside the framework of faujasite or mordenite zeolite in the presence of unredu-ced Ni^{2+} cations (11, 13, 14, 43, 44). Indeed the data reported are very dispersed and con-troversed. But in these conditions metal containing zeolites can act as monofunctional cata-lysts and the zeolite can be considered as inert support for the active metals (23).

The reactions catalysed by nickel zeolite should be clarified in two groups : according to Boudart (45) there are structure insensitive reactions as olefin hydrogenation or paraf-fin dehydrogenation and sensitive ones as hydrogenolysis, methanation...

5.1   Hydrogenation and dehydrogenation reactions

On NiNaY obtained after reduction (12, 46) the degree of conversion in n octane hydroge-nation attained 81 % whereas isolefin and benzene were practically unreactive because the latter cannot penetrate the zeolite cavity where most of the nickel was located. On the contrary, NiNaA reduced at high temperature (large metallic crystallites outside) hydrogenated benzene with 66 % at 180°C.

There is no evidence indicating that the behaviour of metallic species outside zeolite crystals differs from that of the metal supported on other carriers.

Nevertheless in a recent study on the benzene hydrogenation on various NiX (47), it was observed that the Ni°X activity towards benzene hydrogenation is lower than that observed on $Ni°/SiO_2$ in the same conditions. It was available to assume benzene hydrogenation is a metallic structure insensitive reaction but these results lead us to consider that a portion of Ni° was not accessible to benzene molecules by steric hindrance or metal carrier interaction.

Ce^{3+} increases the hydrogenation activity (25). This result can be compared with the data reported for PtCeY (21, 48) and explained by $Ce^{3+}/Ni°$ interaction, or by the additive effect of the electrostatic fiels of the Ce^{3+} cations (21). Ione and al (44) assume the pecularities of the behaviour of the nickel portion fixed in the zeolite cavities are due to the unstable valence state of metal atoms in small clusters. This unstability depends on the degree of interaction between the atoms of metallic clusters and electron acceptor centers of the zeolite framework. These clusters are active for n hexene hydrogenation and inactive for benzene hydrogenation.

In the first case, an electron acceptor center is blocked by a hexene molecule according to

$$Ni^{2+}Z + H_2 \xrightarrow{C_6H_{12}} Ni° + H_2 \quad \overset{\displaystyle -\,C\,-\,C\,-}{\underset{Z}{\diagdown\diagup}}$$

The authors assume a weaker benzene ability to electron-donor.

5.2   Hydrogenolysis and other reactions

Nickel loaded zeolites were active in the reaction of hydrogenolysis of ethane and n hexane (15). The specific activity of reduced nickel decreases with the increasing acidity of the support in order $SiO_2 > Al_2O_3 >$ Silica-Alumina $>$ Y faujasites but the activities remain constant in the faujasite acidity range from NaY to MgY. There are two types of sites of nickel supported by faujasite : type I the most active accounts for 35 % of total surface and is responsible for n hexane hydrogenolysis, type II is responsible for n hexane isomerisation hydrocracking and benzene hydrogenation. Hydrogen sulfide poisons type II sites with one sulfur for every two nickel atoms.

5.3   The bifunctional action of nickel zeolite catalysts in the conversion of toluene were studied (43). The high activity towards the disproportionation of toluene of the reduced nickel zeolite catalysts is attributed to the fact more than 30 % of the nickel ions remain unreduced thereby being the main source of proton acidity. In the presence of hydrogen, the metallic nickel assists the elimination of some unsaturated hydrocarbons which offensive would polymerize and block the active sites.

Nickel and cobalt Y zeolite and mordenite proved to be efficient in the conversion of n hexane with water vapor (49). The 3,24 % Ni Zeolon and 0,76 % NiNaY catalysts were more active than the commercial Ni/Al_2O_3 which contains more nickel (15 %).

Nickel cristallites located on the external surface of the Y type zeolite were investigated for the methanation reaction (50), their turnover number for this reaction was found more than an order of magnitude less than that observed for Ni/Al_2O_3. Ruthenium clusters

inside the supercages were found very active for methanation but exhibited a marked decline
in activity during the reaction, attributed to the build-up of excess carbon which results
from the dissociation of CO. The addition of excess nickel to Ruthenium (1:7 atomic ratio)
brought about a stabilization in activity. NiRu bimetallic clusters were not very dispersed
(2.21 %, but the presence of Ni is believed to moderate the activity of ruthenium providing
more sites for dissociative chemisorption of H_2 (51). There are no systematical studies on
the behaviour of nickel zeolites towards poisons like sulfur, ammonia or coke. No relation
was yet established between the size of nickel particles and the poisoning resistance of
catalysts.

6   CONCLUSION

The conclusions drawn from the works reported up to now do not take into account the fact
that the studied samples are heterodispersed and incompletely reduced. Nevertheless it is
obvious that if the large Ni° particles formed outside the framework do not interact with
the support, the smallest ones inside the zeolite cavities interact strongly with it and are
able to modify the catalytic behaviour of the whole system.

For studying the role played by Nickel aggregates entrapped within the supercages towards
the hydrogenation or hydrogenolysis reactions and their poisoning resistance, it was necessary to start from completely reduced and homodispersed samples. In the first part of this
report, we have seen it was now possible to obtain small Ni° particles entrapped in supercages of X zeolites (Table II). So it was interesting to study the catalytic activity of
these well defined samples towards the two types of reactions -insensitive or sensitive-
and to evidence the interactions of Ni° clusters with another cation like Ce^{3+} or with Lewis
sites. J.F. Tempère reports in the proceedings of this congress (25) the data we have obtained for butane hydrogenolysis. I would only say that the Ni° aggregates with a size lower
than 10 Å are quasi inactive towards these two reactions. Assumptions are made to explain
these results. Probably the ability to obtain small nickel particles into zeolite, involves
a high interaction between the electron acceptor sites and the metal. Conversely to the case
of Platinum in Pt Y, the unstability of small Nickel particles towards the ambient atmosphere makes difficult to evidence these interactions by physical methods like ESCA or EXAFS.

It seems difficult now to use for catalytic application small nickel particles but they
are available to form alloys and bimetallic aggregates with peculiar properties. Besides,
it would be very important to study their selectivity and thioresistance for some catalytic
reactions.

REFERENCES

1 R.Z.C. Van Meerten and J.W.E. Coenen, J. Catal., 37 (1975) 37-43.
2 R.Z.C. Van Meerten and J.W.E. Coenen, J. Catal., 46 (1977) 13.
3 G. Dalmai-Imelik and J. Massardier, Proc. 6th Int. Congress on Catalysis (1976) Chem. Soc
   London Vol 1 pp. 90.
4 G.A. Martin, J. Catal., 60 (1979) 345-355.
5 A. Frennet, L. Degols, G. Lienard and F. Crucq, J. Catal., 55 (1978) 150.
6 M.A. Vannice, R.L. Garten, J. Catal., 56 (1979) 236-248.
7 T.E. Whyte, Catal. Rev., 8 (1973) p. 117.
8 J.A. Rabo, P.E. Pickert, D.N. Stamires and J.E. Boyle, Actes 2eme Cong. Int. Catalyse
   Paris 1960, Vol. 2 (1961) p. 2055.
9 P.B. Weisz, V.J. Frilette, R.M. Maatman and E.B. Hower, J. Catal., 1 (1962) 307.
10 L. Rieckert, Ber. Bunsenges Phys. Chem., 73 (1969) 331.

11 V. Penchev, N. Davidova, V. Kanazirev, H. Minchev and Y. Neinska, Adv. Chem. Ser., 121 (1973) 461-468.
12 N.V. Borunova, L. Kh. Freidlin et al., Zeolity ikh. Sintez, Svoistva i primenenie Nauka M-L (1965) p. 380.
13 Kh. M. Minachev, V.I. Garanin and T.A. Novrusov, Izv. Akad. Nauk. SSSR, Ser Khim, (1973) 330.
14 P. J.R. Chutoransky and W.L. Kranich, J. Catal., 21 (1971) 1-11.
15 J.T. Richardson, J. Catal., 21 (1971) 122-129.
16 W. Romanwski, Roczniki Chemie Ann. Soc. Chim. Polonium, 45 (1971) 427.
17 T.A. Egerton and J.C. Vickerman, J. Chem. Soc. Faraday Trans. I, 69 (1973) 39-49.
18 A.C. Herd and C.G. Pope, J. Chem. Soc. Faraday Trans. I, 69 (1973) 833-839.
19 M.F. Guilleux, D. Delafosse and G.A. Martin, J.A. Dalmon, J. Chem. Soc. Faraday Trans. I, 175 (1979) 165-171.
20 R.A. Dalla Betta, M. Boudart, Proc. 5th Int. Congress on Catalysis 1972, Miami Beach 1 (1973) pp. 329.
21 P. Gallezot, Catal. Rev. Ser. Eng., 201 (1979) 121-154.
22 P. Gelin, Y. Ben Taarit and C. Naccache, J. Catal., 59 (1979) 357-364.
23 Kh. M. Minachev and Ya. I. Isakov, In Zeolite Chemistry and Catalysis (J.A. Rabo ed) ACS Monograph 171 (1976) p. 552.
24 J.B. Uytterhoeven, Acta Phys. et Chemica Szeged Hungary, XXIV 1-2, pp 53-69.
25 J.F. Tempère, this review.
26 E.G. Derouane, J.B. Nagy and J.C. Vedrine, J. Catal., 46 (1977) 434-437.
27 U.N. Ermakov, B.N. Kuznetsov, Kinet. Catal., 13 (1972) 1355 ; Dokl. Akad. Nauk. SSSR, 207 (1972) 644.
28 J. Fournier, Ch. De la Calle, M. Briend and M.F. Guilleux (no published).
29 R.A. Schoonheijdt, L.J. Vandamme, P.A. Jacobs and J.B. Uytterhoeven, J. Catal., 43 (1976) 292.
30 M. Briend-Faure, M.F. Guilleux, J. Jeanjean, D. Delafosse and G. Kjega-Marriadassou, M. Bureau-Tardy, Acta Phys. et Chem. Szeged Hungaria XXIV 1-2 1978 pp. 99-106.
31 M. Briend-Faure, J. Jeanjean, M. Kermarec and D. Delafosse, J.C.S. Faraday I, 74 (1978) 1598.
32 S. Bhatia, J.F. Mathews and N.N. Bakhshu, Acta Phys. et Chem. Szeged Hungaria XXIX 1-2, 1978, pp. 83-88.
33 D. Olivier, L. Bonneviot, M. Richard and M. Che, Proc. (NATO) 2nd Int. Symp. on magnetic Resonance in Colloide and Interface Science, Menton, 1979, in press.
34 M. Briend-Faure, J. Jeanjean, D. Delafosse and P. Gallezot, J. Phys. Chem. (1980)(in press).
35 S. Kjemel, M.F. Guilleux, J.F. Tempère and D. Delafosse (to be published).
36 F. Schmidt, H. Kacirek and W. Gunsser (to be published).
37 M. Che, M. Richard and D. Olivier, J.C.S. Faraday I, 1980 (in press).
38 D. Olivier, M. Richard, L. Bonneviot and M. Che, Actes 32e Réunion Internationale de Chimie Physique, Lyon Septembre 1979, in press.
39 P.A. Jacobs, J.P. Linart, H. Nijs, J.B. Uytterhoeven and H.K. Beyer, J.C.S. Faraday I 73 (1977) 1745-1754.
40 P.A. Jacobs, H. Nijs, J. Verdonck and E.G. Derouane, J.P. Gibson, A.J. Simoens, J.C.S. Faraday I, 75 (1979) 1196-1206.
41 J. Verdonck, P.A.Jacobs and J.V. Uytterhoeven (unpublished results).
42 D. Fargues, F. Vergand, E. Belin, Ch. Bonnelle and D. Olivier, L. Bonneviot, M. Che, 2nd Int. Meeting on the small Particles and Inorganic Clusters, Lausanne, Sept. 1980
43 N. Davidova, N. Peshev and D. Shopov, J. Catal., 58 (1979) 198-205.
44 K.G. Ione, V.N. Rommanikov, A.A. Davidov and L.B. Orlova, J. Catal., 57 (1979) 126-135.
45 M. Boudart, A.W. Adlag, J.E. Benson, N.A. Daugharty, C. Girvin Harkins, J. Catal., 6 (1966) 92-99.
46 P.N. Galich et al., Sb "Neftekhimiya" izd A.N. Turk SSR Ashchabad, 1963, pp. 63.
47 M.F Guilleux, J. Jeanjean, M. Bureau-Tardy and G. Djega-Mariadassou, VIe Symp. Ibero Americano de Catalise, Rio de Janeiro, 1978.
48 F. Figueras, R. Gomez, M. Primet, Adv. Chem. Ser., 121 (1973) 480-489.
49 C.J. Brooks, Adv. Chem. Ser. 102 (1971) 426-433.
50 D.J. Elliott and J.H. Lundsford, J. of Catal., 57 (1979) 11-26.

B. Imelik *et al.* (Editors), *Catalysis by Zeolites*
© 1980 Elsevier Scientific Publishing Company, Amsterdam — Printed in The Netherlands

CHARACTERIZATION AND REACTIVITY OF SMALL METALLIC NICKEL PARTICLES DISPERSED ON X TYPE ZEOLITES.

by G.N. SAUVION, S. DJEMEL, J.F. TEMPERE, M.F. GUILLEUX, D. DELAFOSSE

There have been few studies of the catalytic activity of small metallic nickel particles formed in zeolites, this being due to the difficulty of obtaining a homogeneous, well-dispersed phase for this element.

Most of the work on the catalytic activity of Ni^o zeolite systems has been about particles of heterogeneous, quite often large sizes (> 150 Å). It is understandable that, under these conditions, it is difficult to study the effect of particle size on the catalytic activity.

Recently progress has been made in the preparation of highly dispersed Ni^o on X zeolite (1). It was shown that, by introducing Ce^{3+} ion into the framework (2) or by reduction with atomic hydrogen at low temperatures (3), homogeneous distributions of 7 or 10 Å could be obtained.

In this work we have studied the catalytic activity of metallic nickel particles of sizes between 7 and 30 Å on X zeolite in the butane hydrogenolysis reaction.

EXPERIMENTAL

Our zeolites were prepared by methods described elsewhere (4) (Table 1).

We studied in turn :

a series of $Ni_xNa_yCe_zH_tX$ (samples I, II, III) zeolites differing in the (Ni/Ce) ratio and the acidity,

a sample of $Ni_{10}Ca_{20}H_{12}X$ reduced by atomic hydrogen (sample IV),

a series of Ni/SiO_2 studied for the purpose of comparison.

In most cases, the reductions were carried out in a current of molecular hydrogen (5l/h) at moderate temperatures and in one case (sample IV) in atomic hydrogen at 25°C. The Ni/SiO_2 catalysts were reduced under the conditions recommended by G.A. Martin (5).

The extent of reduction and the metal particle size were determined by magnetic measurements (axial extraction method) at the Institut de Recherches sur la Catalyse at Villeurbanne.

The measurements of the catalytic activity were performed on a dynamic differential microreactor. The reagent gas consisting of a ternary mixture (butane : p = 30 Torr, hydrogen : p = 350 Torr, helium : p = 380 Torr) carefully purified and dried, is sent onto 40 mg of catalyst at temperatures between 200 and 300°C. The gaseous reaction mixture was analysed by gas chromatography. Catalytic activities have been referred to 250°C, the rates being expressed in moles of butane decomposed per second per m^2 of Ni^o. The degree of conversion was kept below 5%. The metal surface areas are expressed by the equation :

$$S = \frac{6 \ 10^4}{8.9 \ d}$$

where d is the metal particle diameter expressed in $\overset{o}{A}$ (10).

RESULTS

On all our catalysts, butane hydrogenolysis involves multiple C-C bond rupture.

Hydrogenolysis is deep and the considerable yields of methane arise both from the hydrogenolysis of propane and ethane. Only traces of isobutane were obtained.

The partial orders with respect to the hydrocarbon and hydrogen were found to be equal to + 0.7 and - 0.7 respectively.

The results concerning the catalytic activity of all the samples are given in Table 1 which suggests :

1) that particles of diameter equal to or less than 10 $\overset{o}{A}$ are inactive,

2) that metallic particles of 30 $\overset{o}{A}$ in diameter are active but less than those supported on silica.

DISCUSSION

Particles of diameter equal to or less than 10 $\overset{o}{A}$.

Firstly it appears evident that the inactivity of 7 $\overset{o}{A}$ particles cannot be attributed to their inaccessibility to the reagent. It is difficult to see how 16 nickel atoms coming from 16 neighbouring supercages could be located inside a sodalite cavity to constitute a 7 $\overset{o}{A}$ aggregate (d = 2.76 $\sqrt[3]{N}$).

These particles (d = 7-10 $\overset{o}{A}$) form on the walls of the supercages. It is possible that for the bigger ones the accessible metallic surface reduces to the four large windows of the supercages (7.4 $\overset{o}{A}$ in diameter) (8). The observed inactivity could be related to an insufficient number of metallic atoms at the accessible surface of these small aggregates. Martin (5) has shown that hydrogenolysis of a saturated hydrocarbon molecules requires a larger number of adjacent nickel atoms : 12 for ethane, 18-22 for butane (6). Now, a simple calculation indicates that 7 $\overset{o}{A}$ nickel particles show less than 16 atoms on the surface and that the window surface can only see no more than 8 nickel atoms. This could explain the inactivity observed for sample I and IV.

Moreover a recent study (7) seems to reveal that nickel particles of small size, like platinum aggregates (8) acquire electron deficient character below a certain size. This seems to be the case of sample IV. In the case of cerium exchanged zeolite (sample I), there could be strong interactions between the Ce^{3+} cation and the electron deficient metal particles, so that the butane molecule can no longer react. Finally, it is not possible to exclude the eventuality of contamination or passivation facilitated by the small particles sizes or even partial reoxidation of the metal by the oxidizing centres (OH, H_2O, Lewis acids) formed during reduction at the zeolite surface.

TABLE 1

Samples	Pre-treatment	α %	dp	S_1	S_2	$r_1\,10^8$	$r_2\,10^8$	E_a	p
I $\quad Ni_8Ce_5H_{40}Na_{15}X$	R 350°C	28	100 % 7 Å	963.1		0.04		39	0.47
IIRD $Ni_9Ce_{5.5}H_2Na_{49.5}X$	RD 350°C	46	65 % 30-25 Å 12 % 15-25 Å 23% <15 Å	421.3	199.7	11.9	25	42	0.51
IIR $Ni_9Ce_{5.5}H_2Na_{49.5}$	R 350°C	20	22 % 30 Å 36 % 15-25 Å 42 % <15 Å	575.3	170.7	12.7	42.9	49	0.38
III $Ni_{22}Ce_6H_1Na_{23}X$	RD 350°C	95	13 % 30 Å 87 % 7 Å	867.1	29.2	0.58	17.3	47	0.46
IV $Ni_{10}Ce_{20}H_{12}Na_{14}X$	H° 25°C	100	100 % 10 Å	674.2		0.19		23	0.37
V $\quad Ni/SiO_2$	R 650°C	100	100 % 25 Å	269.7		38.2		47	0.58
VI $\quad Ni/SiO_2$	R 650°C	100	100 % 60 Å	112.3		200.6		47	0.45
VII $\quad Ni/SiO_2$	R 925°C	100	100 % 150 Å	44.9		16.9		34	0.51

R : dynamic reduction ; RD : dynamic reduction after desorption in vacuum ; α : extent of reduction

dp : particle diameter ; S_1 : total metallic surface area in $m^2 g^{-1}$; S_2 : accessible metallic surface area in $m^2 g^{-1}$

r_1 : rate referred to S_1 in $mol.s^{-1}.m^{-2}$; r_2 : rate referred to S_2 in $mol.s^{-1}.m^{-2}$

E : apparent activation energy in $kcal.mol^{-1}$; p : depth of hydrogenolysis $p = C_1/C_1 + 2C_2 + 3C_3$

Metallic reduction at the zeolite surface.

Sample III ($Ni_{22}Ce_6H_1X$).

Sample III with higher (Ni/Ce) ratio is three times more reduced than sample I. The metallic nickel atoms will have in this case more chance to agglomerate with each other, before interaction with Ce^{3+} cations located in S_I or $S_{I'}$ sites. A bidispersed state is observed with homogeneous particles sizes of 7 and 30 Å. The catalytic activity of sample III is only due to the contribution of 30 Å particles since 7 Å particles are virtually inactive. Under these conditions it appears that sample III is 90 times more active than sample IV with 10 Å diameter particles.

The increase of catalytic activity with particle size can be explained by :

- the better accessibility of the reagent. Most authors think that growth of particles with diameter greater than the supercage diameter is associated with a local loss of cristallinity (8),

- a greater number of adjacent metallic nickel atoms at the surface,

- a lesser extent of metal substrate interactions.

Sample II.

We have also studied a $Ni_9Ce_{5.5}Na_{49.5}H_2X$ zeolite, giving after reduction larger quantities of 30 Å particles, than sample III does. Depending on the pretreatment, smaller particles (with size $<$ 15 Å) will also be obtained and we believe that these particles are inactive. Therefore, the particle distribution appears to be resultant of two families of homogeneous size 27.5 Å and 20 Å for sample IIDR, 30 Å and 20 Å for sample IIR.

These two catalysts are more active than sample III and sample IIR twice as active as sample IIDR. It is tempting to think that 20 Å diameter particles are more active than 30 Å diameter particles. This could explain the differences in activity. However generally speaking, one can observe that for samples I, IIR, IIRD, III and IV, the higher the catalytic activities, the smaller the number of oxidizing centres generated during exchange, desorption or reduction. These results agree with the observations made by Ione (9) where the catalytic activities increase by blockage of the Lewis acids of the zeolite.

CONCLUSION

To summarize we have shown that metallic nickel particles with diameter size $\leqslant 10$ Å located inside the supercages of X type zeolites are inactive or scarcely active. The cause of these low activities is still subject to discussion, but is probably associated with strong interactions between nickel particles and substrate and to the smaller number of adjacent metallic nickel atoms, available at the surface, accessible to the hydrocarbon molecules.

We have also shown that 20 to 30 Å diameter particles dispersed inside the lattice of the zeolite are active. The activities are of the same order of magnitude as those observed with the 25 Å diameter particles supported on silica.

249

REFERENCES

1   D. Delafosse, this review.
2   M. Briend-Faure, M.F. Guilleux, J. Jeanjean, D. Delafosse, G. Djega-Mariadassou and M. Bureau-Tardy, Acta Physica et Chemica, XXIV, 1-2 (1978) pp. 99.
3   D. Olivier, M. Richard, L. Bonneviot and M. Che, to be published.
4   S. Djemel, Thesis, Paris 1980.
5   G.A. Martin, J. Catal., 60 (1979) 345.
6   G.A. Martin and B. Imelik, Surf. Sci., 42 (1974) 157.
7   D. Fargues, F. Vergand, E. Belin, C. Bonnelle, D. Olivier, L. Bonneviot and M. Che, to be published.
8   P. Gallezot, Catal. Rev-Sci Eng., 20, 1 (1979) 121.
9   K.G. Ione, V.N. Romannikov, A.A. Dauydov and L.B. Orlova, J. Catal., 57 (1979) 126.
10 G.A. Martin, Summer School Theorie des Métaux et Catalyse, I, Lyon Villeurbanne, 22 Sept. 1975, C.N.R.S.

Acknowledgements.
   We thank Dr. G.A. Martin for the gift of Ni/SiO_2 samples and his help in carrying out the magnetic measurements.

B. Imelik *et al.* (Editors), *Catalysis by Zeolites*
© 1980 Elsevier Scientific Publishing Company, Amsterdam — Printed in The Netherlands

ETHYLBENZENE HYDROISOMERIZATION ON PtHNaY ZEOLITES

MASAHIRO NITTA[*] and PETER A. JACOBS
Centrum voor Oppervlaktescheikunde en Colloïdale Scheikunde, Katholieke Universiteit
Leuven, De Croylaan 42, B-3030 Leuven (Heverlee), Belgium
([*] on leave from Faculty of Engineering, Hokkaido University, Sapporo 060, Japan)

ABSTRACT

Hydroisomerization of ethylbenzene was carried out in a continuous flow reactor using
PtHNaY zeolites with varying proton and platinum content as catalysts. Detailed product
distributions were determined.

The data obtained qualitatively can be explained using a classic bifunctional
mechanism, the rearrangement of ethylcyclohexyl, methylethylcyclopentyl and dimethylcyclohexyl carbenium ions being rate determining.

Maximum xylene selectivity requires a high number of hydroxyl groups of intermediate
acid strength. Selective o-xylene formation at short contact times is possible via an
energetic favourable pathway if protonated cyclopropane intermediates exist.

1.   INTRODUCTION

A major source of industrial xylene production is the C_8-aromatics cut from reforming
plants. This product stream beside the xylenes contains about 17 % by weight of ethylbenzene. In the so-called "octafining" process, ethylbenzene in a mixed xylene feed is
isomerized towards xylenes. The reactions of major importance in this process are :
hydrogenation of aromatics, isomerization of naphthenes and dehydrogenation of naphthenes.
One of the characteristics of the octafining catalyst is its ability to convert ethylbenzene to xylenes. It seems to be industrial practice to use in the process a platinum
on silica-alumina catalyst with controlled acidity (ref. 1).

Gas-phase xylene isomerization occurs over acidic catalysts. However, ethylbenzene
does not isomerize over acid catalysts but its isomerization proceeds via hydrogenated
intermediates over bifunctional catalysts (ref. 2-4). On Pt-alumina catalysts with
weak acidity, the following reaction sequence seems to exist (ref. 4) :

ethylbenzene $\rightleftharpoons$ ethylcyclohexene $\rightleftharpoons$ 1,2-methylethylcyclopentene
$\rightleftharpoons$ 1,2-dimethylcyclohexene $\rightleftharpoons$ o-xylene

The skeletal rearrangements of tertiary carbocations are rate determining. On a un-
specified zeolite with higher acidity the isomerization of secondary carbocations also
contributes.

In the frame of a better understanding of the action of bifunctional catalysts, the
present work on ethylbenzene isomerization was undertaken. The acidity and metal-loading

of PtHNaY zeolites were gradually changed in order to relate the activity and
selectivity of the catalyst to its hydroisomerizing properties.

2. EXPERIMENTAL

2.1. Materials

Synthetic Y zeolite (56 Al atoms per unit cell on an anhydrous base) was from Union
Carbide Corp., Linde Div. The sample was equilibrated with 0.1 M solutions containing
various amounts of NaCl and NH_4Cl. The NH_4NaY zeolites obtained this way, were exchanged
with the Pt(II)-tetrammine complex. Conventionally prepared DBY (deep-bed calcined)
and USY (ultrastable) zeolites were included as very acidic supports.

The symbols used to identify the different samples, contain two figures : the first
represents Pt content in weight %, the second the percent of the initial Na^+ ions
exchanged with NH_4^+. The sample notation 2PtHY-90, refers to a 90 % exchanged $NaNH_4Y$
zeolite, containing 2 % by weight of Pt and pretreated in the reactor as indicated
below.

2.2. Methods

Catalytic reactions were carried out in a continuous flow reactor with ethylbenzene
saturator. Standard operation conditions were : GHSV = **26**.33 h^{-1}; H_2/ethylbenzene = 75;
catalyst volume = 0.6 ml; size of pellets = 60-80 mesh; catalyst bed diameter = 8 mm.
The catalysts usually were activated in situ at 400°C in a flow of oxygen, followed by
hydrogen reduction for 1 h at the same temperature. Products were analyzed on-line
using two gaschromatographs each equipped with a gas sampling valve. One gaschromato-
graph with a 2m[EHTCP(3%) + bentone 34(7%)] column and TCD detector was used to separate
the three xylenes. On the other GC with a 100 m OV101 capillary column and FID detector,
the remaining products were separated using temperature programming. Sampling is
usually done after 1 h of reaction.

3. RESULTS AND DISCUSSION

3.1. Effect of reaction conditions upon selectivity

At a given contact time, the reaction selectivity changes with temperature as shown
in Fig. 1. It results that a complex reaction network exists, which can be schematized
as follows :

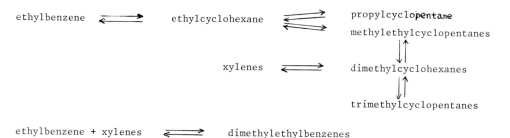

It should be noted that at any reaction temperature, thermodynamic equilibrium exists
between the following components : ethylbenzene and ethylcyclohexane, xylenes and
corresponding dimethylcyclohexanes. This indicates that the hydrogenation-dehydrogenation

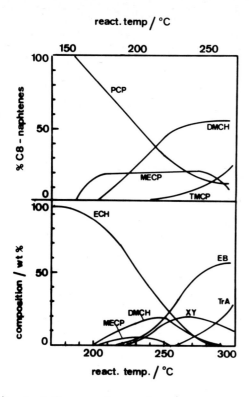

react. temp / °C

ECH = ethylcyclohexane
MECP = methylethylcyclopentanes
DMCH = dimethylcyclohexanes
EB   = ethylbenzene
Xy   = xylenes
TrA = transalkylates of Xy + EB
PCP = propylcyclopentane
TMCP = trimethylcyclopentane

GHSV =26.33 h⁻¹
H_2/EB = 75

Fig. 1. Influence of reaction temperature on ethylbenzene conversion over 2Pt HY-90.

on Pt is far more rapid than the isomerization of naphthenes on the acid sites. Indeed, equilibrium between the methylethylcyclopentanes, dimethylcyclohexanes and trimethyl-cyclopentanes is only approached at high degrees of ethylcyclohexane conversion. It results that the rearrangements among cyclopentyl and cyclohexylcarbenium ions should be rate determining, which agrees with previous work on other supports (ref. 3-4). The formation of MECP prior to DMCH also suggests that isomerization occurs via a successive ring contraction-ring expansion mechanism as proposed earlier (ref. 5).

Table 1 shows at one reaction temperature that the formation of the xylenes and therefore of the corresponding dimethylcyclohexanes is kinetically controlled. Indeed, deviations from thermodynamic equilibrium are pronounced between ethylbenzene and the xylenes. The preferred formation of o-xylene among the xylene-isomers, will be discussed later.

254

TABLE 1.

Distribution of C_8-aromatics (%) formed at 233°C on 2PtHY-90(A) and expected from thermodynamics (B)[1].

	A	B	A	B
ethylbenzene	57.2	3.8	–	–
p-xylene	8.3	23.2	19.7	24.1
m-xylene	22.6	52.6	54.4	54.7
o-xylene	11.9	20.3	25.9	21.2

(1) from ref. 6.

3.2. Effect of Pt-content of HY-90 zeolites

The total xylene yield at a given reaction temperature changes with the Pt loading of the zeolite in a way as should be expected for typical bifunctional behaviour (Fig. 2).

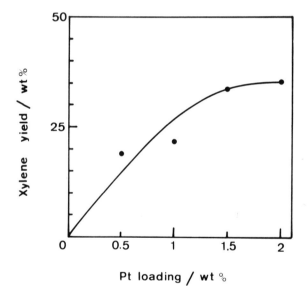

Fig. 2. Influence of Pt content on xylene yield at 270°C. (Reaction conditions are the same as for Fig. 1).

In terms of classic bifunctional behaviour (ref. 7) it is expected that during the rate-determining event ethylcyclohexylcarbenium ions are rearranged first in methylethylcarbenium ions which subsequently are converted in dimethylcyclohexylcarbenium ions. Again by analogy with the bi-functional conversion of paraffins over similar catalysts (ref. 8), even no traces of olefinic intermediates (ethylcyclohexene, dimethylcyclohexenes or branched cyclopentenes) are observed as reaction products.

3.3. Effect of proton content of HNaY zeolites

The increase in xylene yield (at a temperature close to optimum temperature, see Fig. 1) with proton content of the Y zeolite is shown in Fig. 3.

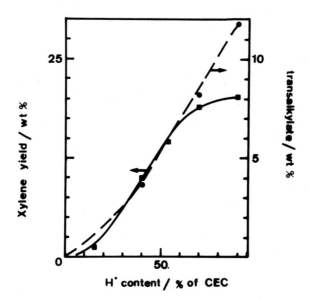

Fig. 3. Influence of proton content on xylene-yield at 270°C and on the extent of transalkylation of the formed C_8-aromatics at 300°C.

At low degrees of NH_4 exchange, a less than proportional increase of the catalytic activity with H^+ content is observed. This is followed by a proportional increase of the activity for both hydroisomerization and transalkylation. At the highest exchange degrees primarily formed xylenes disappear through transalkylation reactions. It is known that starting from 25 % exchange the strong acidity (> 88 % H_2SO_4) in these zeolites increases linearly with the ion exchange degree (ref. 9). These strong acid sites seem to be required for the transalkylation reactions between xylenes and ethyl-benzene. The number of sites in the classes of lower acidity first increase linearly with the H^+ content but at higher exchange degrees level off. It is clear that the rearrangement of naphtenyl carbenium ions follows this behaviour and is already catalyzed by acid sites of lower strength. This also can be derived from Fig. 1. Unfortunately, it is impossible to determine quantitatively the minimum acid strength for this conversion, since the acid strength distribution is given in classes of acid strength which are not narrow enough (ref. 9,10).

It is clear that for conversion of branched-naphtenyl carbenium ions only intermediate acid strength is required. This is confirmed when cracking catalysts such as deep-bed calcined and ultrastable NH_4Y are used as support for Pt. It is general knowledge now that cracking activity for these zeolites increases in the following order :

HY-90 < DBY < USY

It is expected that in this series of Pt-zeolites, selectivity for xylene formation will

decrease in favour of the side reactions catalyzed by stronger acidity, such as trans-
and dealkylation. The expected trend is confirmed by the data of Table 2.

TABLE 2.
Yield of aromatics (%) over Pt-zeolites with strong acidity[1].

Products	1 PtHY-90 (266°C)	1 PtDBY (260°C)	1 PtUSY (269°C)
ethylbenzene	41.7	27.4	12.7
p-xylene	4.7	1.5	3.0
m-xylene	10.3	3.4	7.3
o-xylene	5.1	1.3	2.7
benzene	0.8	1.6	3.9
toluene	0.7	2.0	9.8
polyalkylate	5.1	7.4	22.0

(1) Reaction conditions as for Fig. 1.

3.4. Preferred reaction pathways

The formation of n-propylcyclopentane out of ethylcyclohexane occurs very readily over
2 PtHY-90 (Fig. 1). Also the preferred formation of o-xylene and therefore of 1,2-di-
methylcyclohexane over the same catalyst was already mentioned (Table 1). This is further
illustrated in Fig. 4. When the contact time is decreased, o-xylene gradually becomes a
preferred reaction product (Fig. 4A). When the acidity (number and strength) is decreased,
the same is observed (Fig. 4B), except for the very low conversions (or reaction temperatures). This is also in line with previous observations (ref. 3,4).

Since in no case methylcycloheptane was observed - even not in traces - as reaction
product, isomerisation occurs exclusively via contraction of the 6-ring. If this would
occur via classical carbenium ions, the formation of o-, m-, and p-xylene would require
the formation of primary methylethylcyclopentyl-cations, which event would be rate
limiting. In this way selective o-xylene formation would not be expected (scheme I).
The formation of primary ions is easily avoided via protonated cyclopropane (PCPr)-
species which also is invoked to explain changes in the degree of branching of paraffins
over bifunctional catalysts (ref. 11). If these PCPr-ions are more stable than secondary
carbenium ions the formation of o-xylene is energetically favoured since only PCPr- and
tertiary ions are formed. In case of p- and m-xylene, PCPr- and secondary ions have to
be formed (scheme II). Therefore, the conclusions derived for Pt/alumina (ref. 4) seem
to be of more general validity.

The sudden drop in o-xylene yield (Fig. 4B) at the lowest reaction temperatures is
accompanied by the appearance of relatively high amounts of 1,3- and 1,4-dimethylcyclo-
hexanes. These amounts increase for the low acidity supports. This suggests that a metal
catalyzed direct isomerization of ethylcyclohexane to 1,3 and 1,4-dimethylcyclohexane
occurs to a minor extent at the beginning of the reaction. It is difficult to advance
any reaction mechanism which explains these data. An analogy again exists with n-paraffin
hydrocracking. Indeed, also at the beginning of these reactions small amounts of

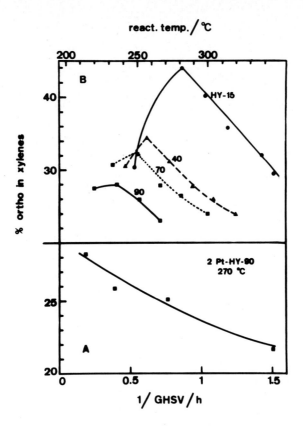

Fig. 4. Preferential formation of o-xylene.

(I)

n-paraffins with smaller carbon number and of hydrogenolytic origin are observed (ref. 8,12). This phenomenon also is not well understood.

(II)

4. CONCLUSIONS

The hydroisomerization of ethylbenzene over PtHNaY zeolites shows that :

- the results can be described in terms of a classic bifunctional mechanism, the rearrangement of following carbenium ions being rate limiting :

ethylcyclohexyl ⇌ methylethylcyclopentyl ⇌ dimethylcyclohexyl

The key-step is the six-ring contraction followed by the five-ring expansion, each being proton catalyzed.

- in order to reach maximum selectivity, acidity of intermediate strength is required.
- selective formation of o-xylene occurs via an energetically favourable reaction pathway if the existence of protonated cyclopropanes with higher stability than secondary carbenium ions is assumed.

ACKNOWLEDGMENTS

M. Nitta acknowledges a research grant from the "Onderzoeksfonds", K.U. Leuven. P.A. Jacobs is grateful to N.F.W.O. (Belgium) for a research position as "Bevoegd-verklaard Navorser". Support from the Belgian Government (Concerted Action in Catalysis) is very much acknowledged.

REFERENCES

1. C.L. Thomas, "Catalytic Process and Proven Catalysts", Academic Press, New York, 1970, pp. 21-21.
2. P.M. Pitts, J.E. Connor and L.N. Leun, Ind. Eng. Chem., 47 (1955) 770.
3. N.S. Gnepp and M. Guisnet, Bull. Soc. Chim. Fr., 5-6 (1977) 435.
4. K.H. Röbschläger and E.G. Christoffel, Ind. Eng. Chem. Proc. Res. Dev., 18 (1979) 347.

5   H. Pines and A.W. Shaw, J. Am. Chem. Soc. 79 (1957) 1474.
6   D.R. Stull, E.F. Westrum and G.C. Sinke, The Chemical Thermodynamics of Organic Compounds, J. Wiley & Sons, New York, 1969.
7   G.A. Mills, H. Heinemann, T.H. Milliken and A.G. Oblad, Ind. Eng. Chem. 45 (1953) 134.
8   M. Steyns, G. Froment, P. Jacobs, J. Uytterhoeven and J. Weitkamp, Erdöll - Kohle - Erdg. - Petrochem. - Brennst. Chem. 12 (1978) 581.
9   D. Barthomeuf and R. Beaumont, J. Catal. 30 (1973) 288.
10   for a review : P.A. Jacobs, Carboniogenic Activity of Zeolites, Elsevier Scientif., Amsterdam, 1977.
11   J. Weitkamp and H. Farag, Acta Phys. Chem. (Hungar.) 24 (1978) 327.
12   P.A. Jacobs, J.B. Uytterhoeven, M. Steyns, G. Froment and J. Weitkamp, Proc. Int. Conf. Molecular Sieves, Naples, June 1980.

B. Imelik *et al.* (Editors), *Catalysis by Zeolites*
© 1980 Elsevier Scientific Publishing Company, Amsterdam — Printed in The Netherlands

CHARACTERIZATION AND HYDROGENATING PROPERTIES OF Y ZEOLITE-SUPPORTED IRIDIUM CATALYSTS.

M. DUFAUX, P. GELIN and C. NACCACHE

Institut de Recherches sur la Catalyse - C.N.R.S.

2, avenue Albert Einstein, 69626 Villeurbanne Cédex

1   INTRODUCTION

A number of past studies have investigated the preparation and the properties of zeolite-supported group VIII metal catalysts. Much of the work in this area, recently reviewed by Gallezot (1), has been carried out on zeolite-supported platinum. The renewed interest on synthesis of hydrocarbons from H_2-CO mixtures has prompted to study the effect of the zeolite support on the dispersion of ruthenium and subsequently on its catalytic behaviour (2,3, 4). Although less extensively studied, the properties of zeolite-supported rhodium have been reported (5,6). In contrast, in our knowledge, not a work on zeolite-supported iridium has been yet published. This investigation was carried out to provide a detailed study on the preparation, structure,chemisorption and catalytic properties of iridium catalysts supported on Na Y zeolite.

2   EXPERIMENTAL

2.1   Materials

The iridium form of Y zeolite was prepared from Na Y zeolite supplied by Union Carbide Linde Division. Cation exchange was effected by stirring for 2 hours Na Y with a solution of chloropentammine iridium dichloride. The exchanged materials were washed with distilled water, filtered and dried at 323 K. The crystallinity of the material was checked by means of powder X-Ray diffraction. Iridium content was determined by atomic absorption spectrophotometry. By this procedure a series of iridium samples, $|Ir(NH_3)_5 Cl|^{2+}_-$ Na Y, loaded with 1.7, 3.4 and 6.2 % by weight iridium, were prepared. Before evaluation of the catalysts, samples were heated in flowing oxygen to 523 K at the rate of 0.5°/min, cooled down and finally reduced in hydrogen at respectively 383, 773 and 923 K.

2.2   Particule size determinations

Transmission electron microscopy (TEM) was used for the determination of the iridium particle size. The thin cutting procedure used in these measurements was similar to that described in (7). Transmission electron micrographs were obtained with a Jeol microscope which permitted resolution to about 0.5 nm. The micrographs were obtained using a magnification, including photographic enlargement, to 10^6.

2.3   Adsorption measurements

Hydrogen, oxygen and carbon monoxide adsorption measurements were performed in a conventional glass vacuum apparatus. Pressures were determined using a Texas Instruments Precision Pressure Gauge. About 0.5 g of Ir-Na Y was placed into a U-Shaped flow through glass cell and subjected to the pretreatment procedure described above. Before adsorption measurements, the reduced samples were outgassed at 723 K. H_2, O_2, CO uptakes were determined at 293 K. Samples were allowed to equilibrate with about 100 Torr gas pressure, and the adsorption isotherm determined between 100 and 200 Torr. Extrapolation to zero pressure provided the total adsorption which occured on the metal portion of the catalyst. The samples were then outgassed at 293 K and a second isotherm (back sorption) was determined which provided, after extrapolation to zero pressure, the reversibly bound H_2, O_2 or CO. The data given in table 1, which represent the strongly bound species, were derived from the difference between the initial isotherm and the back sorption isotherm.

2.4 Infrared studies

The IR spectra were recorded on a Perkin Elmer 521 spectrometer. The powder was pressed into thin discs of 18 mm diameter and weighing about 10-15 mg. The disc was mounted on a disc holder and introduced in a special infrared cell described elsewhere (8) which allowed to undergo the desired treatment.

2.5 Catalytic studies

The hydrogenation of benzene into cyclohexane and styrene into ethylbenzene (EB) and ethylcyclohexane (ECH) were used as test reactions to evaluate the catalytic properties of Ir-Na Y catalysts. The prereduced Ir Na Y sample after being placed in a U-shaped flow-through glass reactor was activated for 2 hours in flowing hydrogen at the same temperature used for its H_2-reduction. N_2-H_2 gas mixture was passed through a saturator kept at 273 K which contained benzene (vapor pressure 31 Torr) or styrene (vapor pressure 1.7 Torr) and then through the catalyst bed. The flow rate was 6 l/h. The hydrogen pressure was 130 Torr, and the reaction temperature 298 K. Analysis of feed and product streams were made by gas chromatography using an Intersmat flame ionization gas chromatograph. The analytical column was a 3 mm x 0.5 m strainless steel tube packed with 10 % $\beta\beta'$ oxidipropionitrile on spherosil XOB-030. The reaction rate was derived from

$$r \text{ mole } h^{-1} g^{-1} = \frac{F}{22.4} \quad \frac{273}{293} \quad \frac{P}{760} \quad \frac{1}{W}$$

where F is the flow rate in l/h, P the pressure in torr of the products and W the catalyst weight in gramme.

3 RESULTS

3.1 Chemisorption studies

The hydrogen and oxygen uptakes are expressed in microatoms per gm of catalyst and carbon monoxide uptakes in micromoles per gm of catalyst. The results are listed in table 1. Comparison of data presented in table 1 leads to the conclusion that for a given iridium loading, the amount of hydrogen and carbon monoxide adsorbed is strongly dependent on the H_2-reduction temperature, while the quantity of oxygen adsorbed is less dependent. It results that the increase of the H_2 reduction temperature produced a decrease of the ability of the iridium catalysts to adsorb hydrogen or carbon monoxide.

TABLE 1

Influence of the H_2-reduction temperature on the CO, H_2 and O_2 adsorption and on the particle size measured by TEM

μ atom g^{-1} Ir	H_2 red. T° K	CO μ mole g$_{cat}^{-1}$	H ads μ atom g$_{cat}^{-1}$	O ads μ atom g$_{cat}^{-1}$	Particule size nm
88 (1.7%)	383	135	142	70	$\leqslant$ 1.
	773	93	104	80	1.
	923	84	100	94	$\sim$ 1.1.
177 (3.4 %)	383	232	232	140	$\leqslant$ 1.
	773	164	206	170	$\sim$ 1.
	923	124	190	174	$\sim$ 1.5
324 (6.2%)	383	420	420	250	$\leqslant$ 1.
	773	300	310	290	1.

3.2 Transmission electron microscopy studies

Electron micrographs were obtained on samples recovered from the adsorption studies. Typical electron micrograph is given in figure 3.1.

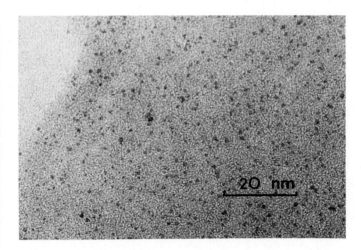

Figure 3.1. Electron micrograph of Ir Na Y H_2-reduced at 773 K.

Evidence from electron micrographs suggests that the temperature of H_2-reduction influenced the number of iridium particles present without modifying significantly the particle size. Indeed all samples showed iridium particles within 0.6-1.0 nm range. A very small percentage of particles with size beyond this range but never exceeding 1.5 nm exists for samples reduced at high temperature. Apparently the temperature of H_2-reduction has no effect on the particle size distribution derived from electron microscopy.

3.3   Infrared studies

The effect of H_2-reduction temperature on CO adsorption was investigated by infrared.

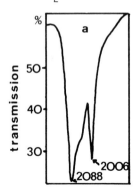 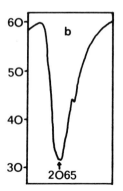

Figure 3.2. Infrared spectra in the carbonyl stretching region of CO adsorbed on Ir Na Y 6.2 wt % Ir
a) H_2-reduced at 383 K   b) H_2-reduced at 773 K.

Figure 3.2a and 3.2b correspond to the infrared spectra in the carbonyl stretching region of CO adsorbed at 293 K on 6.2 wt % Ir-Na Y samples reduced respectively at 383 and 773 K. The CO-infrared spectrum exhibited by the low-temperature reduced sample consists of two intense and narrow bands at 2088 and 2006 cm^{-1} with a shoulder at about 2060 cm^{-1}. In contrast the high-temperature reduced sample shows a strong and relatively broad band around 2065 cm^{-1} with a band at 2006 cm^{-1} very small with respect to the one at 2065 cm^{-1}. Similar results were obtained on the 1.7 and 3.4 wt % iridium. The main feature was that the relative intensities of the doublet at 2088-2006 cm^{-1} with respect to the band at 2065 cm^{-1} decreased along with the iridium loading. Surface reactions between adsorbed carbon monoxide species and oxygen were studied by infrared. Furthermore the effect of methyl iodide adsorption on the iridium-carbon monoxide species was also investigated.

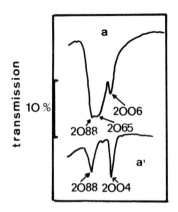 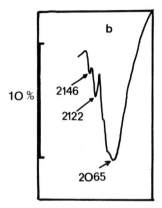

Figure 3.3. Carbonyl infrared spectra of : a) CO adsorbed on Ir-Na Y, 3.4 wt % Ir H_2 reduced at 773 K,   a)   a + O_2 at 373 K   b) a + CH_3 I at 293 K.

Our results, illustrated in figures 3.3a and 3.3b indicate that the doublet at 2088-2006cm^{-1} was almost inchanged upon oxygen adsorption, while the adsorption of CH_3I shifted the doublet at 2146-2122 cm^{-1}. In contrast the single IR band at 2065 cm^{-1} was found insensitive to CH_3I while it showly disappeared when reacting with oxygen at 293 K. The oxygen reaction proceeded at a faster rate at 373 K.

3.4 Catalytic activities

The catalytic activities for the hydrogenation of benzene and styrene of iridium containing zeolites were compared under the reaction conditions outlined above. Since the electron microscopic experiments have indicated that the Ir particles on almost all samples investigated were less than 1.0 nm size it was assumed that the metal dispersion was 100 % making all iridium atoms available for the reaction. Thus to facilitate comparison, the activities were expressed in mmoles of benzene converted per hour and per gramme of metal. The values together with the derived turnover numbers for benzene hydrogenation are given in table 2.

Table 2

Influence of the H_2-reduction temperature on the benzene hydrogenation on Ir Y
(T = 298 K P_{H_2} = 130 torr $P_{C_6H_6}$ = 31 torr).

% Ir	H_2 red T °K	rate mmole h^{-1} g$_{Ir}^{-1}$	N h^{-1}
	383	11	2.2
1.7	773	19	3.8
	923	34	6.6
	383	15	2.8
3.4	773	35	6.8
	923	56	10.7
	383	24	4.6
6.2	773	62	12.0

From this table it is seen that the specific activity of zeolite-supported iridium for the arene reduction is strongly dependent on the temperature of hydrogen reduction and on the metal loading. In general it was found that for benzene hydrogenation and for a given Ir loading the high temperature reduced sample was approximately 3 times more active than the low temperature reduced material. The differences of the specific activity in the metal loading dependence are diminishing with the increase of the metal content. Approximately the same specific activity is found for Ir containing zeolites with 3.4 and 6.2 wt % Ir reduced at high temperature. In table 3 are listed the specific activities of the Ir-containing zeolites for the hydrogenation of C_6H_5-CH = CH_2 into C_6H_5 - CH_2-CH_2 (E B), only the olefinic double bond being reduced, and into C_6H_{11}-CH_2-CH_3, both the aromatic ring and the olefinic bond being reduced. Again it is seen that the zeolite-supported iridium catalysts behave similarly as shown above for the reaction which involves the reduction of the aromatic ring, that is the specific rate of ethylcyclohexane formation is strongly dependent on the H_2-reduction temperature and on the iridium content. In strong contrast the reaction which involves only the hydrogenation of the olefinic double bond is seen almost independent on the H_2-temperature of metal reduction as well almost indifferent on the iridium content.

Table 3

Influence of the H_2-reduction temperature on the hydrogenation of styrene on Ir Na Y
(T = 298 K P_{H_2} = 130 torr P_{Hc} = 1.7 torr)

% Ir	H_2 red T°$_K$	Rate m.mole h^{-1} g$_{Ir}^{-1}$	
		ethylbenzene	ethylcyclohexane
1.7	383	223	3.5
	773	258	8
3.4	383	203	4.4
	773	218	12
6.2	383	160	6.4
	773	176	22.4

4   DISCUSSION

Recent works have pointed out that faujasite-type zeolites are well suited as carriers
for stabilizing very small metal particles. The presence of platinum (1) ruthenium (2,3,4)
and rhodium (5,6) clusters entrapped within the large cavities of the zeolite has been
evidenced by electron microscopy, X-ray diffraction infrared and adsorption studies. A
significant aspect of this work is that NaY zeolite is a good carrier to prepare and stabilize within the zeolite framework well dispersed iridium metal catalyst. Indeed the results
obtained by transmission electron microscopy and reported in table 1 indicate clearly that
in general the size of the iridium particles are less than 1 nm. Furthermore the electron
micrographs of a series of zeolite-supported iridium showed that the shape, the size and
the distribution of the metal particles are almost independent on the temperature of H_2-
reduction and on the metal content. A relative small percentage of iridium particles with
size of about 1.5 nm were observed with those Ir-NaY samples either with the highest metal
content or H_2-reduced at 923 K. The shape of the metal particles appears spherical. It is
well known that one can utilize the average particle size derived from electron micrograph
to estimate the metal dispersion D defined as the ratio of surface to total iridium atoms
(Ir_s/Ir_t = S/S$_t$, (10) where S$_t$ is the iridium surface area when all metal atoms are surface
atoms and S the true iridium surface area for a given sample). Assuming that the area per
iridium atom is a_r = 0.769.10^{-19} m^2 (9) one calculates S$_t$ = 241 m^2 g^{-1}. The metal surface
area of supported iridium is related to the metal particle diameter by the relation
d = 6/ρs, where ρ_{Ir} = 22.4 is the metal density. This relation along with the expression
D = S/St indicates that a metallic dispersion of unity corresponds to a iridium particle
size of about 1 nm and less. Since electron microscopy studies have indicated that zeolite-
supported iridium catalysts are characterized by metal particle size less than 1 nm, it is
reasonable to state that the metal dispersion is unity, which signifies that all iridium
atoms are surface atoms thus available for chemisorption.

Bearing in mind this assumption, the results of chemisorption listed in table 1 allow
to calculate the stoichiometry for CO/Ir$_s$, H/Ir$_s$ O/Ir$_s$ for the various zeolite-supported
iridium catalysts. The values are reported in table 4. This table indicates that the CO
and hydrogen stoichiometries are, for a given iridium content, function of the H_2-reduction
temperature and also for a given H_2-reduction temperature, function of the metal loading.

Table 4.

Stoichiometry for CO, H_2 and O_2 chemisorption on Ir-NaY.

Wt % Ir	H_2-red $T^{\circ}K$	CO/Ir_s	H/Ir_s	O/Ir_s
	383	1.53	1.61	0.80
1.7	773	1.06	1.18	0.91
	923	0.95	1.14	1.07
	383	1.31	1.31	0.79
3.4	773	0.93	1.16	0.96
	922	0.70	1.07	0.98
6.2	383	1.30	1.30	0.77
	773	0.93	0.96	0.90

Furthermore table 4 shows that the CO/Ir_s and H/Ir_s ratios are very close, which suggests that there is a close similitude between the adsorption of CO and H on Ir-NaY samples. Previous chemisorption studies on iridium catalysts had indicated that H/Ir_s ratio may vary within a large range. Anderson (9), Sinfelt and Yates (11), Corro and Gomez (12) have considered a H/Ir_s ratio equal to unity and pointed out that there is a relative good agreement between the particle size determined by electron microscopy or X-ray diffraction and by hydrogen chemisorption using H/Ir_s = 1. In contrast Moroz et al (13) Mc Vicker et al (14) reported H/Ir_s value of about 1.5-1.3 for iridium supported on alumina. Furthermore their data showed that this ratio decreased when the particle size increased. Our results along with those published recently thus indicate that the average hydrogen stoichiometry, in the case of supported small iridium particles, is higher than unity. Since, in general, the experimental H/Ir_s ratios are not integer, one can therefore envisage that the experimental H/Ir_s values correspond to the average values of H/Ir_s = 2 and H/Ir_s = 1, the relative proportion of each type of surface iridium atoms depending strongly on the method of preparation of the catalyst. Hence from the H/Ir_s values and their variations observed in the present work, it may be concluded that the zeolite supported iridium catalysts contain both types iridium atoms, the relative number of those showing H/Ir_s = 2 decreasing with the H_2-reduction temperature and with the iridium content. Similar conclusions were reached in (13) and (14). However while Moroz et al (13) stressed that supported iridium catalysts contain both metal iridium crystallites and ionic iridium, Mc Vicker et al (14) consider that H/Ir_s ratio higher than one, found on Al_2O_3-supported highly dispersed Iridium involves the existence of monoatomically dispersed iridium atoms along with very small metal clusters. Similar considerations are pertinent for carbon monoxide adsorption. The CO/Ir_s stoichiometry given in this work and in (14) suggests that on highly dispersed iridium catalysts monocarbonyl iridium and dicarbonyl iridium species were formed, their formation parallels that of IrH and IrH_2. Further confirmation of the view that two types of surface iridium atoms exist on highly dispersed zeolite-supported iridium catalysts are obtained by examining by infrared the behaviour of these materials towards CO. The IR data have shown that the intensity of the IR bands at 2088 and 2006 cm^{-1} varies in parallel and furthermore that their frequencies were simultaneously shifted upon methyl iodide adsorption. These results indicate that the doublet at 2088-2006 cm^{-1} must be assigned to a single iridium carbonyl species. Since $M(CO)_2$ in a C_{2v} symmetry or $M(CO)_3$ in a C_{3v} symmetry lead to two infrared bands, one can assign the doublet to $Ir(CO)_n$ species, where n = 2 or 3. The IR band at 2065 cm^{-1}, in agreement with (15), is attributed to one CO linearly bonded to iridium metal. This assignment is further confirmed following the $CO-O_2$ results. It is recognized that CO adsorbed on metal surface reacts readily with oxygen. This was found recently for alumina-supported iridium (16). Ourinfrared spectral data have indicated that the band at 2065 cm^{-1} disappeared upon oxygen adsorption, which confirms that the IR band at 2065 cm^{-1} corresponds to CO adsorbed on iridium metal. In conclusion, the infrared results suggest that the adsorption of CO on Ir-NaY samples produces both mono and di-carbonyl species. That conclusion and the quantitative data on CO adsorption presented in table 1 are mutually sustaining and provide support for the occurence of two different types iridium atoms. Their relative contribution is found to depend on the temperature of H_2-reduction and on the iridium content. It is interesting to note that although there is no apparent iridium particle size increase with the temperature of H_2-reduction, the relative number of monocarbonyl species Ir-CO increases as indicated by the development of the IR band at 2065 cm^{-1}. While there is little doubt concerning the assignment of the IR band near 2065 cm^{-1}, the nature of the site on supported iridium catalysts responsible for multiple CO adsorption has not been established conclusively. X-ray diffraction studies on $Ir-Al_2O_3$ showing : $H/Ir_s > 1$, have indicated the existence of an alumina-based spinel structure where iridium-cations are distributed in the octahedral sites (13). The authors concluded that ionic form of iridium coexists with iridium metal crystallites and is responsible for the greater hydrogen adsorption capacity. In contrast Mc Vicker et al (14) suggested that on highly dispersed alumina-supported iridium multiple adsorbate bonding (H_2 or CO) occurs on monoatomically dispersed Ir(0) atoms. The non reactivity of $Ir(CO)_n$ species, entrapped in NaY zeolite, towards oxygen observed in the present work would suggest that iridium in $Ir(CO)_n$ should have a more or less ionic character. In previous works (17-18) we have shown that CO reacts with non-reduced Ir-NaY samples to produce two IR bands on the carbonyl stretching region at 2086 cm^{-1} and 2003 cm^{-1}. These two IR bands were shifted to 2148 and 2100 cm^{-1} following methyl iodide adsorption. These results were interpreted in terms of $Ir(I) (CO)_3$ species generated during the carbonylation reaction. Monovalent iridium carbonyl complex undergoes oxidative addition of methyl iodide to form $(CH_3)Ir(III)I(CO)_3$ with the subsequent (CO) shift at higher frequency (17).

It is clear from the IR spectra shown in figure 3.3 that there is a close similitude between the well identified zeolite-supported $Ir(I)(CO)_3$ (17-18) and the carbonyl iridium species responsible for the IR doublet at 2086 and 2006 cm^{-1}. Because of this analogy it is reasonable to conclude that on H_2-reduced Ir-NaY samples monovalent iridium cations are present which form with CO $Ir(I)(CO)_3$ complexes.

From the evidence presented for NaY zeolite-supported catalysts it may be concluded that on reduced samples a variety of reactive iridium sites existssimultaneously which are isolated monovalent iridium cations and surface iridium atoms on metal crystallites. Their relative importance is likely to be a function of the temperature of H_2-reduction and also a function of the iridium content. The increase of the IR band intensity corresponding to Ir^0 (CO) with the subsequent decrease of those attributed to $Ir(I) (CO)_3$ when the

H_2-reduction temperature increased reflects that Ir(I) cations although relatively stable when entrapped in the zeolite framework are destroyed at higher temperatures and aggregate into small crystallites. The stoichiometry of the hydrogen chemisorption would be $H/Ir_s = 1$ for surface metal atoms and $H/Ir(I) = 2$ for monovalent iridium. Complexes of many group VIII transition metals including Rh(I), Ir(I) were found to bind hydrogen through mechanisms involving the oxidative addition of molecular hydrogen to form hydrido-complexes (9).

$$Ir(I) + H_2 \longrightarrow H-Ir(III) - H$$

On the basis of the iridium surface analysis given above, the oxygen adsorption results may be explained if one considers as in (13) that ionic iridium does not adsorb oxygen. Hence the amount of chemisorbed oxygen would increase when the amount of ionic iridium decreases. This was experimentally observed, the oxygen adsorption involving only surface iridium atoms present on iridium crystallites.

Although most of the published work has established that olefin or aromatic hydrogenation belongs to a class of reactions in which the specific activity of the supported metal is independent of the crystallite size and of the support, recent investigations have demonstrated very conclusively that the specific activity of platinum catalyst in the hydrogenation of ethylene (20), benzene (21), cyclopropane (22) is enhanced when platinum is dispersed within zeolite. However Coughlan et al (2) in their studies on zeolite-supported ruthenium showed that the zeolite -support plays no role in the hydrogenation of benzene. In view of these conflicting conclusions it appears interesting to know if the zeolite support affects the catalytic properties of supported iridium catalyst with respect to other Ir catalysts. According to our experiments over Ir-Al$_2$O$_3$ (3.3 wt % Ir, H_2-reduced at 773 K) where the iridium metal is 100 % dispersed, the specific rate of benzene hydrogenation was found 90 mmoles h^{-1} g_{Ir}^{-1} which corresponds to a turn over number of 18 h^{-1}. The data listed in table 2 indicate that the best turn over number achieved by zeolite-supported iridium is only 2/3 compared with that of alumina-supported iridium. Clearly these results indicate that similarly to Ru-NaY catalysts, the zeolite carrier produces no noticeable effect on the hydrogenating properties of iridium. The striking observation in the present work is the appreciable dependence of the turn over number for benzene hydrogenation on the temperature of H_2-reduction and on the metal content. It is remarkable that the lowest specific activity is found for samples showing the highest H/Ir_s ratios. A similar trend has been shown previously by Aben et al (23) for benzene hydrogenation on alumina-supported platinum. From this work it appeared that highly dispersed platinum was less active than expected. Similar conclusions were reached by Fuentes and Figueras (24). They showed that very small rhodium crystallites supported on alumina were, for benzene hydrogenation, less active than those of moderate size. The specific activity of extremely small metal particles may be altered if one suggests that their electronic properties are different from those of bulk metal or/and if these small crystallites present crystallographic planes different from those encountered on particles of medium size. Let us first focus our attention to the effect of H_2-reduction temperature. Over Ir-NaY the turn over number for benzene hydrogenation was found to increase with the temperature of H_2-reduction. Furthermore this work has well established that for each series of catalysts there is no apparent variation of the iridium particle size while the proportion of the ionic iridium present on the low temperature reduced samples, decreased. Thus it may be suggested that the "reduction temperature effect" observed in this work is due to the presence of ionic iridium stabilized

toward reduction by the zeolite framework. Since isolated ionic iridium species are probably inactive for arene reduction, the TON, listed in table 2, which have been calculated by
assuming that all iridium atoms were efficient for the reaction, are apparent values. The
true TON would be obtained by considering only the iridium surface atoms of the metal crystallites. Since the proportion of ionic iridium decreases with the H_2-reduction temperature
it is clear that the TON should increase. However a possible "surface structure effect"
cannot be excluded. Table 2 shows that for high temperature H_2-reduction, although the
H/Ir_s ratio remain constant, the TON increases with the iridium content. These results are
understood only if one suggests that the benzene hydrogenation is "structure sensisitive"
for metal particles less than 1 nm in size. Thus it must be concluded that although no
conclusive electron microscopy data have been obtained, within the 0.5-1 nm range, with
the increasing reduction temperature and with the increasing metal loading, the smallest
iridium crystallites have the tendancy to aggregate. Thus the relative number of 1 nm
iridium crystallites would increase at the expense of the smallest. The data concerning the
hydrogenation of styrene provide some confirmation of the assumption that the hydrogenation
of benzene may be "structure sensitive" while the hydrogenation of the aromatic ring of
styrene follows the same trend as observed for benzene hydrogenation (the rate of the
hydrogenation of the ring is dependent on the temperature of metal reduction and on the metal content). The TON for the hydrogenation of the olefin double bond in $C_6H_5 - CH = CH_2$
appears to be almost insensitive to the iridium content and to the H_2-reduction temperature.
It is well established that the hydrogenation of olefins is catalyzed by monovalent iridium
complexes (19-25). Hence both ionic iridium species and iridium metal present in Ir-NaY
catalysts are efficient for the total hydrogenation of styrene into ethylbenzene. Since
our results have shown that ionic iridium and zero-valent iridium are surface atoms whatever
are the metal content and the H_2-reduction temperature, the specific activity of Ir-NaY
for the conversion of styrene into ethylbenzene should remain constant.

To summarize, the chemisorption data and the infrared study along with the electron micrographs have indicated that NaY zeolite is suitable to stabilize within the structure both
ionic iridium species and very small iridium clusters. The relative number of these two
species depends strongly on both the H_2-reduction temperature and the iridium loading. The
catalytic measurements provide, with the suggested hypothesis, some evidence that the
hydrogenation of arenes may be "structure sensitive" at least for iridium crystallites which
size is less than 0.5-0.6 nm. Finally in contrast with the results previously established
on zeolite-supported platinum, the specific hydrogenating properties of iridium are not
enhanced when supported on zeolite.

Acknoledgments. We are indebted to Dr. B. Imelik for many useful and stimulating discussions
We are grateful to Mrs C. Leclercq and I. Mutin for the electron micrographs and to
Mr. H. Urbain for running the chemical analysis. C. Naccache thanks G.B McVicker for having
communicated his manuscript before publication.

REFERENCES

1   P. Gallezot, Catal. Rev. Sci. Eng., 20, 1979, 121
2   B. Coughlan, S. Narayana, W.A. Mc Cann and W.M. Carroll, J. Catal., 49, 1977, 97
3   D.J. Elliott and J.H. Lunsford, J. Catal., 57, 1979, 11
4   J.J. Verdonck and P.A. Jacobs, M. Genet and G. Poncelet, J. Chem. Soc. Farad. I 76, 1980, 403.
5   N. Kaufherr, M. Primet, M. Dufaux and C. Naccache, C.R. Acad. Sci., 286c, 1978, 131.
6   M. Primet, J. Chem. Soc. Farad Trans I
7   P. Gallezot, I. Mutin, G. Dalmai-Imelik and B. Imelik, J. Microsc. Spectrosc. Electron., 1, 1976, 1
8   M. Primet, J.C. Védrine and C. Naccache, J. Mol. Catal., 4, 1978, 411
9   J.R. Anderson, J. Catal. 50,1977, 50
10 J.R. Anderson, Structure of metallic catalysts, Academic Press London,1975
11 D.J.C. Yates and J.H. Sinfelt, J. Catal., 8, 1967, 82
12 G. Corro and R. Gomez, React. Kinet. Catal. Lett., 9, 1978, 325
13 E.M. Moroz, S.V. Bogdanov, N.E. Buyanova and O.V. Kovrizhina, Kinet. I. Katal., 19, 1978 1029
14 G.B. Mc Vicker, R.T.K. Baker, R.L. Garten and E.L. Kugler, to be published J. Catalysis
15 L. Lynds, Spectrochim. Acta., 20, 1964, 1369
16 J.L. Falconer, P.R. Wentrcek and H. Wise, J. Catal., 45, 1976, 248
17 P. Gelin, Y. Ben Taarit and C. Naccache, VII Congress on Catalysis Tokyo 1980, preprint B 15
18 P. Gelin, G. Coudurier, Y. Ben Taarit and C. Naccache, Submitted J. Catalysis.
19 R.E. Harmon, S.K. Gupta and D.J. Brown, Chem. Reviews 73, 1973, 21
20 R.A. Dalla Betta and M. Boudart, Proc. 5th Intern. Congress. Catalysis Palm Beach 1972, North Holland Amsterdam, 2, 1973, 1329
21 P. Gallezot, J. Datka, J. Massardier, M. Primet and B. Imelik, Proc. 6th Intern. Congr. Catalysis London 1976, Chem. Soc.,2, 1976, 696.
22 C. Naccache, N. Kaufherr, M. Dufaux, J. Bandiera and B. Imelik, Molecular Sieves 40, 1977, 538, ACS Symposium Series ed. R. Katzer
23 P.C. Aben, H. Van der Eijk, J.M. Oelderik, Proc. 5th Intern. Congress Catalysis, Palm-Beach 1972, North Holland Amsterdam, 1, 1973, 717
24 S. Fuentes and F. Figueras, J. Catalysis, 61, 1980, 442.
25 W. Strohmeir et al, Z. Natur forsch, B.23, 1968, 1377

B. Imelik *et al.* (Editors), *Catalysis by Zeolites*
© 1980 Elsevier Scientific Publishing Company, Amsterdam — Printed in The Netherlands

STUDY OF THE STABILITY TO SULPHUR POISONING OF NICKEL, PLATINUM AND PALLADIUM FIXED ON HYDROCRACKING ZEOLITE CATALYSTS

ECHEVSKII G.V., IONE K.G.
Institute of Catalysis, Novosibirsk, USSR

Processing of sulfided and highly sulfided oils, whose portion in the world's production is continuously growing , necessitates a search for novel catalysts stable to sulphur poisoning .

It is important to establish the nature of factors, determining the activity of hydrocracking catalysts of the sulphur-containing oil fractions. It is also necessary to establish dependences between the nature of the support and the stability of metals fixed thereon to sulphur poisoning. First of all, it is essential to find a parameter which can characterize this stability quantitatively.

In Ref. [1 - 3] the metal stability to sulphur poisoning was characterized by the transformation degree of the sulphur-containing raw material. In this case the activity depends upon a number of factors including dispersity and the metal nature, while the real stability to sulphur poisoning is determined by the kinetics and thermodynamics of the metal-sulphur interaction.

In the present work the dependence of the n-octane hydrocracking rate on the hydrogenation activity of a metal component and that of the metal stability to sulphur poisoning on the nature of the metal and support have been studied.

EXPERIMENTAL

The hydrocracking rate was determined in a flow-circulation system at 350°C, 20 atm, and H_2 : n-octane = 10 : 1. The stability to sulphur poisoning of metals, supported on zeolites, was characterized by the ratio of the rate constants of hexene-1 hydrogenation on pure and sulfided catalysts. The stability coefficient was determined as follows:

$$\lambda = \frac{K^S_{hydr.}}{K^0_{hydr.}} \; 100\%$$

where $K^S_{hydr.}$ is the hydrogenation rate constant of hexene-1, containing 0.05 % sulphur; $K^0_{hydr.}$ is the hydrogenation rate constant of "Chemically pure" hexene-1 determined in a flow-circulation system at 150°C and

15 atm.

Alumina, silica, klinoptilolite, mordenite and Y-type zeolite both in decationated and dealuminated forms were used as supports. The metals were supported on zeolites by the ion exchange technique.

EXPERIMENTAL RESULTS

Fig. 1 shows the dependence of hydrocracking rate of n-octane on the hydrogenation activity of a metal component. Pd, Mo, Pt, Ni, Ru, Re were used as hydrogenating components.

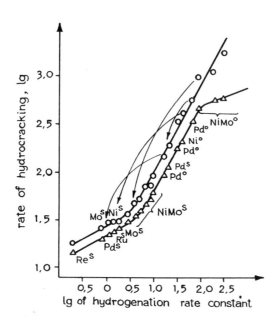

Fig. 1. n-octane hydrocracking rate versus hydrogenation activity ($K^o_{hydr.}$) of the metal component in zeolite catalysts
1 - the overall reaction rate
2 - the formation rate of C_3-C_5 paraffins.

The introduction of 0.05% sulphur into n-octane decreases both the hydrogenation rate of hexene-1 and the hydrocracking rate of n-octane. All the data lie on the same curve independently of the metal nature and sulphur presence. The common dependence of the hydrocracking rate on the activity of a metal component in hydrogenation reaction for both sulfided and nonsulfided catalysts indicates that the overall hydrocracking rate on zeolite catalysts is determined only by the level of activity of the metal component. This result is in agreement with the concept of the hydrocracking mechanism on zeolites, involving the paraffin dehydrogenation as a rate-determining step [4, 5]. Hence, the problem of creating of hydrocracking catalysts on zeolites stable to sulphur poisoning is reduced to fixing a metal with a high hydro-dehydrogenation activity in the sulfided state. As reported in [1 - 3] this may be reached by two ways: i) fixing on a support of a highly dispersed metal component; that is structural promoting, or ii) using of supports effecting the metal electron state   ; that is electron promoting. In the former case the coefficient α, which characterizes the hydrogenation rate fall before and after the sulfuration, should be constant, in the latter case, it should depend upon the support nature.

TABLE 1

The hydrogenation activity and the stability to sulphur poisoning of nickel-containing catalysts. $T_{reac.}$ = 200°C.

No	Sample	Na %	%Me	$K^o_{hydr.}$ mmole l/g	$K^s_{hydr.}$ Me sec.	α %
1	Ni-black	–	100	0,20	0,020	7,6
2	Ni/O$_2$	–	60	1,50	0,060	3,6
3	NiaY	4,0	5,4	14,5	0,5	3,7
4	NiHY-1	0,4	8,0	9	0,34	3,7
5	NiHY-2	0,4	12,0	7,4	0,27	3,6
6	NiH-klinoptilolit	–	3,0	15,00	0,550	3,8
7	NiH-mordenite	0,6	2,9	12,0	0,440	3,7
8	NiH-mordenite	0,6	3,2	33,70	1,300	3,9

As is seen in Table 1, the nickel black is two times more stable to sulphur poisoning than nickel in the supported state. The stability of the latter to the influence of sulphur compounds depends neither on the support type, nor on the sodium concentration therein.

The coefficient of the stability to sulphur poisoning of nickel, fixed on mordenite, is the same as other catalysts have, but its hydrogenation activity in the sulfided state is essentially higher.

Platinum-containing catalysts (Table 2) are two times more stable to sulphur poisoning than nickel catalysts. The stability of all the platinum-containing samples studied is also independent of the support nature, metal content in the catalyst and sodium content in the zeolite. Moreover, no difference has been found between the coefficients of the stability to sulphur poisoning for the platinum black and the supported metal.

The stability of palladium catalysts to sulphur poisoning (Table 3) is the same for all the samples in study. The only exception is palladium, supported on alumina and silica characterized by somewhat higher value of stability coefficient. Thus, from all the above one may conclude that the change in the support acidity by varying the decationation and dealumination degree of zeolite (up to module 6), the use of various types of zeolites, variations in calcination temperature of samples in oxygen don't affect the stability to sulphur poisoning.

TABLE 2

The hydrogenation activity and the stability to sulphur poisoning of platinum-containing catalysts. $T_{reac.}$ = 200°C.

No	Sample	%Na	%Me	$K^o_{hydr.}$ mmole.l g.Me sec	s m^2	$K^s_{hydr.}$ mmole.l g.Me sec	d%
1	P-black	—	100	0,4	3	0,03	8,1
2	P/Al$_2$O$_3$	—	1,0	66,5	480	5,70	8,6
3	P/SiO$_2$	—	1,0	31,3	230	3,10	9,9
4	PNaY	5,9	3,3	4,2	30	0,5	9,9
5	P HY-1	0,9	0,5	22,2	160	2,20	9,9
6	P HY-2	0,9	1,0	15,0	110	1,1	7,2
7	P HY-impregnation	0,9	1,0	11,0	80	0,9	8,0
8	P- aH-klinoptilolit	—	1,5	21,0	150	1,5	7,4

TABLE 3

The hydrogenation activity and the stability to sulphur poisoning of palladium containing catalysts. $T_{reac.}$ = 200°C

No	Sample	%Na	%Me	$K^o_{hydr.}$ mmole.l g.Me sec	s m^2	$K^s_{hydr.}$ mmole.l g.Me sec	d%
1	Pd-black	—	100	2,0	5	0,11	5,5
2	Pd/Al$_2$O$_3$	0,05	1,0		390		8,5
3	Pd/SiO$_2$	—	1,0	76,5	190	7,60	9,9
4	PdNaY/300°,O$_2$/	3,22	1,6	41,6	100	2,3	5,6
5	PdNaY/500°,O$_2$/	3,22	1,6	53,0	130	2,3	4,4
6	Pd HY-1	0,9	0,6	103,0	250	5,10	4,9
7	Pd HY-2/300°,O$_2$/	0,8	1,2	50,0	120	2,9	5,9
8	Pd HY-2/500°,O$_2$/	0,8	1,2	44,0	110	2,4	5,4
9	Pd HY-3	0,7	3,1	28,0	70	1,6	5,6
10	Pd HY-dealum. SiO$_2$/Al$_2$O$_3$ = 6,1	0,6	1,0	125,0	305	5,10	4,1
11	PdNaH-klinoptilolit	—	1,0	128,0	320	7,30	5,7
12	PdH-mordenite	0,6	0,6	168,0	410	8,0	4,7

At the same time, the hydrogenation activity of metals fixed on the zeolite acid sites, mainly on mordenites, is higher both in the initial and sulfided states. This latter may be due to a higher dispersity of the metal, fixed on these zeolite supports.

In Ref. [1 - 3] it is supposed, that in decationated and dealuminated Y-type zeolites, the platinum and palladium species are transformed to the electrondeficient state as a result of their interaction with the zeolite acid sites. This is equivalent to the supposition that the acid sites of zeolites are electron promoters of metals. As evidenced by our results, such an interaction with acid sites of support (even if it does occur) does not increase the metal stability to sulphur poisoning.

As shown in [5,6], the metal electron deficiency, caused by the interaction with the support, decreases in the presence of reaction media due to the interaction in the system reagent-metal. The metal-support interaction takes place only in the absence of the reaction medium components. It is likely that the interaction with the sulphur containing reagents must determine the total state of the supported metal even on the most acid zeolites.

Hence, we may conclusively state that the increase in the support acidity leads to the growth of the metal stable dispersity, but does not make the catalyst more stable to sulphur poisoning.

Thus, the zeolite support is unlikely to be an electron promoter of metals, but it favours the distribution of the metal in high degree of dispersity and, therefore is likely to be a structural promoter.

REFERENCES

1   R.A. Della Betta, M. Boudart, Proc. 5-th Intern. Congress on Catalysis, 2, 1972, 1329 pp.
2   J.A. Rabo, V. Shomaker, P.E. Pickert, Proc. 3-ird Intern. Congress on Catalysis, Amsterdam, 2, 1965, 1264 pp.
3   M.V. Landau, V.Ya. Kruglikov, N.V. Goncharova, Kinetika i Kataliz, 17 (1976) 1281.
4   P.N. Kuznetsov, D.M. Anufriev, K.G. Ione, Kinetika i Kataliz, 19 (1978) 1520-1526.
5   V.N. Romannikov, K.G. Ione, L.V. Orlova, P.A. Zhdan, Proc. IV-th Soviet-French Seminar on Catalysis, Tbilisi, 1978.
6   K.G. Ione, V.N. Romannikov, A.A. Davydov, L.B. Orlova, J. Catal., 57 (1979) 126-135.

B. Imelik *et al.* (Editors), *Catalysis by Zeolites*
© 1980 Elsevier Scientific Publishing Company, Amsterdam — Printed in The Netherlands

SULFUR RESISTANCE OF MODIFIED PLATINUM Y ZEOLITE

TRAN MANH TRI, J. MASSARDIER, P. GALLEZOT and B. IMELIK
Institut de Recherches sur la Catalyse, C.N.R.S. - 2, avenue Albert Einstein
69626 VILLEURBANNE Cédex.

ABSTRACT

The sulfur resistance of different modified PtNaHY zeolites has been investigated for n-butane conversion. Addition of Ce or Mo to PtNaHY zeolites leads to catalysts showing an enhanced sulfur resistance. On PtNaHY and PtCeY zeolites, it has been evidenced that sulfur resistance is related to the electronegative character (e.c.) of platinum. In PtMoY zeolite, the electronic state of Pt is not changed and the sulfur resistance has been attributed to a synergetic effect due to Pt-Mo associations.

1   INTRODUCTION

Poisoning of Pt catalysts by sulfur is an important problem in a number of catalytic processes. Rabo et al. (ref. 1) were the first to point out the sulfur resistance of platinum in Y zeolite and previous experiments in this laboratory have also evidenced that the sulfur resistance of Pt is strongly dependent on the electronic structure of metal (ref. 2). Therefore, it seems attractive to prepare more suited metal catalysts by altering their electronic properties, which may induce both enhanced activity of structure sensitive reactions such as hydrogenolysis and an increased sulfur resistance. Such modifications of electronic properties for a given metal may be induced when the particle size and/or the particle environment including the support are modified. On platinum Y zeolites, homogeneous states of dispersion in a large range of particle size may be obtained and the properties of the support may be adjusted by changing the charge compensating cations. In this work, the sulfur resistance of different modified Pt Y zeolites has been investigated in the n-butane hydrogenolysis. The influence of the particle size and of the addition of cerium and molybdenum on the poison sensitivity of the catalysts is discussed.

2   EXPERIMENTAL

The PtNaHY zeolites are obtained by ions exchanging a NaY zeolite (Linde SK 40) in ammonia solutions of $PtCl_2$ (refs. 1, 3). The platinum zeolites were treated either by heating at 650 K under O_2 followed by reduction at 600 K under H_2 which leads to 1 nm Pt particles fitting in the zeolite supercage (1 nm PtY zeolite) or by heating at 820 K under O_2 and reduction under H_2 at the same temperature which gives rise to 2 nm PtY zeolites occluded in the zeolite bulk (ref. 4). The PtCeY zeolites were prepared in the same way as the PtNaY zeolites from CeY zeolites already obtained by ion exchange with water solutions of $Ce(NO)_3$. The PtMoY zeolites are issued from PtY containing 1 nm Pt particles left in contact at 360 K with a known amount of $Mo(CO)_6$ vapor which are then decomposed under H_2 at 580 K. Well dispersed Pt/SiO_2 catalyst (60 % dispersion) has also been studied. The different catalysts are presented in Table 1.

Reactions were carried out at about 600 K in a conventionnal flow reactor operating in the differential mode at low conversion. Samples were reactived in flowing H_2 at 600 K for about 3 h or 16 h with PtMoY. H_2S (1000 ppm of H_2S in H_2) was introduced on the catalyst by successive pulses of about 5 cm^3 either at room temperature or near the reaction temperature (600 K).

3   RESULTS

As observed from the electron microscope photographs, the presence of Ce^{3+} ions exchanged before reduction favors the formation of smaller particles with a size distribution centered at 0.8 nm instead of 1 nm for the PtNaY zeolites. At the reverse, the decomposition at 580 K of $Mo(CO)_6$ left in contact with 1 nm PtY zeolite leads to 2 nm particles which remain occluded in the zeolite bulk.

The reaction of n-butane (nC_4) on Pt catalysts gives hydrogenolysis and isomerization with the respective turnover frequencies N_H and Ni. The poisoning of catalysts is carried out by flowing 5 cm^3 of diluted H_2S under H_2 (3 $l.h^{-1}$) during 0.25 h. It is assumed that the whole amount of poison is left on the metal since the poison adsorption energy on SiO_2 or Al_2O_3 is significantly lower than on metal (ref. 5). As a matter of fact, the adsorption of S on platinum was monitored in a previous study (ref. 2) by the shift of ν_{CO} frequency of coadsorbed CO.

The relative activity N_H/N_H^o, N_H^o and N_H being respectively the hydrogenolysis activity for unpoisoned and poisoned catalysts, has been plotted versus the number of H_2S molecules introduced on the catalysts at room temperature (see Fig. 1) or at about 600 K (see Fig. 2). The decrease of activity is linear for the first doses but an inflexion occurs subsequently The effect of a concentration C of the poison on the activity is represented by $N_H = N_H^o$ (1-αC) up to the inflexion. According to Maxted (ref. 6), the α coefficient allows the comparison of the relative sensitivity of each sample to H_2S poisoning. It is observed that the poisoning coefficient α decreases in the sequence : 2 nm PtY > 2 nm Pt/SiO_2 > 1 nm PtY > 0.8 nm PtCeY > 2 nm PtMoY (see Table 1). Moreover, it may be noticed that the poisoning efficiency at room temperature is slightly weaker than the poisoning at higher temperature, probably because of a partial desorption of the sulfur by the hydrogen treatment during the increase of the temperature up to the reaction temperature (600 K).

The influence of sulfur poisoning on the selectivities of the various catalysts has also been determined. The nC_4 hydrogenolysis involves three reactions : $nC_4 \rightarrow C_1 + C_3$ (1), $nC_4 \rightarrow 2 C_2$ (2) and $C_3 \rightarrow C_2 + C_1$ (3) (turnover frequencies N_1, N_2, N_3 respectively). Table 2 gives the ratios Ni/N_H (isomerization selectivity) and Ne + N_3/N_H (selectivity in ethane formation) before and after poisoning. It is noteworthy that no significant change in selectivity occurs.

4   DISCUSSION

The poison sensitivity of each sample, except for the particular PtMoY catalyst, is strongly dependent on the metal particle size : the smaller the particle size, the higher the sulfur resistance. It is also strongly dependent on the particle environment, since acid sites and multivalent cations increase the resistance to poison. This has already been observed for benzene hydrogenation, the 1 nm PtY zeolite completely poisoned by H_2S or

TABLE I
Nomenclature, particle size and α coefficient of the differents catalysts.

Sample	Particle size (nm)	$\alpha^{(1)}$	$\alpha^{(2)}$
10.9 % Pt Na HY	1	6.2	8.7
10.9 % Pt Na HY	2	10	
5 % Pt 1 % Ce Y	0.8	4.5	6.3
10.9 % Pt 3.5 % Mo Y	2	2.2	2.4
6 % Pt/SiO$_2$	2	7.2	

α is expressed in number of Pt atoms poisoned by 1 molec. of H$_2$S.
(1) H$_2$S is adsorbed at room temperature
(2) H$_2$S is adsorbed at reaction temperature

TABLE II
Relative selectivities of the different catalysts

Sample	N_2+N_3/N_H			Ni/N_H		
	a)	b)	c)	a)	b)	c)
1 nm 10.9 % Pt Na HY	0.28	0.25	0.3	0.15	0.16	0.14
2 nm 10.9 % Pt Na HY	0.20	0.20		0.35	0.45	
0.8 nm 5 % Pt Ce Y	0.29	0.27	0.26	0.07	0.04	0.1
2 nm 10.9 % Pt 3.5 % Mo Y	0.45	0.48	0.46	0.03	0.03	0.03
2 nm 6 % Pt/SiO$_2$	0.33	0.36		0.53	0.5	

a) fresh catalyst
b) H$_2$S poisoned catalyst : H$_2$S adsorption at room temperature
c) H$_2$S poisoned catalyst : H$_2$S adsorption at reaction temperature

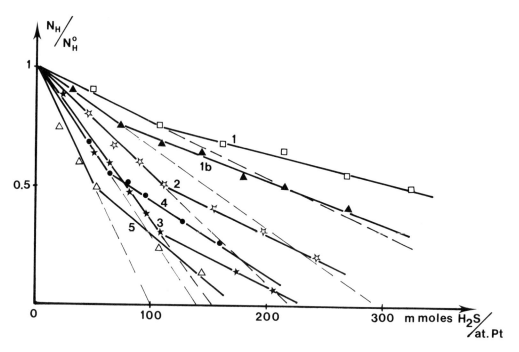

Fig. 1. Sulfur poisoning of the catalysts, after H_2S adsorption at room temperature : curves (1) : 2 nm Pt Mo Y, (2) 0.8 nm Pt Ce Y, (3) 1 nm Pt Na HY, (4) Pt/SiO$_2$ (5) 2 nm Pt Na HY. Curve 1b correspond to 2 nm Pt Mo Y with poisoning reported to mole H_2S/ at (Pt + Mo).

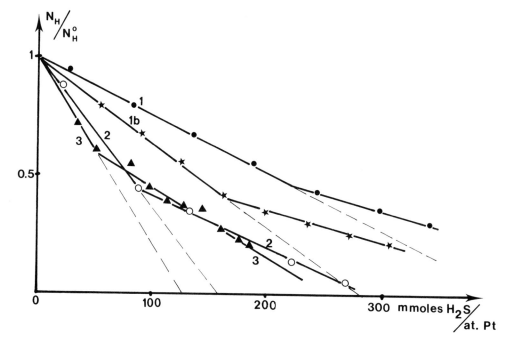

Fig. 2. Sulfur poisoning of the catalysts after H_2S adsorption near reaction temperature : curves (1) 2 nm Pt Mo Y, (2) 0.8 nm Pt Ce Y, (3) 1 nm Pt Na HY. Curve 1b as for Fig.1.

thiophen were partially regenerated at lower temperatures than the catalysts with higher particle size (ref. 2). This dependency of the sulfur resistance can be related to the electronic properties of platinum. Thus the electronic structure of platinum in the different catalysts has been investigated by X-ray absorption edge spectroscopy (refs. 7, 8). It was shown that the electrophilic character (e.c.) of platinum follows the sequence PtCeY > 1 nm PtNaHY > 2 nm PtNaHY (ref. 7). On the electrophilic platinum, there are less electrons available for the charge transfer from Pt toward the electronegative sulfur atoms issued from the H_2S decomposition on the metal. This leads to a weaker bond between S and Pt and an enhanced sulfur resistance on the sample with the most electrophilic platinum (PtCeY). This interpretation still holds if part of the poison is adsorbed on the carrier. Indeed, on the same carrier, the most sulfur resistant samples are those with the larger e.c. Moreover, Pt/SiO_2 and 2 nm PtNaHY zeolite have about the same particle size and similar poisoning curves whereas the nature of the support is very different. Therefore it seems that sulfur adsorption is restricted on the metal at least up to the inflexion point of the poisoning curves. This is in agreement with, the high difference in the adsorption energy of sulfur on the platinum and on the carrier.

With respect to PtMoY zeolites, the greater resistance to H_2S poisoning cannot be attributed to the e.c. of platinum since X-ray absorption edge spectroscopy (refs. 7, 8) has shown that the electronic structure of platinum is not changed. Similar behaviour has been observed on $Ni-Mo/Al_2O_3$ where the sulfur resistance is increased for methanation by the addition of MoO_3 (ref. 9). This has been explained by a greater sulfur adsorption capacity of Mo containing catalysts and the possible participation of sulfided molybdenum sites in the catalytic mechanism. The enhanced activity for nC_4 hydrogenolysis observed on PtMoY zeolite (ref. 7) and the sulfur resistance seems due to a Pt-Mo synergism.

The change in the slope of the poisoning curve is observed on the whole set of catalysts. It can be explained by the presence of a second type of active sites which would be less sensitive to poisoning and less active. Herington and Rideal (ref. 10) have shown by calculations based on random adsorption on a model lattice that the inflexion point can be accounted for on the basis of the requirement of a set of adjacent unoccupied surface atoms, necessary for the poison and/or for the reactant adsorption. Indeed the adsorption of H_2S (refs. 11, 12) and of saturated hydrocarbons on metals (refs. 13, 14) points to the requirement of multiple adsorption sites. Also the question arises whether the active sites remaining on the surface are modified by the sulfur adsorbed on the neighbouring atoms. The fact that the selectivities of the different reactions are not modified by H_2S adsorption indicates that the sites are not drastically changed. However, additionnal measurements of the kinetic orders and of the activation energies must be performed on catalysts poisoned to different extent to ascertain this conclusion.

5   CONCLUSION

The H_2S poisoning of PtNaHY and PtCeNaY zeolites is strongly dependent on the electrophilic character of platinum. The possibility to change this e.c. by the modification of the support by ion exchange allows to prepare more sulfur resistant catalysts. On the Pt-Mo zeolite, the enhanced sulfur resistance cannot be explained by the electronic properties of Pt, it is more probably due to a synergetic effect of the association Pt-Mo. The formation of active molybdenum sulfide as assumed on $Ni-MoO_3/Al_2O_3$ (ref. 9), where a similar

284

effect has already been mentionned would explain this higher sulfur resistance.

REFERENCES

1 J.A. Rabo, V. Schomaker and P.E. Pickert, Proceedings of the 3rd International Congress on Catalysis, North-Holland, Amsterdam, 2, 1965, 1264.
2 P. Gallezot, J. Datka, J. Massardier, M. Primet and B. Imelik, Proceedings of the 6th Int. Congress on Catalysis, Chemical Society, London, 2, 1977, 696.
3 P. Gallezot, Catal. Rev. Sci. Eng., 20, 1979, 121.
4 P. Gallezot, I. Mutin, G. Dalmai-Imelik and B. Imelik, J. Microsc. Spectroscop. Electron. 1, 1976, 1.
5 R.W. Glass and R.A. Ross, J. Phys. Chem., 77, 1973, 2571 and 2576.
6 E.B. Maxted, Adv. in Catalysis, 3, 1951, 129.
7 Tran Manh Tri, J. Massardier, P. Gallezot and B. Imelik, 7th Int. Congress on Catalysis, Japan, in press.
8 P. Gallezot, R. Weber, R.A. Dalla Betta and M. Boudart, Z. Naturforsch., 34 A, 1979, 40.
9 C.H. Bartholomew, G.D. Wheaterbee and G.A. Jarvi, J. Cat., 60, 1979, 257.
10 E.F.G. Herington and E.K. Rideal, Trans. Faraday Soc., 40, 1944, 505.
11 I.E. Den Besten and P.W. Selwood, J. Cat., 1, 1963, 93.
12 J.M. Saleh, C. Kemball and M. Roberts, Trans. Faraday Soc., 57, 1961, 1771.
13 A. Frennet, G. Lienard, A. Crucq and L. Degols, J. Cat. 53, 1978, 150.
14 G.A. Martin, J. Cat., 60, 1979, 345.

B. Imelik *et al.* (Editors), *Catalysis by Zeolites*
© 1980 Elsevier Scientific Publishing Company, Amsterdam — Printed in The Netherlands

EFFECT OF METAL DISPERSITY ON THE ACTIVITY OF ZEOLITE CATALYSTS CONTAINING TRANSITION METALS

N.P.DAVIDOVA, M.L.VALCHEVA, D.M.SHOPOV
Institute of Organic Chemistry, Bulgarian Academy of Sciences,
Sofia III3, Bulgaria

INTRODUCTION

In the present work, the effect of the dispersity of metal particles on the catalytic activity is studied on the reaction of toluene disproportionation on nickel-zeolite catalysts. The latter are obtained by ion exchange (NiCaY and their mixed forms - CrNiCaY and PbNiCaY), by deposition (NiO/CaY), or by mechanical mixing of the constituents (NiO+CaY). The metal dispersity is varied:

- by introduction of Cr(III) or Pb(II) in NiCaY,
- by varying the sequence of ion introduction - Pb(II) in NiCaY, or Ni(II) in PbCaY,
- by varying the manner of ion exchange - consecutive or similtaneous introduction of the ions,
- by varying the way of preparation - ion exchange, daposition, or mechanical mixing of the constituents.

The changes in the dispersity of the metal particles in the zeolite intracristalline structure are followed by the change in the size of the unit cell (X-ray data). The parameter of the unit cell and the average size of the nickel particles are determined in air-dry state, reduced state, state of maximal catalytic activity and after deactivation. The X-ray date for the parameter of the unit cell of samples in different states provide information with respect to the changes taking place during Ni(II) reduction, the mutual effects between the cations in the zeolite framework and the possible changes during the interaction of the active sites with the substrate.

EXPERIMENTAL

The catalysts are obtained on the basis of a synthetic zeolite of Y type with molar ratio $SiO_2/Al_2O_3 = 5$. The Ca-form of the zeolite (CaY) is obtained by ion exchange with Ca(II) from a 4 N solution of $CaCl_2$. The Ni-form (NiCaY) is prepared by a further ion exchange of the Ca-form with Ni(II) from a 0,I N solution of $Ni(NO_3)$. The mixed forms (CrNiCaY and PbNiCaY) are obtained by ion exchange of NiCaY with Cr(III) and Pb(II) from 0,I N solutions of the corresponding nitrates or by ion exchange of

PbCaY with Ni(II) and simultaneous introduction of the Ni(II) and Pb(II) in CaY. The sample NiO/CaY is obtained by deposition of NiO on the calcium form. The mechanical mixtures (NiO+CaY) and (Cr_2O_3+NiCaY) are prepared by mechanical mixing and long grinding of the components. The composition of the catalyst samples is presented on Table I.

TABLE I

Composition of the examined samples

Sample	Composition (wt %)					
	Ca(II)	Ni(II)	Cr(III)	Pb(II)	NiO	Cr_2O_3
CaY	6.6	–	–	–	–	–
NiCaY	5.7	2.3–2.9	–	–	–	–
CrNiCaY	5.2	2.3–2.9	0.4–I.0	–	–	–
PbNiCaY	5.6	2.3–2.9	–	0.8–3.0	–	–
NiPbCaY	5.6	2.3–2.9	–	0.8–3.0	–	–
(NiPb)CaY	5.6	2.3–2.9	–	0.8–3.0	–	–
NiO/CaY	6.6	2.3–2.9	–	–	2.9–3.8	–
NiO+CaY	6.6	2.5–7.5	–	–	3.2–7.5	–
Cr_2O_3+NiCaY	5.7	2.3–2.9	0.7–2.5	–	2.0–7.5	2.0–7.5

The catalytic studies are carried out in a flow system at atmospheric pressure. Pretreatment of the samples, including the the reduction is performed in a hydrogen flow at 450°C for 2 hours (after the temperature is slowly elevated to this value). The conditions of the catalytic experiments are: temperature 330°C, space velocity of I.I h^{-I} and ratio hydrogen:toluene = IO.

The samples for the X-ray analysis are prepared in the following way: after reduction ("red" state), in the state of maximal catalytic activity ("cat" state) and after deactivation ("deact" state) the samples are cooled and passivated with argon.

X-ray analysis is performed with U = 40 kV and CuK$_\alpha$ irradiation. The parameter of the unit cell and the average size of the metal particles are determined as described in [1].

RESULTS AND DISCUSSION

The study of the catalytic activity of nickel-containing samples reveals that when Ni(II) is reduced to metallic state only on the zeolite surface (Fig.I, curves I and 2), only dealkylation takes place and the catalyst is soon deactivated. When Ni(II) ions are present in the zeolite structure (NiCaY), the major part of them is reduced to metallic nickel, which partially migrates and agglomerates on the zeolite surface as nickel crystallites, sized about 300–500 A and partially remains in the intracrystalline structure in form of clusters [2]. The formation of the clusters can be verified using the change of the unit cell parameter in the transition from air dry to reduced state.

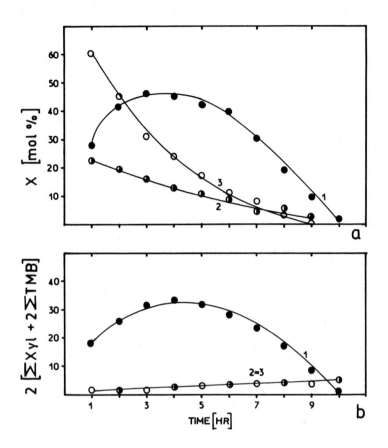

Fig.I. Total conversion extent of toluene (a) and the sum of the disproportionation products (b) as a function of time for the samples:
I. NiCaY; 2. NiO/CaY; 3. NiO+CaY (5 wt % Ni)

The deactivation of the catalysts is primarily related to the reaction
of dealkylation, which, in the cases of samples obtained by deposition, or
by mechanical mixing, takes place on the metallic nickel deposited on the
zeolite surface after reduction. Since dealkylation is a strongly exothermal reaction, in the very beginning of the conversion the temperature is
highly increased, as a result of which additional agglomeration of the
metal particles and a rapid coke formation take place.

The increased selectivity and stability of NiCaY effected by the introduction of Cr(III) and Pb(II), or by the mechanical mixing with Cr_2O_3
(Fig.2) is related to the higher dispersity of the nickel particles both
in the intracrystalline strructure and on the surface [3-5].

The effect of varying the sequence of ion introduction and of varying
the way of ion exchange on the dispersity of metal particles in the zeo-

288

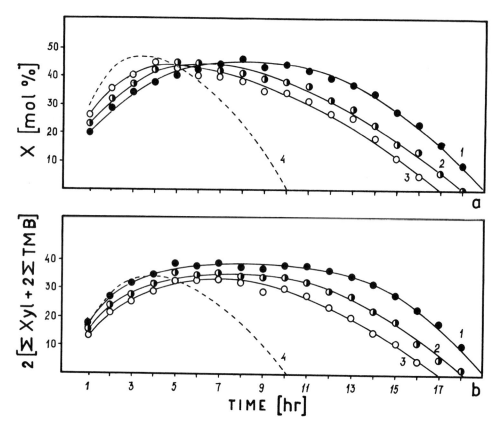

Fig.2. Total conversion extent of toluene (a) and the sum of the disproportionation products (b) as a function of time for the samples:
I. CrNiCaY; 2. PbNiCaY; 3. NiCaY+Cr$_2$O$_3$; 4. NiCaY .

lite intracrystalline face and the influence of these factors on the activity and particularly on the selectivity and stability of the catalytic
action is presented for the Pb containing samples on Table 2 and on Fig.3.

The data presented on Table 2 reveal that the size of the unit cell of
NiCaY significantly increases after reduction due to the formation of metallic clusters in the intracrystalline structure. There is no appreciable
change in this parameter in the mixed forms (CrNiCaY, PbNiCaY and NiCaY+
Cr$_2$O$_3$), which points to a lowered reducibility of Ni(II) and/or a hampered
agglomeration of the metallic nickel particles, as in [6].

The introduction of Ni(II) in the CaY under conditions of reduction
and simultaneous dehydration leeds to an increase in the parameter of the
unit cell, whereas the introduction of Cr(III) and Pb(II) leeds to a decrease. After introduction of Cr(III) and Pb(II) in NiCaY, however, the
parameter of the unit cell remains constant. The observed increase in the
parameter of the unit cell for NiCaY is attributed to the presence of metal particles in the zeolite inner face in the form of "grapes", the size

TABLE 2

Parameter of the unit cell in air dry state, in reduced state, in state of maximal catalytic activity and in deactivated state for the examined samples

Sample	Parameter of the unit cell ($\overset{o}{A}$)			
	Air dry state	"red" state	"cat" state	"deact" state
NiCaY	24.66	24.79	24.77	24.78
CrNiCaY	24.69	24.65	24.66	24.64
PbNiCaY	24.68	24.73	24.73	24.72
NiPbCaY	24.66	24.68	24.65	24.64
(NiPb)CaY	24.67	24.65	24.62	24.6I
NiO/CaY	24.63	24.64	24.66	24.67
NiO+CaY	24.63	24.63	24.65	24.66
Cr_2O_3+NiCaY	24.66	24.68	24.72	24.74

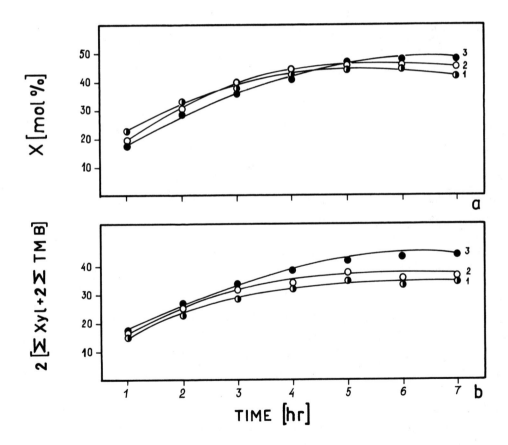

Fig.3. Total conversion extent of toluene (a) and the sum of the disproportionation products (b) as a function of time for the samples:

I. PbViCaY; 2. NiPbCaY; 3. (NiPb)CaY

of which in some cases exceeds that of the large cavities, without breaking the structure [7,8] . These "clusters of grapes" are formed by agglomeration of separate blocks sized about IO Å from the vicinal pores. No
noticeable increase in the parameter of the unit cell is observed for the
mixed forms, which can be explained with the reduced capacity for the formation of large "grapes" of metallic nickel in the inner parts of the
framework due to hampered migration and agglomeration. Similar effect on
the dispersity of metal particles obtained on reduction of Ni containing
zeolites have some other cations, such as Ce(III) [6].

A simultaneous effect of deceleration of the migration and agglomeration of the metal on the zeolite surface is observed (Fig.4), which results
in an increased hydrogenating capacity of the metal (Ni(0) + H_2) and
farther limiting of the coke formation.

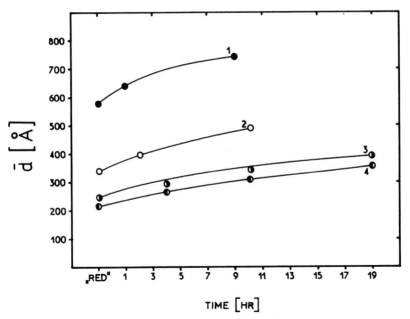

Fig.4. Dispersity of the nickel particles (Å) as a function of time
(hours) for the samples: I. NiO+CaY; 2. NiCaY; 3. NiCaY+Cr_2O_3 and
4. CrNiCaY.

There is no change in the size of the nickel particles in the intracrystalline structure during the reaction and after deactivation, as it
can be seen from the constant size of the unit cell of the samples in the
transition from "red" to "cat" and "deact" states (Table 2). However, the
size of the nickel particles on the zeolite surface increases with time
(Fig.4). This effect is most pronounced for the sample NiO+CaY, which
can be explained with the intense course of the dealkylation and the rapid
coking of the catalyst.

The deactivation of the samples is undoubtedly connected with the coke deposition, leeding to the appearance of a band about 1590 cm^{-I} in the IR spectrum, characteristic for coke-like products [9]. The band intensity increases with time, i.o. in the transition from "cat" to "deact" state (Fig.5a). The amount of coke deposited for a given time period on the mixed forms (Fig.5b) is less than that on NiCaY, which correlates with the observed increase in the selectivity and stability of the mixed forms.

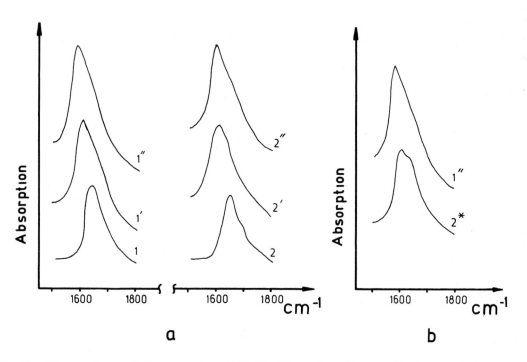

Fig.5. IR spectra of the samples NiCaY (I) and CrNiCaY (2) in "red" state (I,2); "cat" state (I',2'); "deact" state (I'',2") (a) and 8 hours from the beginning of the reaction (b).

On the basis of the obtained results it can be concluded that the dispersity of the metal particles in the intracrystalline structure and on the surface of zeolite catalysts containing transition metals, affects the coke formation processes and provides possibilties for their limitation. Particular attention must be pay to the effect of samples preparation on the dispersity of the metal and on the influence of this factor on the activity and particularly on the selectivity and stability of the catalytic action.

REFERENCES

I   G.B. Bokij and M.A. Poraj-Koshiz, X-Ray Analysis, I, Moscow, 1974,
     44I pp.
2   J.B. Uytterhoeven, Acta Physica et Chemica, 24 (1978) 53-69.
3   N.P. Davidova, N.V. Peshev, M.L. Valcheva and D.M. Shopov, Acta
     Physica et Chemica, 24 (1978) II3-II7.
4   N.P. Davidova, N.V. Peshev, M.L. Valcheva and D.M. Shopov, Mechanism
     of the Catalytic Reactions, I, Moscow, 1978, II pp.
5   M.L. Valcheva, T.P. Aleksandrova, N.P. Davidova and D.M. Shopov,
     Proceedings of the fourth International Symposium on Heterogeneous
     Catalysis, Varna, 1979, 397 pp.
6   M. Briend Faure, M.F. Guilleux, J. Jeanjean, D. Delafosse, G. Diega
     Mariadassou, M. Bureau-Tardy, Acta Physica et Chemica, 24 (1978)
     99-I06.
7   D.A. Agnevskij, V.N. Kvoshopkin, M.A. Kipnis, Supported Metal Catalysts
     for the Conversion of Hydrocarbons, Proceedings of the Allunion Confe-
     rence, Novosibirsk, 1978, 89 pp.
8   G. Dalmai, B. Imelik, C. Leclercq, I. Mutin, J. Microsc. (France),
     20 (1974) I23.
9   D. Eisenbach, E. Gallei, J. Catalysis, 56 (1979) 377.

METAL-ZEOLITES : TRENDSETTERS IN SELECTIVE FISCHER-TROPSCH CHEMISTRY

PETER A. JACOBS

Centrum voor Oppervlaktescheikunde en Colloïdale Scheikunde, Katholieke Universiteit
Leuven, De Croylaan 42, B-3030 Leuven (Heverlee), Belgium

ABSTRACT

The thermodynamic and kinetic limitations upon carbon number distribution in Fischer-
Tropsch synthesis have been discussed. Schulz-Flory polymerization schemes impose
severe limitations upon this selectivity. Deviations from Schulz-Flory kinetics by
secondary reactions on dual-component-catalysts (classical CO reduction function mixed
with a shape selective acidic zeolite) improve the selectivity mainly in the gasoline
carbon number range. The concept of a particle size effect in Fischer-Tropsch chemistry
was derived from the results on RuY zeolites. Increased selectivities are the result
of primary effects at the surface of the active metal. A new polymerization scheme was
developed (Extended-Schulz-Flory model) which fits every possible product distribution
in terms of metal particle size distribution and chain growth possibility. In this
scheme, the kinetic limitations upon product selectivity are much less severe. The
model is consistent with a new mechanism for chain propagation.

1.   NEW INCENTIVES FOR RESEARCH ON SYNTHESIS GAS CONVERSION

The exponential growth of petroleum consumption after world war II has led to a
complete dependence on this source of energy in most sectors of economy and daily life.
Therefore, the shock of the 1973 oil embargo and the ever increasing crude oil prices
since then, have had a profound impact on the western world's way of thinking on
matters of energy and basic chemicals supply. Solid fossil fuels, and particularly
coal, are far more abundant long-term resources than oil and gas (ref. 1) and will have
to be used to an increased extent for the production of fuels and chemicals.

Basically four types of processes can transform coal into fuels or into feedstocks,
which subsequently can be treated by chemical industry : pyrolysis, liquefaction,
solvent extraction and gasification. It seems that synthesis gas processes overcome a
number of drawbacks encountered with the other methods (ref. 2). Moreover, in-situ
coal gasification seems to be near cost effectiveness (ref. 3). Furthermore, if the
probable price development for coal and crude oil is assumed, synthesis gas at competitive
prices is to be expected from coal gasification plants by the early 1990s (ref. 4). This
would form the basis for replacement of at least part of the naphta for olefin and
aromatics production, provided selective processes for synthesis gas conversion are
available at that time. Meanwhile new processes converting syngas to glycol (ref. 5),
acetic acid, ethanol, acetaldehyde (ref. 6,7) have been anounced. The technology for
methanol synthesis is widely applied and can be used in connection with the Mobil

methanol-to-gasoline technology, which in this way constitutes a unique pathway from coal to high quality gasoline (ref. 8).

Classical Fischer-Tropsch (FT) chemistry for the moment only industrially practised at the SASOL plants, is far less selective (ref. 9). The product of such a plant consists of a broad spectrum of hydrocarbons between C_1 and C_{30} and is mainly composed of linear paraffins. The use of such an effluent for fuels and chemical feedstocks requires lengthy and expensive separation procedures. Therefore, improvement of the FT selectivity seems to be a highly timely research goal, of major economical importance.

In the present work, it is aimed to review briefly the advances of FT chemistry achieved during the second research wave (1970-1980). Emphasis will be placed on the processes in which zeolites intervene either directly of through a secondary action. It will be shown that in order to be able to improve reaction selectivity, renovative concepts on the nature of the FT catalysts and the reaction mechanism had to be developed. Product selectivity will be considered mainly in terms of carbon number distribution, irrespective of the chemical nature of the products in each carbon number class. Indeed, provided that the required chemical changes at the catalysts surface can be made, the new concepts derived here are valid irrespective of the chemical nature of the products.

2. STATE OF THE ART IN FISCHER-TROPSCH CHEMISTRY

2.1. Constraints upon FT selectivity imposed by thermodynamics

Typical synthesis reactions of hydrocarbons out of synthesis gas (SYN-gas) are :

$$(2n+1)H_2 + nCO \rightleftharpoons C_nH_{2n+2} + nH_2O \tag{1}$$

$$2nH_2 + nCO \rightleftharpoons C_nH_{2n} + nH_2O \tag{2}$$

Two important side reactions are the Boudouard and the watergas shift (WGS) reactions respectively :

$$2CO \rightleftharpoons CO_2 + C \tag{3}$$

$$CO + H_2O \rightleftharpoons CO_2 + H_2 \tag{4}$$

Using available thermodynamic data, standard free energy changes per carbon atom can be calculated for reactions (1) and (2) in the temperature range of interest (150-450°C) (ref. 10,11). It follows that :

- the equilibrium conversion will increase with pressure; for catalysts with low activity, the conversion will eventually be limited at high reaction temperatures.
- paraffins are preferred over olefins for a given chain length and over the whole temperature range (alcohols are less stable than the corresponding paraffins).
- methane is always the preferred alkane.
- the stability of paraffins decreases with increasing chain length (for alcohols the opposite is true).
- only the olefin selectivity is temperature dependent : light olefins are the preferred products only at high reaction temperatures.

Therefore, the selective synthesis of either higher paraffins or lower olefins will totally rely upon the catalyst. Also if relatively high amounts of methane have to be avoided, high catalyst selectivity is a prerequisite.

2.2. Kinetic limitations upon FT selectivity

All proposals for a FT mechanism describe hydrocarbon formation by a polymerization scheme. In such scheme, one carbon at the time is added to the growing chain on the catalyst surface and it is assumed that the growth constants are independent on carbon number. Such a simple chain growth (SCG) scheme was first advanced by Anderson (ref. 12,13). Recently a simplified form of the SCG scheme was presented as the so-called "Schulz-Flory (SF) polymerization" mechanism (ref. 14). Mathematically this is translated as follows :

$$W_n = n \, \alpha^{n-1} \, (1-\alpha)^2 \qquad (5)$$

W_n is the product weight fraction of carbon number n; α is the chain growth probability, which can be derived from a plot of log W_n/n against n.

A high number of published FT product distributions can be fairly accurately described by this formalism (Table 1).

TABLE 1.
FT product distributions described by SF kinetics.

Catalyst			Products	α	Applicable in carbon number range	Ref.
Support	Metal	Promotor				
kieselguhr	Co	ThO_2	hydrocarbons	0.54	C_5-C_{13}	15
-	Fe	K_2CO_3	hydrocarbons	0.51	C_1-C_{14}	16
-	1Fe+0.1Cu	$K_2O+K_2SiO_2$	hydrocarbons	0.46	C_2-C_{11}	13
Al_2O_3	Fe_3O_4	K_2O	alcohols	0.36	C_1-C_6	14
-	Fe_3O_4	-	alcohols	0.40	C_1-C_6	17
silica	Ru	-	hydrocarbons	0.90	C_1-C_{12}	18
garnierite	Ni	-	hydrocarbons	0.30	C_1-C_4	19
(benzene)	$W(CO)_6$	$AlCl_3$	side chain in alkylbenzenes	0.17	C_1-C_5	20

It follows that distributions for several classes of products obtained in various conditions over the most common active metals are consistent with the SF formalism. Deviations at the lower carbon numbers are most pronounced over Co. An entirely satisfactory explanation for these deviations have never been advanced (ref. 21). A number of implications follow from such kinetic behaviour :

- the product distribution is determined only be one physical parameter = the chain growth probability.
- when the concentration of one component is determined, all others are fixed. In this way, Dry (ref. 9) for Sasol experimental data plotted the selectivities for various product cuts against methane or hard wax selectivity and encountered maxima in these curves. The maximum selectivity from SF kinetics is compared to these data in Table 2. Except for the C_2 fraction the agreement is again excellent.

TABLE 2.

Maximum selectivity W_n^m (%) for FT products according to SF kinetics and from experimental data (ref. 9).

Product	C_2	C_3	C_6-C_{12} gasoline	$C_{13}-C_{18}$ diesel	$C_{24}-C_{35}$ medium wax
calc.	29.6	18.8	43.5	20.9	22.2
exp.	17	17	40	18	22

- for any carbon number a theoretical maximum can be calculated together with the corresponding optimal polymerization degree (see Fig. 1). The maximum selectivity decreases exponentially with the product carbon number.

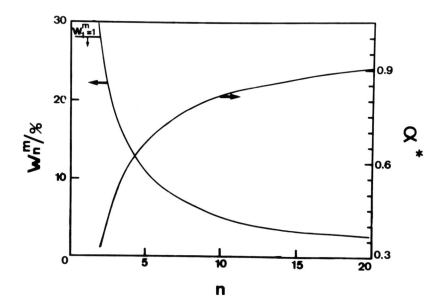

Fig. 1. Maximum yield (W_n^m, weight %) and optimal growth probability (α^{*}) of FT synthesis according to SF kinetics.

The main conclusion is that further improvement of the FT selectivity for fuels or chemical feedstocks (light olefins, aromatics or $C_{15}-C_{20}$ α-olefins for detergents use) will require catalysts which circumvent SF kinetics. This can be done when primary reaction routes are shifted or when secondary effects are exerted such as cracking, hydrogenolysis, steric exclusion, diffusion limitation or a combination of them.

3. NON-SCHULZ-FLORY BEHAVIOUR EXERTED BY SECONDARY REACTIONS : USE OF TWO-COMPONENT CATALYSTS

The primary FT products out of SYN-gas and obtained according to SF kinetics, can be tranformed over a selective catalyst, physically mixed with the FT active solid. This is the new MOBIL approach, in which a CO reduction function is mixed with a shape selective zeolite of the ZSM-type (ref. 21-24). In this way, the selectivity for high-quality gasoline of the acid ZSM zeolite is obtained through secondary conversion of primary FT hydrocarbons or alcohols. The maximum gasoline yield of 40 % predicted by SF kinetics in this way is increased up to 60 % (ref. 26). If the zeolite component of such catalyst is an ultrastable zeolite, a typical cracking product spectrum, rich in aromatics is obtained (ref. 27).

Pertinent data for dual-component catalysts are shown in Table 3. It is a necessity that the activity of both components is tuned. The growth probability on the FT component has to be sufficient, which requires low reaction temperatures, low H_2/CO ratios and high pressures. The cracking activity of the zeolite component should be high enough in order to observe an important secondary effect. This requires reaction temperatures preferentially above 275°C. Typical isosynthesis catalysts (ZrO_2 and ThO_2) operating above 400°C and under a high pressure of SYN-gas (ref. 28) are preferred FT components. Also iron-based FT catalysts can be used. Ru-catalysts are less suitable since rather low synthesis temperatures (below 250°C) are required to avoid high methane yields. Table 2 shows that for Ru and Fe, the secondary selectivity imposed by the HZSM-5 zeolite (high yield of aromatics in the C_5^+ fraction) is less pronounced.

In isosynthesis (ref. 28), high yields of isobutane and isopentane are obtained. In the presence of HZSM-5 zeolite, the product becomes a high quality gasoline (C_5-C_{11}). Other zeolites can also be used. In any case, the FT product will reflect the typical cracking or shape selective pattern exhibited by each particular zeolite. The major influence of the zeolite upon the apparent growth probability of the FT component is given in Fig. 2. The ZSM-5 zeolite strongly reduces the growth probability above C_7. This is for a major part due to secondary shape selective reactions in this zeolite.

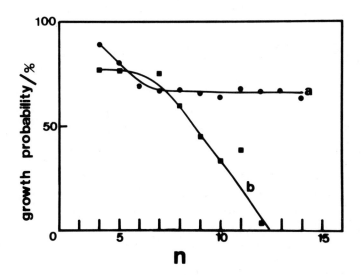

Fig. 2. Growth probability at different carbon numbers for a typical FT liquid effluent over Fe (a) and over a dual component catalyst (b) consisting of the same FT catalyst and a HZSM-5 zeolite (data derived from ref. 26).

298

TABLE 3.

Effect on product selectivity of zeolite addition to a FT catalyst.

No.	1	2	3	4	5	6	7	8	9[a]	10[b]
Catalyst	$1ZrO_2$ + $1SiO_2$	$1ZrO_2$ 1HMOR	$1ZrO_2$ + 1HZSM	$1ThO_2$ 1HZSM	$1ThO_2$ + 1 Erionite	$1ThO_2$ + 1REY	$0.05RuC_2$ + 0.95HZSM	0.22Fe + 0.78HZSM	Zn + Cr + HZSM	HZSM-5
React. temp., °C	427	427	427	427	427	427	294	316	427	371
Press., atm.	83	83	83	84	84	84	77	14	83	1
H_2/CO	1	1	1	1	1	1	2	2	1	–
CO conv. (%)	14	10	19.6	11.8	11.3	5.3	90	83	57	100
CO_2 (wt %)	5.8	1	4.3	–	–	–	3.8	–	39.4	–
Prod. distrib. (wt %)										
C_1	24.8	44.9	1.6	17.3	13.7	42.8	26.1	38	3.9	1.0
C_2	4.7	9.1	6.8	28.2	49.6	17.1	3.4	14	13.1	1.1
C_3	2.5	2.5	6.0	26.9	20.9	10.9	3.2	7	22.9	17.2
C_4	9.5	4.6	0.8	20.0	14.4	26.7	7.1	8	15.5	25.6
C_5^+	24.0	34.8	84.8	8.9	1.4	2.5	60.4	33	44.5	55.1
MeOH + Me_2O	34.5	4.1	0	0	0	0	0	0	< 0.1	75.0
% arom. in C_5^+	46.0	94.5	99.8	42.0	0	68.0	19.1	28	75.6	75.0
Ref.	21	21	21	22	22	22	23	23	21	25

[a] ZnO, 16, Cr_2O_3, 44, HZSM, 40 %.
[b] methanol as feed.

Most of the selectivity changes observed upon tuning the two components by changing operational parameters are understood by the separate action of both components. Illustrative data are available (ref. 23). The following observations are consistent with FT behaviour :

- increase of CO conversion and (C_1+C_2) yield with pressure
- decrease in C_5^+ and increase in C_1+C_2 content with increasing reaction temperature, decreasing space velocity and increasing H_2/CO molar ratio.

The increased yield of C_5^+ aromatics at higher reaction temperatures and lower space velocities can be explained by the action of the ZSM-5 zeolite.

The flexibility of such processes, in casu of the SYN-gas to gasoline conversion can be increased by the use of two separate reactors containing a classic FT catalyst and a ZSM-type zeolite respectively. Ample data illustrate this (ref. 29). However, the results of table 2 (no. 1 and 3) show that upon mixing the isosynthesis with the ZSM component, CO conversion is increased and the C_1 yield suppressed. A logical explanation of these synergistic effects is that a "drain-off" action is exerted by the zeolite to overcome thermodynamic equilibrium (ref. 21). Intermediates formed on the FT component are intercepted by the zeolite and converted to components inert to further chain growth (ref. 25,26). These low methane yields are not found for Fe and Ru mixed with ZSM-5. This is understood if for methane formation and chain growth two discrete species are responsible. Indeed, methane formation over Fe and Ru seems to occur preferentially via surface carbides while chain growth occurs via a = CHOH surface intermediate (see later). The latter one seems to be spilled over to the zeolite and stabilized there. Methane once formed by the surface carbide mechanism is inert to both catalyst components. The exact mechanism remains obscure but the intermediate could be the same as for the methanol decomposition on HZSM-5 as suggested from the very similar product distribution (Table 3).

Dual-component catalysts consisting of a methanol synthesis function and an acid zeolite are also advantageous. In this way the primary synthesis product is partially dehydrated to dimethylether. Consequently, a higher SYN-gas conversion is reached with a gas less rich in hydrogen (ref. 30). This broadens the operation conditions of the methanol synthesis catalyst. If zeolite ZSM is used again as acidic component a useful hydrocarbon mixture is obtained at once (ref. 31).

4.   NON-SF BEHAVIOUR DUE TO PRIMARY EFFECTS

Zeolites Y with Ru (ref. 32) and Fe (ref. 33) in the inner cages, also are very efficient in limiting the size of the FT products. The deviations from SF kinetics are analogous to those reported for Fe/ZSM (Fig. 2). The growth probability suddenly drops around C_9-C_{10} (ref. 32). The RuY and Fe/ZSM catalysts are distinctly different in two aspects :

- the metal loading is lower in case of Ru : 2-6 wt % against 25 % for the two-component catalysts.
- Ru metal is encaged inside the Y zeolite crystals, while for the Fe/ZSM system only physical mixing is reached.

A detailed study of the Ru-Y zeolite system resulted in the development of renovative concepts in FT chemistry which will allow to design tailor-made selective FT catalysts. This will be developed in the following paragraphs.

4.1. Ruthenium metal encaged in zeolite Y

The behaviour of Ru in zeolite Y proved to be vital to the interpretation of FT results. A detailed and complete description of the parameters controlling the physical state of Ru in faujasites has been published (ref. 34). By a combination of physical and chemical techniques the following information was obtained :

i.   Decomposition of ruthenium(III)hexammine exchanged in zeolite Y in vacuo or in an inert atmosphere leads to autoreduction :

$$Ru(NH_3)_6^{+3} \longrightarrow Ru^{\circ} + 3N_2 + 3H^+ + 3/2H_2 \tag{6}$$

ii. Metal clusters obtained in this way are pyrophoric and accommodated in the super-cages.

iii. After reduction in the presence of water a bimodal metal particle size distribution is found : the smaller particles are inclosed in the supercages, the larger ones (3-4 nm) are accommodated in holes in the zeolite crystal, allegedly created by the hydrolyzing action of water upon the aluminosilicate structure.

iv. Metal migration out of the zeolite crystal occurs under influence of oxygen.

v.   Temperature programmed oxidation (TPO) of the Ru-metal phase is a convenient technique to determine the metal particle size distribution of Ru in the zeolite : larger particles are oxidized at increasing temperatures, Ru in the supercages (1.4 nm) is oxidized at room temperature, the 2.0-4.0 nm clusters in the crystal holes between 150 and 350°C.

4.2. FT carbon number distribution on Ru-Y zeolites

In order to reach maximum selectivity at a given carbon number in the SF kinetic regime, operation conditions and catalysts characteristics have to be such that α is optimum. For a RuNaY-30 zeolite (30 % of the exchange capacity is Ru^{3+}) with average particle size of 4 nm, the influence of reaction conditions is shown in Fig. 3.

For the first two product distributions (a and b), a typical SF distribution with α = 0.3 and 0.6 respectively, is obtained. When reaction conditions are changed as to increase α (c and d), deviations from SF kinetics become more and more important and finally a pronounced selectivity for C_8-C_{10} hydrocarbons is obtained. Therefore it is important on these catalysts to operate at maximum growth probability in order to be able to observe deviations from SF behaviour. The origin of these deviations also has to be different from those obtained for the dual-component catalysts.

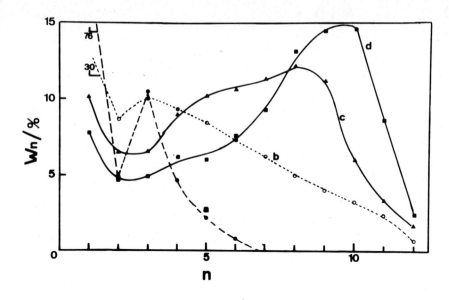

Fig. 3. Influence of reaction conditions on FT carbon number distribution over RuNaY-30; a, 270°C, H_2/CO = 2.5, 15 atm.; b, 250°C, H_2/CO = 2, 15 atm.; c, 200°C, 15 atm., H_2/CO = 2; d, 200°C, 25 atm., H_2/CO = 1.5.

4.2. FT particle size effect on Ru-metal in faujasite-type zeolites

Other size limitations of FT products at high growth probabilities can be obtained on zeolites with increased steam stability (Fig. 4).

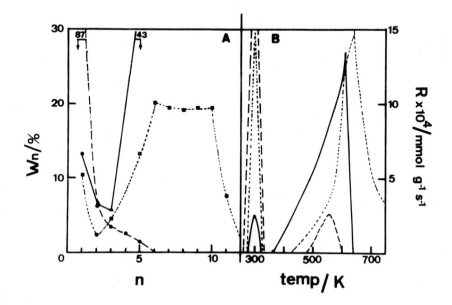

Fig. 4. FT product distribution (A) and TPO (B) curves on 1, ... Ru(20)NaY; 2, ----- Ru(14)La(80)NaY; 3, ——— Ru(14)LaY. (FT conditions : H_2/CO = 1.5; 200°C; 15 atm.). Ref. 35, 36.

The TPO curves show that the Ru-particle size distribution is narrow and approaches mono-dispersion. The average size decreases when the steam stability of the zeolite increases. At the same time the FT product distribution is cut off at shorter carbon numbers. Particle sizes of 1.5, 2.5 and 4.0 nm limit the size of FT products at a chain length of 10, 5 and 1 carbon numbers respectively.

It can be excluded that these deviations from SF kinetics are due to secondary effects of the zeolite :

i.   secondary hydroisomerization and hydrocracking do not occur since at all carbon numbers, the relative isomer distribution on Ru(40)NaY with residual acidity (see equation 6) is very similar to the one on Ru on silica (non-acidic support) (ref. 32).

ii. diffusion cannot explain the differences in FT products of Fig. 5 since porosity of the support, reaction conditions and reaction rate are identical in the three cases.

iii. the effect of secondary cracking on the catalyst terminating at C_5 was determined experimentally (ref. 35). 1-Decene added to the feed in reaction conditions only underwent secondary isomerization.

Therefore, the limit imposed by the Y-zeolite on the size of the FT products is the result of the particle size of the Ru metal, which in its turn is determined by the steam stability of the zeolite or by the pore dimensions of other structures would be used.

4.3. FT synthesis : a "demanding" reaction

It was found (ref. 36) that the average Ru particle size not only influences the maximum carbon number but also the turnover number for CO disappearance. Fig. 5 shows that this number increases smoothly with the average particle size, while the maximum size of the FT products is linearly related to the metal particle size.

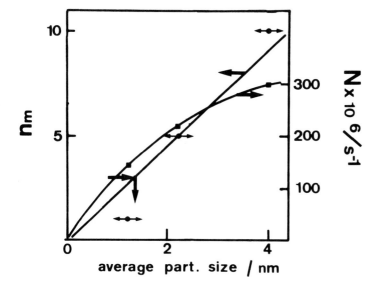

Fig. 5. Relation of the turnover number for CO disappearance (N) and the maximum size of the FT product (n_m) with the average particle size for RuY zeolites (data from ref. 35,36).

Using these data, the turnover numbers can be recalculated as a number of mole of FT product of maximum size desorbed per unit time from the particles having the average size of the near-monodisperse distribution. For the RuY zeolites an average value of 1.07 ± 0.1 is obtained. So independent of metal particle size, one mole of FT product of a given carbon number is synthesized on each particle per unit time. Its carbon number is determined by the size of the metal particle.

4.4. Particle size effect in bimetallic zeolites

When large Ru particles are alloyed with Cu or Ni, smaller surface ensembles are expected. The effect of particle size upon FT product size should be easily visualized in this way. On the other hand, the FT method at high growth conditions should be a gentle method to obtain information upon particle size distribution and in the case of multimetallic systems upon phenomena as surface dilution and metal mixability. Data are published which illustrate this (ref. 37). In Fig. 6 the FT product distributions over Ru(40)Cu(10)Y and Ru(40)Ni(10)Y are shown.

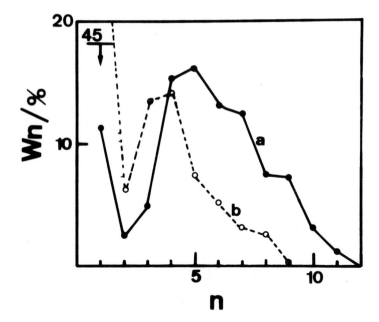

Fig. 6. FT product distribution over a, Ru(40)Cu(10)Y reduced at 500°C (conv. 2.4 %; C_{12}^+ : 1.8 %) and b, Ru(40)Ni(10)Y reduced at 500°C (conv. 11.1 %; C_{12}^+ : 0 %) (ref. 37). (Reaction conditions : Fig. 5).

In both cases the product size is restricted but less abruptly than for the RuNaY samples (Fig. 4). The CO conversion on an equal amount of catalyst also has decreased (30 % on original sample). This is indicative of metal sintering out of the zeolite crystal resulting in the formation of large particles (10-20 nm) (ref. 34). The very low yield of C_{12}^+ (no large particles) can only be reconciled with the decreased activity if surface dilution of Ru by Ni or Cu has occurred and almost no segregation of the two metals exists. The Ru ensembles have an average diameter between 1.7 and 2.5 nm

(Fig. 6 shows that C_4-C_5 hydrocarbons are synthesized over the mentioned particle sizes). The Ni ensembles diluting the Ru surface are active as methanation sites.

4.5. Mathematical formulation of the particle size effect in FT chemistry : the EXTENDED SCHULZ-FLORY (ESF) model

The following concepts are at the basis of this new model (ref. 38) :

i.   the size of a surface ensemble of FT active atoms imposes a strict maximum upon the chain length of the products formed on it (Fig. 6).

ii. the reaction mechanism can be described by a polymerization scheme at the surface of the active particle, the number of polymerization steps is limited by the first condition.

iii. at any moment only one molecule is growing on each particle (section 4.3.).

iv. a linear relationship exists between particle size and the maximum chain length of the hydrocarbon formed on it (Fig. 6).

All this can be quantified when a function for the particle size distribution is available. For non-zeolitic supports the usual skewed Gaussian distribution is used (ref. 39). For metals encaged in zeolite crystals, the first half of an ordinary Gauss curve proved satisfactory. Using numerical methods, the amount of a given carbon number is calculated on each particle. Using the particle size distribution curve, the hydrocarbon distribution in terms of carbon numbers can then be calculated (ref. 38).

The method allows to fit all published carbon number distributions : the minor deviations from SF kinetics as well as the strongly deviating distributions. This is illustrated in Table 4. The product distributions are now dependent upon three parameters : the chain growth probability, the average particle size and the width of the particle size distribution.

In every case very reasonable values of the parameters used to fit the experimental products are obtained. This now explains the bimodal product distributions reported for sulphur-treated Co on Kieselguhr catalysts (ref. 44). Definitely two distributions of particle sizes are present, one probably being formed by sulphur dilution of the Co surface.

It is also clear that not only zeolite-based FT catalysts can produce selectively FT product cuts. The only condition is that a very narrow particle size (or surface ensemble size) distribution is present. On the other hand, only when high chain growth probabilities can be realized, deviations from SF kinetics on such catalysts are observed. The latter of course can be achieved by a proper choice of reaction conditions.

In the ESF model, selectivities of 100 % can be obtained for each carbon number, provided that mono-disperse systems can be prepared. A more realistic prediction of maximum selectivity is given in Fig. 7. Under these conditions selectivity decreases gradually from 57 % at C_2 towards 47 % at C_{15}. If appropriate zeolite frameworks, proper metal loading techniques are used so as to form metal particles of identical sizes in the zeolite cages, this optimum selectivity should easily be approached.

TABLE 4

Parameters in the ESF method used to obtain a confident fit of the FT carbon number distributions.

Catalyst (% metal)	Deviations from SF	Gauss[a] curve	$\bar{D}^b$ (nm)	σ^c	α^d	Ref.
Ru/alumina (1)	(C_9-C_{10}) : max.	skewed	4.0	1.56	0.95	40
Co/alumina (2) (pores : 6.5 nm)	C_4 : max.	skewed	1.8	1.44	0.89	41
id. (pores : 30 nm)	$C_{14}-C_{15}$: max.	skewed	5.4	2.10	0.89	41
RuLaY (1.2)	C_5 : max.	1st half	2.2	1.00	0.85	35,36
RuNaY (2.1)	C_{10} : max.	1st half	4.4	1.80	0.77	35,36
Co/Kieselguhr	C_2-C_5 : too low	skewed	4.0	1.80	0.79	15,17
Mn/Fe precip. oxides	C_3 : max.	skewed	1.2	1.40	0.90	42
$HFe_3(CO)_{11}^-/Al_2O_3$	C_5 : max.	skewed	2.2	1.80	0.90	43

a, shape of the particle size distribution curve; b, average particle size; c, square root of the variance of the particle size distribution; d, chain growth probability of FT products.

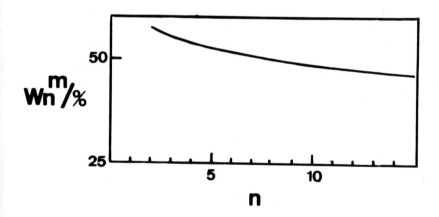

$W_n^m/\%$

Fig. 7. Optimum FT product selectivity (W_n^m) at different carbon numbers using the ESF model, assuming the particle size distribution found in zeolites (1st half of a Gauss curve with $\sigma = 1$) and high growth probability ($\alpha = 0.95$).

5. NEW EVIDENCE FOR THE FT MECHANISM

The ESF model is true, irrespective of the detailed reaction mechanism. The only growth scheme which has to be excluded is the classical insertion mechanism of CH_2 entities in the metal-alkyl chains. New evidence on the nature of the surface intermediates has been found in the transient reaction conditions before the catalyst has reached steady state operation (ref. 45). The following new observations show that methane formation and chain growth occur via distinct surface intermediates :

i. on a clean FT active metal surface, CO_2 and CH_4 are the primary products.

ii. the FT products obtained over a surface pre-covered with ^{13}C are not enriched in ^{13}C except for C_1.

iii. when in transient conditions the CO_2 concentration decreases, higher hydrocarbons are formed.

iv. water addition accelerates the third phenomenon.

v. phenomenon iii is faster, when the catalyst has higher WGS activity (or generally lower support acidity).

This suggests that methane formation occurs via surface carbide, which is generally accepted now (ref. 46). The intermediate for chain growth is most probably oxygen containing. This explains why only on typical FT active metals C-O-H bands were found, but never on methanation catalysts as Ni (ref. 47). A more thorough argumentation on all this can be found in the original paper (ref. 45). The chain of events in FT chemistry is now presented as follows :

$$CO \rightleftharpoons C_{ads} + O_{ads} \tag{7}$$

$$C_{ads} + 2H_2 \rightleftharpoons CH_4 \tag{8}$$

$$O_{ads} + H_2 \rightleftharpoons H_2O \tag{9}$$

$$C_{ads} + H_2O \rightleftharpoons (CHOH)_{ads} \tag{10}$$

$$2(CHOH)_{ads} + H_2 \rightleftharpoons (C_2H_4OH)_{ads} + H_2O \tag{11}$$

CO dissociation is a slow step in the succession of events, since at high growth rates oxygenated species are abundant on the surface (ref. 46). The metal particle size effect can be understood if hydrocondensation of oxygen-containing C_1 species occurs as follows :

This gives 1-alcohols or 1-olefins as final products depending upon support acidity or the hydrogenating power of the metal (ref. 45). The intermolecular dehydration of CHOH species should also be more rapid than the intramolecular chain dehydration or the subsequent product desorption.

6. CONCLUSIONS

With the new view on the Fischer-Tropsch selectivity and mechanism presented here, the expectations can revive that highly selective catalysts for industrial use can be manufactured. However, to achieve this goal a series of practical problems at the catalyst manufacturer-level have to be overcome : the preparation of a quasi-monodisperse metal particle size distribution has to be realized and stabilized against hydrothermal working conditions.

The encagement of these particles in a stable manner in zeolite cages seems the most likely route to prepare selective FT catalysts. Perfect hydrothermal stability of the zeolite framework is a prerequisite. The limited amount of metal which on a weight basis can be stabilized this way, could impose a limit on the overall activity, although in the highly exothermic synthesis, this may turn out to be advantageous.

ACKNOWLEDGMENTS

The author acknowledges a research position as "Bevoegdverklaard Navorser" from
N.F.W.O. (Belgium). I am indebted to Dr. H.H. Nijs which was at the basis of most of the
experimental work and of much of the ideas. The continuous interest of Prof. J.B.
Uytterhoeven in this work is also acknowledged. I am grateful to the Belgian Government
(Diensten Wetenschapsbeleid) for the material help in the realization of this work.

REFERENCES

1   P.H. Spitz, Prepr. Div. Petrol. Chem. ACS, 21 (1976) 562-572.
2   B.M. Harney and G.A. Mills, Hydrocarb. Process., Febr. (1980) 67-71.
3   Anonym., Hydrocarb. Process., Dec. (1979) 17.
4   L. Bornhofen, Hydrocarb. Process., March (1980) 34J-P.
5   K. Weissermel and H.J. Arpe, Industrial Organic Chemistry, Verlag Chem. 1978,
     p. 135-136.
6   ibid., p. 158.
7   I. Wender, Catal. Rev. Sci. Eng. 14 (1976) 97-129.
8   J.J. Wise and S.E. Voltz, ERDA Contract No. E(49-18)-1773.
9   M.E. Dry, Ind. Eng. Chem., Prod. Res. Dev. 15 (1976) 282-286.
10   D.R. Stull, E.F. Westrum and G.C. Sinke, The Chemical Thermodynamics of Organic
     Compounds, J. Wiley, 1969.
11   Von E. Christoffel, I. Swijo and M. Baerns, Chem. Zeit., 102 (1978) 19-23.
12   H.H. Storch, N. Columbic and R.B. Anderson, The Fischer-Tropsch and related
     syntheses, Wiley 1951, p. 585-591.
13   R.B. Anderson, Catalysis, 4, P.H. Emmett, Ed., Reinhold, 1956, p. 585-591.
14   G. Henrici-Olivé and S. Olivé, Angew. Chem. Int. Ed., 15 (1976) 136-142.
15   H. Pichler and G. Krüger, Herstellung Flüssiger Kraftstoffe aus Kohle, Metro-Druck
     (Bonn) 1973.
16   A.W. Weitkamp, H.S. Seelig, N.J. Bowman and W.E. Cady, Ind. Eng. Chem. 45 (1953)
     343-347.
17   H. Pichler, H. Schulz and M. Elstner, Brennst. Chem. 48 (1967) 78-85.
18   H.H. Nijs, P.A. Jacobs and J.B. Uytterhoeven, J.C.S. Chem. Comm. (1979) 180-181.
19   P.A. Jacobs, H.H. Nijs and G. Poncelet, J. Catal. (1980) to be published.
20   G. Henrici-Olivé and S. Olivé, Angew. Chem. 91 (1979) 83-84.
21   C.D. Chang, W.H. Lang and A.J. Silvestri, J. Catal. 56 (1979) 268-273.
22   C.D. Chang, W.H. Lang and A.J. Silvestri, U.S. Patent 9,086,262 (1978).
23   W.O. Haag and T.J. Huang, U.S. Patent, 9,157,338 (1979).
24   C.D. Chang, W.H. Lang and A.J. Silvestri, Belg. Patent 463,711 (1974).
25   C.D. Chang and A.J. Silvestri, J. Catal. 47 (1977) 249-259.
26   P.D. Caesar, J.A. Brennan, W.E. Garwood and J. Ciric, J. Catal. 56 (1979) 274-278.
27   W.H. Seitzer, U.S. Patent 4,139,550 (1978).
28   H. Pichler and H.H. Ziesecke, The Isosynthesis, Bureau of Mines Bulletin (1950)
     488.
29   See U.S. Patents (1977) 4,041,094; 4,041,096; 4,044,063; 4,045,505; 4,049,741;
     4,052,477; 4,046,831; 4,046,829; 4,046,830; 4,059,648 assigned to Mobil Oil
     Corporation.
30   J.C. Zahner, U.S. Patent 4,011,275 (1977).
31   C.D. Chang, W.H. Lang, A.J. Silvestri and R.L. Smith, U.S. Patent 4,096,163 (1978).
32   H.H. Nijs, P.A. Jacobs and J.B. Uytterhoeven, J.C.S. Chem. Comm. (1979) 180-181.
33   D. Ballivet-Tkatchenko, G. Coudurier, H. Mozzanega and I. Tkatchenko, Fundam. Res.
     Homog. Catal., in the press.
34   J.J. Verdonck, P.A. Jacobs, M. Genet and G. Poncelet, J.C.S. Faraday I 76 (1980)
     403-416.
35   H.H. Nijs, P.A. Jacobs and J.B. Uytterhoeven, J.C.S. Chem. Comm. (1979) 1095-1096.
36   H.H. Nijs, P.A. Jacobs, J.J. Verdonck and J.B. Uytterhoeven, Proceedings Int.
     Conf. Zeolites, Naples, June 1980.
37   H.H. Nijs, P.A. Jacobs, J.J. Verdonck and J.B. Uytterhoeven, Proceed. Int. Symp.
     "Small Metal Clusters", Lyon, Sept. 1979, Elsevier Scientif., 1980.
38   H.H. Nijs and P.A. Jacobs, J. Catal. 1980, in the press.
39   J. Anderson, Structure of Metallic Catalysts, Academic Press, 1975, p. 364.
40   J. Madon, J. Catal. 57 (1979) 183.
41   D. Vanhove, P. Mucambo and M. Blanchard, J.C.S. Chem. Comm. (1979) 605.
42   H. Kölbel and D.T. Tillmetz, Belg. Patent 837,628 (1976).

43   D. Commereuc, Y. Chauwin, F. Hughes and J.M. Basset, J.C.S. Chem. Comm. (1980) 154–155.
44   R.J. Madon and W.F. Taylor, Adv. Chem. Ser. 178 (1979) 93–111.
45   H.H. Nijs and P.A. Jacobs, J. Catal., to be published.
46   V. Ponec, Catal. Rev. Sci. Eng. 18 (1978) 151.
47   G. Blyholder and L.D Neff, J. Catal. 2 (1963) 183.

B. Imelik *et al.* (Editors), *Catalysis by Zeolites*
© 1980 Elsevier Scientific Publishing Company, Amsterdam — Printed in The Netherlands

Y-ZEOLITES LOADED WITH IRON CARBONYL COMPLEXES : ACTIVITY AND SELECTIVITY IN CO + H_2 CONVERSION

D. Ballivet-Tkatchenko, G. Coudurier and H. Mozzanega

Institut de Recherches sur la Catalyse - C.N.R.S. - 2, avenue Albert Einstein

69626 Villeurbanne Cédex

1   INTRODUCTION

Thermodynamic considerations and experimental evidence indicate that by using CO + H_2 mixtures as reactants one should be able to produce a broad range of molecules including alcohols, acids, aromatic hydrocarbons, alkanes and alkenes with either linear or branched skeleton. This situation leads to the major problem aera : CO/H_2 synthesis-product selectivity. Today one of the main objectives is the development of new catalyst systems which promote the more desirable products such as low molecular weight hydrocarbons. To this end we have to focus our efforts on the determination of the composition and the structure of the active catalyst on the atomic scale.

The aim of our study is to prepare better defined catalytic systems by combining components with well defined structures and properties such as zeolites and molecular complexes of transition metals. From the point of view of reactivity and catalysis hope is still high that this type of adducts will be efficient as/or more efficient than classical heterogeneous catalysts. The interaction between the support and the complex and the redox behavior of these partners are important factors which will influence the type of catalyst formed. Earlier works from this laboratory (refs. 1, 2) have shown that the adsorption of certain transition-metal carbonyls into an HY-zeolite framework and subsequent thermal desorption lead either to ions (Mo, Fe) or to metal (Re, Ru). However the oxidation reaction can be suppressed by using a non-acidic zeolite i.e. the NaY-type. We report here the behavior of such $Fe_3(CO)_{12}$-NaY adducts upon thermal treatments and their reactivity in the Fischer-Tropsch synthesis.

2   EXPERIMENTAL SECTION

2.1   Materials

The NaY faujasite was supplied by Linde Co. (SK 40 Sieves). A conventional exchange with NH_4Cl provides a NH_4Y sample (unit cell composition : $(NH_4)_{46}Na_{10}Al_{56}Si_{136}O_{384}$). Heating this sample for 15 h in oxygen and 3 h in vacuo (10^{-5} torr) at 350°C leads to the hydrogen form HY.

$Fe_3(CO)_{12}$ was prepared according to ref. 3.

The $Fe_3(CO)_{12}$-Y adducts are prepared under argon atmosphere with the zeolites previously heated in vacuo at 350°C. The impregnation of the support is performed either from pentane solution at 25°C or from dry mixing of the carbonyl and the zeolite to avoid any complications from the solvent. In this last preparation the sample stands in vacuo for 24 h at 60°C in order to favour the sublimation of the carbonyl into the pores of the zeolite. The

TABLE 1

Carbonyl Stretching Frequencies for Iron Compounds and Adducts with $AlBr_3$, HY and NaY

Carbonyl compound	$\nu_{CO}(-CO)$, cm^{-1}	$\nu_{CO}(\rangle CO)$, cm^{-1}
$Fe_3(CO)_{12}$ [a]	2058 s, 2053 s, 2036 s	1871-1862 w, 1823 m, br
$Fe_3(CO)_{12}$ [b]	2097 w, 2056 s, sh, 2039 s, 2018 s, sh, 2006 sh, 1987 sh	1855 w, 1825 mw
$Fe_3(CO)_{12}$ [c]	2103 w, 2046 vs, 2023 mb, 2013 sh	1867 w, 1838 mw
$Fe_3(CO)_{12}/AlBr_3$ [d]	2124 w, 2081 ms, 2070-2008 s, b	1922 mw, 1548 s
$Fe_3(CO)_{12}/HY$ [e]	2112 mw, 2056 s, 2030 m, 1985 sh, 1950 m	1795 mw, 1760 m
$Fe_3(CO)_{12}/NaY$ [f]	2112 mw, 2060 s, 2045 s, 2025 s, 1995 s, 1965 s, 1945 s	1800 mw, 1770 m
$Fe_3(CO)_{12}/NaY$ [g]	2080 m, 2045 s, 2015 s, 1950 s, b, 1910 sh	
Fe film + CO [h]	1980 s, 1900 sh	

[a] In argon matrix, ref. 5   [b] In KBr pellets, this work   [c] n-hexane solution, this work   [d] ref. 4   [e] (1 % Fe) ref. 2

[f] (1 % Fe)   [g] sample decomposed at 200°C and readmission of CO (300 bar) at 25°C   [h] ref. 6.

amount of iron anchored is determined by chemical analysis. The loadings correspond to 3-12 Fe atoms per unit cell.

2.2 Infrared and volumetric experiments

The procedure has been described in ref. 2.

2.3 Catalytic experiments

The runs are performed in a static reactor (300 ml) for 15 h under an initial 20 bar pressure with a sample weight leading to 0.4 mg-atom of iron. Neither the unloaded zeolites nor the $Fe_3(CO)_{12}$ are active in $CO + H_2$ conversion in our experimental conditions.

The products are analysed by gas chromatography usually on five different columns in order to detect CO, H_2, CO_2, H_2O and C_1 to C_n hydrocarbons (alkanes, alkenes, alcohols).

3 RESULTS AND DISCUSSION

Since the $Fe_3(CO)_{12}$-Y adducts are new catalyst precursors it seems desirable to describe in more detail the fate of $Fe_3(CO)_{12}$ when anchored in a zeolite and the influence of the different process variables on activity and selectivity. Finally the particularity of these systems will be stressed.

3.1 Interaction and thermal behavior of $Fe_3(CO)_{12}$ anchored in Y-zeolites

Upon adsorption of the iron carbonyl on the zeolite no CO is evolved. The $Fe_3(CO)_{12}$-NaY adduct presents $\nu(CO)$ bands in the ir spectrum which are not centered at the same frequencies than those of $Fe_3(CO)_{12}$ taken alone. Terminal and bridging carbonyl ligands are still present but the ν (bridging CO) are shifted towards lower values ($\Delta\nu \simeq 70$ cm^{-1}). The same behavior is found with a $Fe_3(CO)_{12}$-HY adduct and the spectrum is more informative due to the presence of ν_{OH} bands. Only the 3640 cm^{-1} band decreases while a broad one develops near 3530-3520 cm^{-1}. This result points out the presence of the carbonyl complex in the zeolite supercage and the interaction is depicted as acid-base interactions according to the similarities found with the $Fe_3(CO)_{12}$-AlBr$_3$ adduct in which AlBr$_3$ is linked to the oxygen atoms of the bridging carbonyls (ref. 4). The same $\nu(CO)$ frequency shifts are observed (Table 1) so that with the zeolite systems the cation (H^+ or Na^+) acts as a Lewis-acid and the oxygen atom of the bridging carbonyl as a Lewis-base.

A progressive heating of the $Fe_3(CO)_{12}$-NaY adduct at temperatures higher than 60°C promotes the decarbonylation which is complete at 200°C under vacuum. The intensities of the $\nu(CO)$ bands decrease monotoneously without any frequency shifts. Readmission at room temperature of carbon monoxide as a molecular probe restores $\nu(CO)$ bands which are centered at different frequencies. Five maxima are easily distinguished and the spectrum thus obtained is quite different from that recorded with a metallic iron film (Table 1). The number of bands and their frequencies lie in the range observed for Fe(0) molecular clusters (ref. 7). This similarity leads us to propose that the decarbonylation of the $Fe_3(CO)_{12}$-NaY adduct yields high dispersed metallic particles which upon CO atmosphere form discrete iron carbonyl species. These species loss their CO very easily as a vacuum treatment at 120°C removes the $\nu(CO)$ bands.

3.2 Fischer-Tropsch reaction : the prerequisite to get Fe-Y catalyst precursors

$Fe(CO)_{12}$-NaY and -HY adducts were the starting materials. They are either totally decar-bonylated at 200°C under vacuum or not before the run.

The $Fe_3(CO)_{12}$-HY adduct exhibits no catalytic activity in the temperature range studied (200-300°C). During the decarbonylation under vacuum, several stoichiometric reactions take place as evidenced by mass spectrometry and infrared. The water-gas shift reaction, the hydrogenation of CO_2 and the oxidation of Fe(0) into Fe(II) species by the zeolite protons account for the formation of CO_2, H_2O, CH_4, higher hydrocarbons, H_2 and Fe^{2+} (ref. 8). The sample thus obtained, a Fe^{2+}-HY, is also inactive in the catalytic CO + H_2 conversion as standard Fe^{2+}-NaY and Fe^{3+}-NaY exchanged ones. Therefore $Fe^{2+,3+}$-Y zeolites are not catalyst precursors and the protons oxidize the Fe(0) moieties even under the CO + H_2 reducing atmos-phere as the non-decomposed $Fe_3(CO)_{12}$-HY adduct is also inactive.

On the opposite the $Fe_3(CO)_{12}$-NaY adduct is active. A non-decomposed sample exhibits a significant activity at 230°C whereas that for the decarbonylated one appears at 200°C. As IR experiments show an increase in the stability of the $Fe_3(CO)_{12}$ units upon heat-treatment under CO atmosphere so that total CO evolution is effective only at 230°C, the catalyst is, thus, certainly not $Fe_3(CO)_{12}$. The cluster has to be transformed into species which form less stable bonds with carbonyl ligands upon CO re-adsorption.

3.3 Effect of the reaction parameters on activity and selectivity

A non-decomposed $Fe_3(CO)_{12}$-NaY adduct (4 % Fe) was used and effects of the inlet CO/H_2 ratio and reaction temperature on CO, H_2 conversions and selectivity were studied at cons-tant initial pressure (20 bar) and reaction time (15 h). In all experiments H_2O, CO_2, alkanes and alkenes (up to C_{12}) are produced. Linear alcohols are also detected to a minor scale. It should be emphasized that the results were intended to demonstrate qualitative trends rather than quantitative kinetic data with this typical catalyst. Moreover the high CO conversion levels achieved in the present work will not be the limiting factor to observe side reactions.

At increasing reaction temperatures (230-350°C), the product selectivity is shifted towards C_1-C_4. The alkene/alkane ratio declines at higher reaction temperatures whereas the branched-/linear-alkane ratio increases as CO_2 formation. These observations are entirely consistent with the behavior of the classical Fischer-Tropsch catalysts.

A reaction temperature of 250°C was used to study the other reaction parameters. Increase in the CO/H_2 ratio increases the consumption of CO whereas the percent converted decreases. At the opposite the H_2 conversion (%) increases but the consumption of H_2 is constant. In fact this means that higher molecular weight products are formed under low hydrogen pressure and the quantitative analyses of the products effectively show a decrease in C_1-C_3 yield and a subsequent increase in $C_3{}^+$ hydrocarbons (Table 2). The selectivity for CO_2 remains constant which apparently indicates that CO_2 is a primary product in the Fischer-Tropsch synthesis. We have checked that its hydrogenation and the water-gas shift are not signifi-cant reactions. The alkene/alkane ratio greatly varies with (i) CO/H_2 and (ii) the chain-length. An increase in hydrogen pressure increases the alkane production as alkene hydroge-nation is a secondary reaction which takes place with Fischer-Tropsch catalyst and in this work, with the zeolite system. For a constant CO/H_2 inlet, $C_2^=/C_2$ ratio is consistently much lower than $C_3^=/C_3$ while $C_4^=/C_4$ is complicated by the existence of the butene isomers (Table 2).

TABLE 2

Effect of CO/H_2 ratio on CO and H_2 conversions, $C_2^=$-$C_4^=$ formations and CO_2, C_1-C_3 selectivities (expressed as mole percent of CO converted into the desired product).

CO/H_2	CO conv. %	H_2 conv. %	CO_2 %	CH_4 %	C_2 %	C_3 %	$C_2^=/C_2$ %	$C_3^=/C_3$ %	$C_4^=/C_4$ %
1/4	73	39	10	20.6	9	9.9	-	0.8	-
1/2	64	57	9.5	7.4	4.4	7.7	4.5	45	16
1/1	48	66	10.4	5.2	2.8	5.3	7.1	47	19.5

Although ethylene is more readily hydrogenated than the other alkenes it can also participate to the formation of higher molecular weight hydrocarbons.

If the Fischer-Tropsch synthesis is performed with ethylene as a co-reactant, the significant changes in selectivity are found for CO_2, C_3 and for the i-C_4/n-C_4 ratio. The changes observed for CH_4 and $C_3^=/C_3$ ratio can also be attributed to the CO/H_2 modification (1/3.5 instead of 1/4) if one takes in account the hydrogen consumed for ethylene hydrogenation (Table 3). It appears that as CO_2 seems to be a primary product (see above), the

TABLE 3

Effect of ethylene as a co-reactant on CO, H_2 conversions, $C_3^=$ formation and CO_2, C_1-C_4 selectivities (expressed as mole percent of CO converted into the desired product). Initial pressure : 20 bar, $C_2^=/CO/H_2$ = 1/1/4, reaction temperature : 250°C.

co-reactants	CO conv. %	H_2 conv. %	CO_2 %	CH_4 %	C_3 %	$C_3^=/C_3$ %	iC_4/nC_4 %
$CO + H_2$	73	39	10	20.6	9.9	0.8	4.8
$CO + H_2 + C_2^=$	73	48	7.6	11	13.2	16.7	1.6

decrease for its selectivity in the presence of ethylene indicates that CO consumption now occurs in part through a reaction producing no CO_2 and more hydrocarbons, especially C_3. This reaction could involve the insertion of CO into a metal-ethyl bond which is a well-known and easy step with soluble transition metal complexes. Then ethylene participates to the Fischer-Tropsch reaction but the presence of CO is essential for chain growth as an ethylene + H_2 feed mainly gives ethane. This result is in agreement with the observations of Pichler et al. (ref. 9). [14]C ethylene is incorporated into higher hydrocarbons and acts as an initiator of the chain growth.

3.4   Chain-length distribution

The formation of C_n hydrocarbons from CO + H_2 involves a chain-growth mechanism. The molecular distribution will depend on (i) the propagation/transfer rate ratio and (ii) the side reactions. It is obvious that hydrogenation, isomerisation and hydrogenolysis can occur according to the metal, the support and the experimental conditions. From the point of view of mechanism, these secondary reactions are masking the primary growth process and render difficult the improvement of product selectivity. Therefore one has to look for new catalyst compositions and to clear up the selectivity parameters.

A non-decomposed $Fe_3(CO)_{12}$-NaY adduct (4 % Fe) provides a molecular weight distribution which is reported in figure 1, curve 1. An exponential decrease of the mole percent of CO

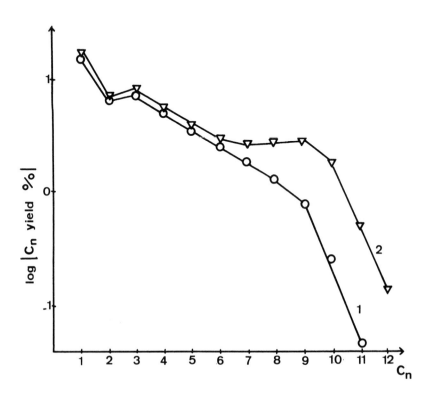

Fig. 1. Effect of different iron-NaY zeolite on the product distribution in the Fischer-Tropsch reaction. CO/H_2 = 1/4, reaction temperature : 250°C, initial pressure : 20 bar, reaction time : 15 h.
(1) $Fe_3(CO)_{12}$-NaY adduct (4 % Fe). (2) Fe-NaY sample (4 % Fe).
The C_n yield corresponds to CO converted into C_n/initial CO amount in percent.

consumed to form C_n is observed from C_1 to C_9 with a consistently lower value for C_2. Hydrocarbons higher than C_9 are only in trace amounts. This distribution is independent of the CO/H_2 ratio and of the reaction temperatures (230-300°C, CO/H_2 = 1/4). Two questions arise from this peculiar chain-length distribution : (i) what is the value of the chain-growth probability α and (ii) is this selectivity relevant to the iron-zeolite systems ?

The chain-growth rate can be expressed, for example, by the Schulz-Flory distribution law which is called, in Polymer Chemistry, the "most probable distribution" (refs. 10, 11). The α value is 0.58 and 0.60 if respectively calculated from the slope and the ordinate inter-cept of Figure 1. This is in good agreement with the data reported for conventional iron catalysts (ref. 10). Nevertheless the predicted maximum achievable for the C_2-C_4 fraction is 56 % which is much higher than that found in our run (28 %). This discrepancy casts some doubt on the mechanism meanings of the Schulz-Flory equation in our conditions. The corre-lation is :

$$\log \frac{m_n}{n} = \log \frac{(1-\alpha)^2}{\alpha} + n \log \alpha$$

where m_n is the weight fraction of each carbon number fraction, n is the carbon number and α is the probability of chain growth. It is worth recalling that this one-parameter equation is operative (i) when a linear addition polymer is formed by a constant rate of initiation, monomer concentration invariant, transfer to solvent but not to monomer and termination by disproportionation or (ii) when a linear condensation polymer is formed by assuming equal reactivity of all chain ends or (iii) when a linear condensation polymer is formed by allowing the units to interchange in a random manner or (iv) when a low molecular weight linear polymer is formed from a higher molecular weight linear polymer by random scission (ref. 12). It is obvious that these conditions are not full filled in the standard Fischer-Tropsch synthesis : (i) solvent is absent, (ii) C_2 chains are more readily converted than the others and (iii) (iv) no kinetic control by the catalyst is involved.

In order to assess the special distribution reported in Figure 1, curve 1, an other Fe-NaY catalyst precursor was prepared so as to get metallic particles on the external surface of the zeolite. This is performed by heating the $Fe_3(CO)_{12}$-NaY adduct (4 % Fe) in vacuum from 25° to 250°C during 1 h and further evacuation at 250°C for 15 h. Particles of 200-300 Å in diameter are thus obtained. Higher CO, H_2 conversions and CO_2 selectivity (13 %) are found with the Fe-NaY sample. Figure 1, curve 2 reports the chain-length distribution. A drastic change occurs in the C_6-C_{10} region. These hydrocarbons are more favoured and the chain growth probability is not constant over the C_1-C_{12} range. It can be concluded from these experiments that the $Fe_3(CO)_{12}$-NaY precursor induces a peculiar selectivity in the hydrocarbon chain length. As the $Fe_3(CO)_{12}$ moieties are located in the supercages, this hydrocarbon distribution can be achieved by a cage effect as the C_9 length fits with the supercage diameter. Parallel studies by Jacobs et al. (ref. 13) on Ru-Y samples and Blanchard et al. (ref. 14) on $Co_2(CO)_8$-Al_2O_3 samples point out the same dependence effect of the porosity of the inorganic matrix on the upper limit of the hydrocarbon chain. Nevertheless as the particle size of the catalyst can be limited by the pore diameter in which it is trapped, one has to get a better understanding of the reaction mechanism in order to determine the primary effect.

We have found that the chain-length selectivity can be altered by the addition of an acidic support, i.e. HY zeolite, to the $Fe_3(CO)_{12}$-NaY sample in the autoclave. Figure 2 reports the hydrocarbon distribution obtained (curve 1) together with that for the $Fe_3(CO)_{12}$-NaY taken alone (curve 2). The presence of the HY zeolite leads to an increase in the C_4-C_5 fraction with a concomitant decrease of the higher hydrocarbons. Moreover isoalkanes are the major products (iC_4/n-C_4 = 2 instead of 0.05). The acidity of the added zeolite promotes cracking of the hydrocarbons and therefore modifies the Fischer-Tropsch distribution.

Cracking experiments performed under hydrogen pressure with the HY show than n-octane is transformed into C_2 (traces), C_3, C_4 (iC_4/n-C_4 = 5), C_5 and C_6 whereas n-butane is transformed into C_2 (traces), C_3, i-C_4, C_5 and C_6 but, in this latter case, the activity is quite low.

316

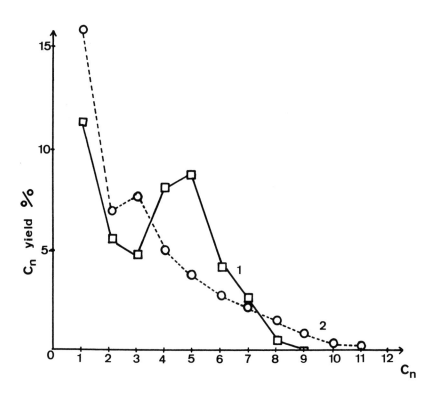

Fig. 2. Modification of the product distribution in the Fischer-Tropsch reaction by addition of an HY zeolite. CO/H_2 = 1/4, reaction temperature:250°C, initial pressure : 20 bar, reaction time : 15 h.
(1) $Fe_3(CO)_{12}$-NaY (4 % Fe) + HY. (2) $Fe_3(CO)_{12}$-NaY (4 % Fe).
The C_n yield corresponds to CO converted into C_n/initial CO amount in percent.

4   CONCLUSION

The results described in this paper show that molecular iron complexes are valuable candidates to introduce Fe(0) particles into zeolites provided that no acidity is present.

According to this procedure iron-zeolites are catalyst precursors in the Fischer-Tropsch synthesis. They behave as conventional ones if one considers the effect of temperature and CO/H_2 ratio upon the C_1-C_3, C_3^+ fraction and alkene/alkane ratio. The originality of the $Fe_3(CO)_{12}$-NaY adduct lies in the upper limit of the chain-length of the hydrocarbons formed. Hydrocarbons higher than C_9 are only produced in trace amounts. The key parameter for selective Fischer-Tropsch synthesis appears to consist in the formation and stabilization of small aggregates which can be afford by trapping them in the pores of an inorganic matrix. Nevertheless the selectivity can be shifted towards C_4-C_5 in promoting side reactions such as cracking by the addition of acidic supports.

REFERENCES

1 P. Gallezot, G. Coudurier, M. Primet and B. Imelik, Molecular Sieves II, ACS Symp. Ser. n° 40, 1977, pp. 144-155.
2 D. Ballivet-Tkatchenko and G. Coudurier, Inorg. Chem. 18 (1979) 558-564.

3 W. Mc Farlan and G. Wilkinson, Inorganic Synthesis VIII, Mc Graw Hill, 1966, pp. 181-183.
4 J.S. Kristoff and D.F. Shriver, Inorg. Chem., 13 (1974) 499-506.
5 M. Poliakoff and J.J. Turner, J. Chem. Soc., Chem. Commun. (1970) 1008.
6 G. Blyholder and M.C. Allen, J. Amer. Chem. Soc., 91 (1969) 3158-3162.
7 P. Chini, G. Longoni and V.G. Albano, Adv. in Organomet. Chem. 14, Academic Press, 1976, pp. 285-344.
8 D. Ballivet-Tkatchenko, G. Coudurier, H. Mozzanega and I. Tkatchenko, Fundamental Research in Homogeneous Catalysis, Vol. 3, Plenum Press, 1979, pp. 257-270.
9 H. Pichler and H. Schulz, Erdöl u. Kohle-Erdgas Petrochemie, 23 (1970) 651.
10 G. Henrici-Olivé and S. Olivé, Angew. Chem. Int. Ed. Engl. 15 (1976) 136-141.
11 R.J. Madon, J. of Catalysis, 57 (1979) 183-186.
12 L.H. Beebels Jr, Molecular Weight Distributions in Polymer, Interscience Publish., 1971, pp. 7-8.
13 H.H. Nijs, P.A. Jacobs and J.B. Uytterhoeven, J. Chem. Soc., Chem. Commun., (1979) 180-181, ibidem (1979) 1095-1096.
14 D. Vanhove, P. Makambo and M. Blanchard, J. Chem. Soc., Chem. Commun., (1979) 605.

B. Imelik *et al.* (Editors), *Catalysis by Zeolites*
© 1980 Elsevier Scientific Publishing Company, Amsterdam — Printed in The Netherlands

INFLUENCE OF PLATINUM CONTENT ON THE CATALYTIC ACTIVITY OF PtHY and PtHM ZEOLITES.

F. RIBEIRO[+], Ch. MARCILLY[x], M. GUISNET[+], E. FREUND[x], H. DEXPERT.[x]
+ Université de Poitiers   x IFP Rueil Malmaison France

INTRODUCTION

In the course of a more extensive work dealing with the preparation and the catalytic properties of platinum HY zeolite and platinum H-mordenite (ref. 1 - 4), we have been lead to study the changes of activity and selectivity of these solids for the hydroisomerization of n-hexane, as a function of their platinum content. It is assumed (ref. 5 - 10) that the rate of isomerization of n-paraffins such as n-hexane or n-heptane first increases with an increasing platinum content, then reaches a plateau for a value of approximately 0.5 - 1%wt. The classical bifunctional mechanism states that the metallic function corresponding to this plateau is sufficient for the hydrogenation - dehydrogenation equilibria between paraffins and olefins to be reached ; the reaction rate is then determined by the skeletal isomerization of carbocations on the acidic function and is independent of the metal content.

The purpose of the present work is to examine if the preceding conclusions are still valid for high platinum contents, when an adequate preparation procedure is used so as to insure a close contact between the acidic and metallic functions.

EXPERIMENTAL

The two PtHY and PtHM series have been prepared from the corresponding ammonium zeolites by using the technique of ion exchange with competition as described in reference 1 ($Pt(NH_3)_4^{2+}$ is used, the competing ion is NH_4^+). The zeolites are first diluted in an alumina gel (67 % wt), extruded and calcined at 773 K for 4 hours, with a stream of wet air for Y zeolite, dry air for mordenite. After ion exchange, the extrudates are calcined at 773 K for 2 hours, then reduced with hydrogen 2 hours at 723 K for the PtHY series, 8 hours at 778 K for the PtHM series. The platinum contents vary from 0 to saturation.

Zeolite crystallinity is estimated from X-ray diffraction intensities and nitrogen adsorption at 77 K. The metallic phase dispersions are measured by H_2-O_2 titration and high resolution electron microscopy. The acidic function is studied by IR spectroscopy (pyridin desorption at increasing temperatures). The coke contents are determined by elemental microanalysis (sensitivity $\sim$ 0.2 % wt).

The test reaction (hydroisomerization of n-hexane) is done under the folowing conditions : total pressure 3 MPa, H_2/nC_6 = 4. All catalysts are first aged for 16 hours at 573 K. Activities are measured at 533 K, temperature for which conversions are less than 10 %.

RESULTS

1 Catalytic results

(i) <u>Aging period</u>. Figure 1 shows the evolution of overall conversion, isomerization and cracking for the PtHY and PtHm series respectively.

320

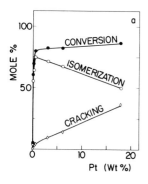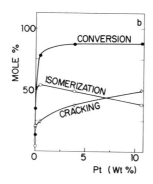

Fig. 1. Variation of the total conversion, isomerization and cracking as a function of platinum content (a) PtHY zeolite, (b) PtH-mordenite.

A platinum content increase causes a rapid increase of overall conversion, up to about 0.6 % platinum wt ; a plateau is then reached. Cracking increases and reaches high values (36 and 50 % respectively for the saturated PtHY and PtHM). Isomerization goes through a maximum occuring for low platinum contents (about 0.6 % wt). The distribution of cracked products varies markedly with platinum content, as is shown in figure 2 which gives the ratio $r = C_1 + C_2/C_3$ as a function of platinum content ; r varies from a low value to 1 and reaches a plateau at 3 - 4 % platinum wt for the PtHY series. The variations of r are similar for the PtHM series. However, r remains less than 1 (0.80 at saturation).

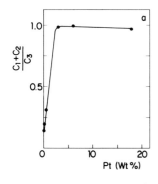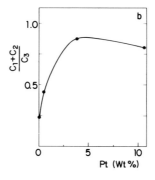

Fig. 2. Variation of the molar ratio $C_1 + C_2/C_3$ as a function of platinum content (a) PtHY zeolite, (b) PtH-mordenite.

(ii) <u>Kinetic measurements (after aging)</u>. The isomerization rate measured at 533 K after the aging period shows a complex variation (see Fig. 3.) : it increases rapidly at low platinum contents, reaches a maximum and then decreases at high metal contents. The maximum isomerization rate is higher and corresponds to a higher metal content for the PtHM series.

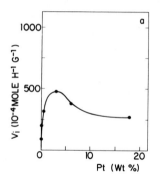 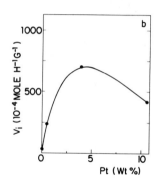

Fig. 3. Variation of the isomerization rate as a function of platinum content (a) PtHY zeolite, (b) PtH-mordenite.

2 Physico-chemical characterization of catalysts

(i) Zeolite crystallinity. B.E.T. areas of fresh and used catalysts are shown in Table 1.

TABLE 1

B.E.T. areas of catalysts.

Sample	Pt (% wt)	S (fresh)	S zeolite (fresh)	S (coked)	S zeolite (coked)
HY	0	382	724	303	489
HPtY	0.01	378	714	364	672
	0.03	384	732	375	705
	0.17	-	-	365	675
	1.0	390	754	355	645
	2.08	370	713	352	658
	6.7	340	673	335	657
HM	0	335	467	-	-
HPtM	0.005	348	506	-	-
	0.16	330	453	350	513
	1.34	340	495	340	495
	3.67	340	521	315	443

These areas are referred to the initial carrier (mixture zeolite + alumina). We have indicated in this table the computed values of areas of zeolitic phases, assuming a value of 210 m^2/g and 270 m^2/g for the alumina in the case of zeolite Y and mordenite respectively. No significant area decrease is observed, including the used catalysts, except for the HY catalyst. An X-ray diffraction study confirms that crystallinity has not significantly changed after reduction and reaction.

(ii) Dispersion of the metallic phase. Values obtained from H_2-O_2 titration and electron microscopy are listed in Table 2.

TABLE 2
Metallic phase dispersion and localization.

Sample	Pt (% wt)	d_1	d_2	D	D_v	D_S^1	D_S^2
HPtY	0.01	2.0	-	-	-	-	-
	0.17	1.8 - 2.0	2.8	12.8	13.7	15.5	3.6
	1.0	2.5	3.4	10.2	10.5	10.4	3.3
	2.08	-	5.2	-	-	-	-
	6.7	4.5	4.9	9.0	9.8	9.5	3.0
HPtM	0.16	1.5	2.2	10.6	10.6	10.6	3.4
	1.34	2.0	1.8	6.1	6.9	5.6	1.8
	3.67	2.2	2.0	4.8	5.4	3.9	1.2

Crystallite diameter : d_1 electron microscopy, d_2 H_2-O_2 titration.
Average distance (in projection) : D experimental, D_v; D_S computed (zeolite crystals D_S^1 500 Å, D_S^2 5000 Å). All distances in nm.

Figure 4 is an example of a micrograph obtained for a large crystal of HPtY zeolite saturated with platinum.

400Å

Fig. 4. Influence of platinum content on the catalytic activity of PtHY and PtHM zeolites.

The data of Table 2 suggest the following comments :
- the dispersions are markedly below 100 %, even at low platinum contents. Thus, high dispersions are not obtained with the simple activation procedure used in this work.

The dispersions are higher in mordenite than in Y zeolite.

- the dispersions decrease when the metal content increases. However, the metallic area remains an increasing function of platinum content and reaches high values for the saturated zeolites (12 m^2/g zeolite for Y zeolite).

(iii) <u>Localization of the metallic phase</u>. From the electron micrographs, the average distance, or rather, the projection of the average distance on the plane of the micrograph between platinum crystallites can be estimated and compared with the theoretical distances computed according to two different hypotheses :

- the platinum crystallites are uniformly spread inside the zeolite crystals (distance in projection : D_v) ;

- the crystallites are located on the external surface of the zeolite crystals (in this case, the zeolite crystal dimensions are needed)   (distance in projection : D_S).

The data of Table 2 clearly show that platinum crystallites are located inside the crystals, in agreement with previous observations (ref. 11 - 12), in spite of the fact that their average diameter is larger than the zeolite cage dimensions.

(iiii) <u>Coke content</u>. The coke contents of the PtHY and PtHM series are shown as function of platinum content in Figure 5.

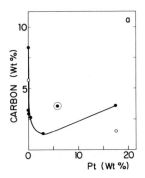
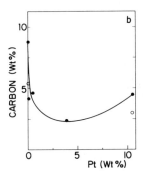

Fig. 5. Variation of the coke content as a function of platinum content (a) HPtY zeolite, (b) HPt-mordenite. ● stripped with hydrogen ○ desorbed under vaccum.

For both series, a curve with a minimum located at a metal content of about 1 % wt is obtained (the sample PtHY 5.8 % platinum wt has been coked under more severe conditions than the other samples).

These coke contents are obtained for the used catalysts, which have been stripped for 30 minutes under 3 MPa of hydrogen at 523 K, cooled under nitrogen, and then quickly analyzed. A second set of measurements has been carried out for some samples after a vaccum treatment (10^{-4} Torr) at 573 K. The values thus obtained are also given in Figure 5. A noticeable decrease in coke content is apparent : the coke can be partially desorbed. This is indirectly confirmed by tentative microanalysis by electron energy loss spectroscopy in a scanning transmission electron microscope : the samples are heavily contaminating.

(iiiii) Study of acidity. It has been carried out for the HPtY series (fresh and used catalysts). The main results are :

    - strong Brönsted acidity is more important for the saturated zeolite than for HY, before reaction ;

    - there is no noticeable difference of acidity (Brönsted + Lewis) among used catalysts ;

    - strong Brönsted acidity (corresponding to pyridine still adsorbed at 573 K) disappears almost completely after coking.

DISCUSSION OF RESULTS

We do not find the expected result : occurrence of a plateau for the hydroisomerizing activity above a certain (small) platinum content. To interpret the decrease in activity at high platinum contents, two different hypothesis can be put forward :

    - modification of the acidic function under the influence of platinum ;

    - increase of coking and modification of the nature of coke at high platinum contents.

Let us consider the first hypothesis. We do find an increase of strong Brönsted acidity with increasing platinum content for the fresh catalysts ; however, no significant difference is apparent among used catalysts, at least with the method used.

The second hypothesis is supported by the curves of Figure 5 : the maximum isomerizing activity corresponds to the minimal coke content. Furthermore, results obtained during the aging period show that selectivity for isomerization steadily decreases as the platinum content increases (and cracking increases).

The high value of $C_1 + C_2/C_3$ ratio implies that, at high platinum contents, hydrogenolysis on the metal becomes more important than acid cracking (the phenomenon is more marked on zeolite Y, the acidity of which is weaker) (ref. 13 - 16). It seems logical to admit that hydrogenolysis is the measurable manifestation of a metallic activity differing from simple hydrogenation - dehydrogenation reactions. The metallic phase (of very high surface at high platinum contents) would produce coke precursors, and indirectly poison the acidic function.

To sum up, at low metal contents, the metal influence is beneficial : it slows down the formation of coke on the acidic function ; hydrogenolysis and formation of coke precursors on the metal are negligible. At high metal contents, this beneficial influence is counterbalanced by the formation of coke originating from the metal.

These conclusions lead to complicate the scheme proposed by the classical bifunctional mechanism, as the activity of the metallic phase cannot be reduced to a simple hydrogenating - dehydrogenating function.

REFERENCES

1 F. Ribeiro, Thesis, Université de Poitiers, France, 1980.
2 Ch. Marcilly and F. Ribeiro, Rev. Inst. Fr. Pétr. 34 (3) (1979) 405 - 428.
3 Ch. Marcilly, F. Ribeiro and G. Thomas, C.R. Acad. Sc. Paris 287 C (1978) 431 - 434.
4 M. Guisnet, Ch. Marcilly and F. Ribeiro, J. Cat. to be published.
5 F. Chevalier, M. Guisnet and R. Maurel, Proc. Intern. Congr. Catalysis 6th London the Chemical Society London 2 (1976) 478.
6 F. Chevalier, Thesis, Université de Poitiers, France, 1979.
7 F. Chevalier, M. Guisnet and R. Maurel, C.R. Acad. Sc. Paris 282 C (1976) 3 - 5.

8 A.P. Bolton, M.A. Lanewala and P.E. Pickert, J. Cat. 8 (1967) 95 - 97.

9 P.B. Weisz, Adv. Catalysis 13 (1962) 137 - 189.

10 C. Gueguen, M. Guisnet, G. Lopez and G. Perot, Proc. of the symp. on zeolites, Szeged Hungary (1978) 207 - 213.

11 J. Datka, P. Gallezot, B. Imelik, J. Massardier and M. Primet, Ref. 5, 696 - 707.

12 P.A. Jacobs and J.J. Verdonck, J.C.S. Faraday I 76 (1980) 403 - 416.

13 H. Matsumoto, Y. Saito and Y. Yoneda, J. Cat. 22 (1971) 182 - 192.

14 J.L. Carter, J.A. Cusumano and J.H. Sinfelt, J. Cat. 20 (1971) 223 - 229.

15 C. Herrera, Thesis, Université de Poitiers, France, 1977.

16 A.M. Gyul'maliev, I.I. Levitskii and E.A. Uddl'tzova, J. Cat. 58 (1979) 144 - 148.

B. Imelik *et al.* (Editors), *Catalysis by Zeolites*
© 1980 Elsevier Scientific Publishing Company, Amsterdam — Printed in The Netherlands

KINETIC OF WATER-GAS SHIFT REACTION ON Y ZEOLITE

A. L. LEE, K.C. WEI, T.Y. LEE, J. LEE

Institute of gas technology, 3424 South State Street, Chicago, Il, 60616 USA

ABSTRACT

   The kinetics of water-gas shift reaction on Ni-Mo-Y zeolite has been studied. The intrinsic reaction rate was obtained experimentally in a Carberry type Continuous-Stirred-Tank-
Reactor (CSTR). The gas-solid film resistance was eliminated by high speed rotation of the
pellets and the pore diffusion resistance was minimized by selecting pellets of small size
($\sim$1.4 mm). The reaction temperature and pressure ranged from 176° - 400°C and 4-69 bars
respectively. The rate is kinetics controlled at low temperatures (176° -232°C) and pore
diffusion controlled at higher temperatures (260° -400°C). The rate equation was found to be
Lanqmuir-Hinshel wood type with two parameters. A possible reaction mechanism was proposed
and discussed. Also, the catalyst was found of possessing excellent resistance to H_2S and
NH_3 poisoning. Catalytic deactivation by carbon deposition or poisoning of phenol and ben-
zine was investigated and a deactivation model was established which described the activity
decay accurately. Moreover, Y zeolite was demonstrated to be superior than other catalysts
in reactivity under identical test conditions.

1 INTRODUCTION

   Water-gas shift reaction is widely used for the manufacture of synthesis gas. Although
it has been studied extensively over promoted iron oxide catalysts but there seems little
information on the zeolite supported metal catalysts. Furthermore, in spite of considerable
data available on the kinetics of the reaction but little agreement on either the precise
form of the rate equation or the value of the rate constants or their activation energies.

   Moe (1) gave an empirical correlation of the water-gas shift reaction on chromia promoted iron oxide catalysts for industrial design application. He also gave a general discussion of the effects of temperature, pressure and optimal design of a reactor. A statistical
experiment design approach to study the kinetics of this reaction was reported by Hulbert
and Vasan (2). A qualitature discussion of the reaction mechanism was also presented by them
Bohlbro (3,4) compared the kinetics obtained with and without minor quantities of alkali
present in the catalyst. He also found that, over a wide range of conditions, the data could
be satisfactorily correlated by empirical power law form. Mars et al. (5) published a com-
perative study of fourteen iron-based catalysts containing different amounts of chromia
promoter and concluded that the activity per unit surface area was approximately the same
which indicated that the chromia was acting as a stabilizer.

   Recent emphasis on coal gasification, liquefaction and combustion revitalizes new efforts
and strives to understand this important reactor better. It is the intention of this study
in water-gas shift reaction on ions exchanged zeolite for the following conceivable advantages, namely :

o high reactivity and stability.

o less sintering degradation.

o high acid resistance.

o less carbon deposition at severe conditions.

With these in mind, the kinetics of water-gas shift reaction on a Union Carbide Ni-Mo bi-ion exchanged zeolite Y was investigated in a Spinning Basket Constant Volume Reactor (Carberry type Continuous-Stirred-Tank-Reactor). A new intrinsic rate equation and a possible reaction mechanism were elucidated from the experimental results. The rate constant and its dependence on temperature as well as the catalyst deactivation by carbon deposition and poisoning were also investigated. The superiorty of the Y zeolite over other catalysts in a prolong period of test under identical conditions was demonstrated in this paper.

2   EXPERIMENTAL

A brief description of the catalyst, equipment and experimental procedure are given below.

2.1   Catalyst

The catalyst was furnished by Union Carbide Corp. It is a Y zeolite with certain percents of Ni and Mo cations. Due to the contract agreement, the detail information of the catalyst is not allowed for release.

2.2   Equipment

A magnetic drive constant volume (300 CC) high pressure reactor was used for the study. Surrounding the reactor was an electric furnace with temperature controller and recorder. A simplified diagram of the reactor is shown in Fig. 1.

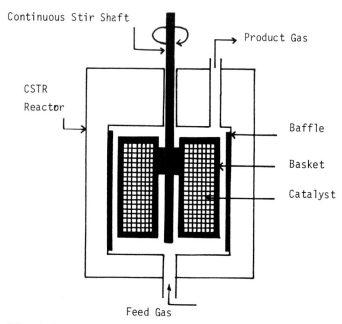

Fig. 1 Continuous-Stirred-Tank-Reactor

The four thin baskets of catalyst holders were fixed on the central rotating shaft of the reactor. A preheater for raising the temperature of the reactants to the reaction tempera- ture was placed at the front end of the reactor. Regulators were used to adjust the gas flow rate and the front end pressure. At the outlet of the reactor, a back pressure regula- tor was installed for maintaining the reactor pressure. A thermal couple was inserted into the vicinity of the reaction zone in order to monitor and control the reaction temperature. The preheated reactants were flowing in from the bottom of the reactor and out from the top. Flow rates were measured by calibrated rotameters or manometers placed in before and after the reactor. In addition, a wet test meter was connected at the rear end for double check the constant flow rate in a steady state continuous operation. Gas chromotograph and infra- red analyzers (for CO and CO_2) were used to analyzed the gas composition and occasionally confirmed by mass spectrometer. The complete set-up has been reported by Lee (6) elsewhere.

2.3   Procedure

Approximately 10 grams of 1.4 mm size catalyst pellets were placed in the four sample baskets. Temperature, pressure and flow rate of feed gas (CO, N_2) were established first. Apredermined amount of water was metered into the preheater and mixed well with the feed gas before entering the reactor. Gas chromatograph, infrared analyzers were activated to monitor the gas composition continuously. The steady state continuous flow rate was measu- red by rotameter (or manometer) and wet test meter. Normally a space velocity of $80 cm^3$/unit volume catalyst/min. was used. Material balances on carbon, hydrogen and oxygen were always ascertained for each of the run. All the runs were under isothermal and isobaric conditions.

3   RESULTS AND DISCUSSION

The temperature and pressure of reaction ranging from 176° to 395°C and 4-69 bars res- pectively were studied. The gas-solid film resistance was eliminated by the rapid spinning of the pellets in the baskets and the small pellets ($\sim$1.4 mm) were used to minimize the pore diffusion (intraparticle) effect. The resulting kinetics were summarized below :

3.1   Intrinsic reaction rate

A general form of Langmuir-Hinshelwood was used to correlate the experimental data,

$$Y_{CO} = \frac{k_1 P_{CO} P_{H_2O}}{1 + k_2 P_{CO} + k_3 P_{H_2O} + k_4 P_{CO} P_{H_2O}} \qquad (1)$$

A multiple regression analysis was applied to the reciprocal form of equation (1). Statistical analysis indicated that k_2 and k_3 were insignificant in the correlation, hence the resulting intrinsic rate was an equation with two parameters :

$$Y_{CO} = \frac{k_1 P_{CO} P_{H_2O}}{1 + k_4 P_{CO} P_{H_2O}} = \frac{K_1 0_e^{-Ea/RT} P_{CO} P_{H_2O}}{1 + k_4 P_{CO} P_{H_2O}} \qquad (2)$$

In general, a multiple correlation coefficient of 0.95 was obtained. A typical compari- son of predictions calculated by equation (2) and experimental data at 343°C and various total pressures is shown in Figure 2.

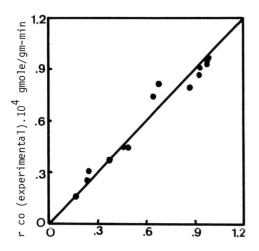

r co (model).10^4 gmole/gm-min

Fig. 2 Comparison of Predictions with Experimental Data

It can be seen that an excellent agreement between them was achieved.

An Arrhenius plot of the intrinsic reaction rate vs. temperature is shown in Fig. 3.

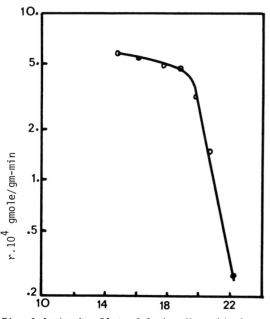

$1/T.10^4$, K^{-1}

Fig. 3 Arrhenius Plot of Carbon Monoxide Conversion Rate

It appears that the water-gas shift reaction is controlled by kinetics at low temperatures (between 176° -232°C) and possibly by pore diffusion (intraparticle diffusion) at higher temperatures (260° -400°C). The transition zone was observed between 232° -260°C. The least square fit of k_1, activation energy E_a in the two zones are listed in Table 1.

TABLE 1

Rate Constants and Activation Energies

	Zone 1 (260° -400°C)	Zone 2 (176° -232°C)
k_i	8.55×10^{-14}	1.34×10^{-5}
E_a	$0.95^{Kcal}/\text{g-mole}$	$20.5^{Kcal}/\text{g-mole}$
k_4	4.91×10^{-11}	4.91×10^{-11}

The high activation energy of 20.5 kcal/g-mole in zone II support the proposed kinetics and diffusion controlling step in zone I and zone 2 respectively.

3.2 Reaction Mechanism

A hypothetical reaction mechanism for the water-gas shift reaction based on the obtained rate equation is

$$CO + Y \underset{k_A}{\overset{k_A}{\rightleftarrows}} CO \cdot Y \quad \text{(Adsorption)}$$

$$H_2O + Y \underset{k_B}{\overset{k_B}{\rightleftarrows}} H_2O \cdot Y \quad \text{(Adsorption)}$$

$$CO \cdot Y + H_2O \cdot Y \underset{k_S}{\overset{k_S}{\rightleftarrows}} CO_2 \cdot Y \vdots H_2 + Y \quad \text{(Reaction)}$$

$$CO_2 \cdot Y + H_2O \cdot Y \underset{k_D}{\overset{k_D}{\rightleftarrows}} CO_2 + H_2 + Y \quad \text{(Desorption)}$$

$$CO + H_2O \rightleftharpoons CO_2 + H_2 \quad \text{(overall)}$$

If desorption is the rate controlling step and

$$K = \frac{K_A K_B}{K_D} \quad K_S \gg K_A \text{ and } K_B$$

and K_A and $K_B \ll 1$

where $K_i = \frac{k_i}{k_i}$, i = A, B, S, D

then equation (2) can be deduced again,

$$Y_{CO} = \frac{KS_a k_D P_{CO} P_{H_2O}}{1 + K_S Kp_{CO} P_{H_2O}} = \frac{k_1 P_{CO} P_{H_2O}}{1 + k_4 P_{CO} P_{H_2O}}$$

where S_a = total number of active sites.

This result gave certain validity to the proposed mechanism.

3.3 Catalyst Deactivation

The catalyst deactivation is expressed in a separate form normally,

$$-Y_d = f_1(T) f_2(C) f_3(a)$$

or in a more specific form,

$$-Y_d = -\frac{da}{dt} = k_0 e^{-E_d/RT} C^\alpha a$$

where   T = Temperature

      a = activity = $\dfrac{\gamma}{\gamma_0}$

      t = time

      α = exponential constant

      γ = rate of reaction at time t

      c = concentration

Equation (3) was used to correlate the deactivation of catalyst. Experiments were carried out with feed gases containing 0.1, 0.4 and 1.0 percents phenol. The best values obtained by least square fit of equation (3) were

$k_o = 4.44 \times 10^{-5}$

$E_d = -1.33 \ Kcal/g\text{-}mole$

$\alpha = 0.1813$

A comparison between predictions (equation 3) and experimental results was shown in Fig. 4.

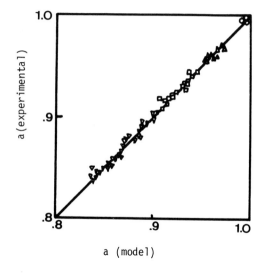

Fig. 4 Comparison of Calculated and Experimental Values on Catalytic Deactivation

$C_{\phi OH}$ mol %

water-free

 ○ 0.0   ▲ 0.1   ◘ 0.4   ▼ 1.0

Again an excellent agreement was achieved.

Froment and Bishoff (7) described the activity decay via carbon deposition, coking or fouling on the catalyst pores and surface which could show mass transfer rate or even prevent reactants getting into the active sites. The low deactivation energy obtained in this study indicated this was a strong possibility of causing the activity decay. Instead of phenol, benzene was used the same phenomenon of decline in catalytic activity was observed.

A high steam concentration appears to suppress the carbon deposition. At a optimal temperature, by keeping a high steam to phenol (or benzene) ratio, we could have a good possibility of avoiding carbon deposition and maintaining catalytic activity in the water-gas shift reaction (Lee (6)).

A maximum of 2 mole % H_2S and 0.3 mole % NH_3 have been blended in the feed gas separately for investigation of the poison effects on catalyst. No appreciable decline in conversion rate of CO was observed. Thus, it was concluded that the Ni-Mo-Y zeolite possessed super resistance to sulphur and ammonia poisoning.

3.4 Comparison with other catalysts

A cobalt-molybdenum catalyst from Shell Oil Company, another also Co-Mo catalyst from Girdler Chemical, Inc., both of them are on aluminium support were tested for comparison. The feed gas was dosed in 0.75 mole % H_2S, 0.15 mole % NH_3, 1.0 mole % benzene, 0.03 mole % phenol, 10.0 mole % CO, 50 mole % steam and balanced of by N_2. The carbon monoxide reaction was monitored for a long period of time for the three catalysts -Union Carbide, Girdler Chemical and Shell Oil. Results of the prolong test at 395°C and 65 bars are plotted in Fig. 5, and the Y zeolite showed the highest rate of conversion.

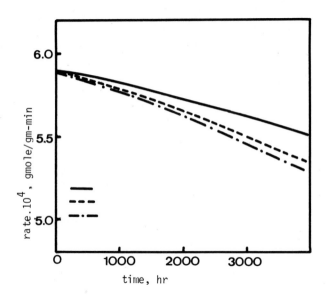

Fig. 5 Comparison Study of Catalyst Activity at T=395 c and P=65 bars
    Union Carbide
    Girdler Chemical
    Shell Oil

REFERENCES

1 J.M. Moe, Chem. Eng. Prog., 53 (1962) 33
2 H.M. Hulburt,and Vasan, C.D.S., A.I.Ch.E., 7 (1961) 143
3 H. Bohlbro, J. of Catalysis, 3 (1964) 207.
4 H. Bohlbro, Acta Chem. Scand., 5 (1961) 502
5 P. Mars, M.S. Gorgels, and P. Zwietering, Actes Internationale Congres Catalyse 2e, (1960) 2429
6 A.L. Lee, "Catalyst Selection and Evaluation" paper presented at the Center for Professional Advancement, East Brunswick, N.J. March 13-16, 1979.
7 G.F. Froment and K.B. Bischoff, Chem. Eng. Aci., 16 (1961) 189.

334

NOMENCLATURE

k = rate constant
K = equilibrium constant
γ = reaction rate g-mole/g(catalyst) - min
p = pesssure   newtons/M2
T = temperature κ
t = time
$E_{a(d)}$ = activation or deactivation energy
γ = catalyst
S_a = number or concentration of active sites
C = concentration mole %

a = activity = $\dfrac{\gamma}{\gamma_0}$

k^0 = Arrhenius' constant (pre-exponential constant)

B. Imelik *et al.* (Editors), *Catalysis by Zeolites*
© 1980 Elsevier Scientific Publishing Company, Amsterdam — Printed in The Netherlands

THE INFLUENCE OF CALCIUM IONS ON THE PROPERTIES OF NICKEL FAUJASITE CATALYSTS
FOR THE HYDROGENATION OF CARBON MONOXIDE

N.JAEGER, U.MELVILLE, R.NOWAK, H.SCHRÖBBERS, G.SCHULZ-EKLOFF
Research Group Applied Catalysis, University of Bremen, D-2800 Bremen 33

1. INTRODUCTION

The investigation of the catalytic methanation of coal via synthesis gas receives in-creasing attention in view of the dependence of industry and households on natural gas on the one hand and diminishing natural resources on the other hand. Also the development of active and stable catalysts with low metal content and optimum dispersity of the active metal is still one of the major objectives in catalytic research, the more so in view of the increasing prices for catalytically active metals.

Besides the widely used conventional nickel catalysts, which have been studied exten-sively (ref. 1), also nickel zeolites have been found to be active in the methanation re-action (ref. 2). Investigations on reduced nickel faujasite catalysts have shown that the reducibility and the resulting catalytic activity is depending on Ni^{2+} location, nature and concentration of other elements, pretreatment and reduction conditions (ref. 3,4). The postulation of Egerton and Vickerman (ref. 5) that tetrahedrally coordinated Ni^{2+} might be less readily reduced than the octahedrally coordinated Ni^{2+} in sites SI, has found atten-tion, although a clear experimental evidence for this hypothesis is still missing. In the following the influence of calcium ions on the activity of reduced nickel faujasite catalysts for the hydrogenation of carbon monoxide is reported.

2. EXPERIMENTAL

2.1 Preparation of the catalysts

The NaX zeolite was prepared by hydrothermal synthesis. Ion exchange was carried out in solutions of $Ca(CH_3COO)_2$ and $Ni(CH_3COO)_2$. The extent of ion exchange was determined by atomic absorption spectroscopy (Table 1).

TABLE 1

Composition of the samples used for the catalytic measurement

Abbreviation	Composition	Remarks
NiCaX 3.1	$Ni_{8.6}Ca_{20.8}Na_{27.1}X$	Consecutive exchange of Ca(3x) and Ni(1x)
NiCaX 1.1	$Ni_{10.1}Ca_{10.7}Na_{44.5}X$	Consecutive exchange of Ca(1x) and Ni(1x)
NiCaX 1.2	$Ni_{20.3}Ca_{8.9}Na_{27.6}X$	Consecutive exchange of Ca(1x) and Ni(2x)
NiCaX 5	$Ni_{9.6}Ca_{10.6}Na_{45.5}X$	Simultaneous exchange of Ca and Ni(1x)
NiX (57)	$Ni_{24.6}Na_{36.8}X$	
NiX (17)	$Ni_{7.6}Na_{70.8}X$	

336

2.2 Carbon monoxide hydrogenation measurements

The catalysts were activated in argon ($420^{\circ}C$, 16h, heating rate: 5°/min) and reduced with hydrogen ($300^{\circ}C$, 25h, 1bar). The reaction was carried out in a fluidized bed reactor (1 bar, CO/H_2 = 3/7, space velocity: 300 h^{-1}) at 250, 300 and $350^{\circ}C$. The reaction products were analysed by capillary gas chromatography.

2.3 Characterizations of the catalysts

(i) X-ray analysis. All catalysts were examined by X-ray diffraction in the hydrated form following the ion exchange, in the dehydrated reduced form and following the full catalytic reaction cycle at all three temperatures. Diagrams in Fig. 1 were obtained by photometric evaluations of films.

(ii) Susceptibility measurements. Magnetic susceptibility measurements of Ni^{2+} and Ni containing zeolite catalysts were carried out by the Faraday method at room temperature and field strength up to 12 $\times 10^3$ Oe. All catalysts were examined in the hydrated form following the ion exchange, in the dehydrated form and in the reduced form prior to and after the full reaction cycle at all three temperatures.

(iii) Thermal analysis. The DTA measurements were carried out with a thermobalance (Linseis, Selb) using alumina as reference and platinum crucibles. The catalysts were analysed following the ion exchange and after activation in argon ($420^{\circ}C$). The samples (about 50 mg) were subjected to a standard hydration procedure prior to the temperature programmed dehydration (5°/min).

3. RESULTS

3.1 Catalytic activities

TABLE 2

Hydrocarbon yields (mg (hydrocarbon)/g (nickel)·h) and product distributions (weight % of C_1 - C_{6+}) of the carbon monoxide hydrogenation reaction at 1 bar, CO/H_2 = 3/7 and 300 h^{-1} space velocity

Catalyst	$250^{\circ}C$				$300^{\circ}C$				$350^{\circ}C$			
	Yield after		Products after		Yield after		Products after		Yield after		Products after	
	1h	6h	1h	6h	1h	6h	1h	6h	1h	6h	1h	6h
NiCaX 3.1	3	3	68	71	27	68	96	96	134	100	99	98
			13	13			3	3			1	2
			12	10			<1	<1			-	<1
			7	6			<1	<1			-	-
			-	<1			-	<1			-	-
			-	-			-	-			-	-
NiCaX 1.1	4	4	64	70	82	103	93	95	164	70	98	97
			11	9			4	4			2	3
			16	13			2	1			<1	<1
			9	7			<1	<1			-	-
			<1	<1			-	-			-	-
			-	-			-	-			-	-
NiCaX 1.2	11	19	71	76	83	131	97	98	213	89	99	98
			10	9			2	2			1	2
			12	10			1	<1			-	<1
			6	4			<1	-			-	-
			1	1			-	-			-	-
			-	<1			-	-			-	-

Catalyst	250°C				300°C				350°C			
	Yield after		Products after		Yield after		Products after		Yield after		Products after	
	1h	6h	1h	6h	1h	6h	1h	6h	1h	6h	1h	6h
NiCaX 5	14	27	77	72	209	214	93	91	499	520	99	99
			9	8			4	6			1	1
			9	10			2	2			-	<1
			4	6			<1	<1			-	-
			1	4			<1	<1			-	-
			-	<1			-	-				
NiX (57)	32	41	79	78	199	210	96	95	342	348	100	100
			10	10			3	4			-	-
			7	8			<1	<1			-	-
			3	3			<1	<1			-	-
			1	1			-	-			-	-
			<1	<1			-	-			-	-
NiX (17)	53	46	68	67	242	242	95	96	786	544	100	99
			9	9			4	3			<1	1
			13	12			1	<1			-	-
			7	7			<1	<1			-	-
			3	4			-	-			-	-
			<1	1			-	-			-	-

The results given in Table 2 show, that the catalysts markedly differ in activity (hydrocarbon yield referred to the nickel content) and stability. With about 0.5 g of the most active catalyst ($\hat{=}$ 100 mg nickel) nearly one half of the carbon monoxide was converted to methane at 350°C.

3.2 Characterization of the catalysts

(i) X-ray diagrams. Fig. 1 depicts two typical sequences of diagrams obtained for catalyst NiCaX 3.1 and NiCaX 5. Curves a,b,c were obtained for the hydrated.samples following the ion exchange, for the reduced samples and for the catalysts following the full catalytic reaction cycle at 350°C respectively.

The reduced samples show an increased background in the neighbourhood of the Ni 111 signal, however, in no case a clear Ni 111 signal could be observed at this point. After the full catalytic reaction cycle, the Ni 111 signal could be observed only in the case of NiCaX 3.1 even though NiCaX 5 has the same Ni content. The two samples are representative for different sintering behaviour depending on the preparation conditions of the catalysts.

(ii) Magnetic moments of Ni^{2+} ions and ferromagnetic susceptibilities. The experimentally determined magnetic moments of the Ni ions are listed in Table 3 for the hydrated and dehyhrated samples prior to reduction.Based upon the simplified assumption of essentially two symmetries for the coordination of Ni ions within the dehydrated zeolite lattice
- octahedral coordination to lattice oxygen in SI positions in the hexagonal prisms
- tetrahedral coordination to lattice oxygen and remaining water resp. OH groups in SII, SII' and SI' positions,
the number of Ni ions in both symmetries has been calculated from the experimental results

338

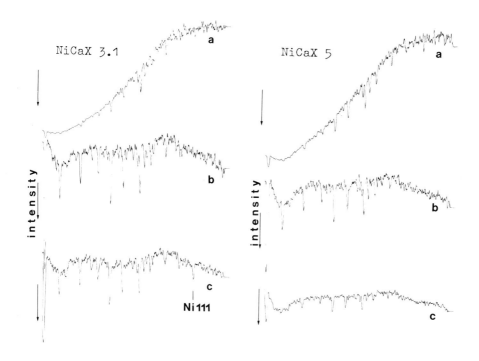

Fig.1. X-ray diagrams for sample NiCaX 3.1 and NiCaX 5: a) hydrated b) dehydrated and c) following full catalytic reaction cycle at 350°C.

using $n_{eff} = 3.2$ and $n_{eff} = 4.6$ for Ni complexes with O_h and T_d symmetry respectively (ref. 5).

For the hydrated samples most of the Ni^{2+} ions should be localized in the supercages in the form of hexaquocomplexes according to the experimental magnetic moments. Even though the assumption neglects the likely distortion of the assumed symmetries in all positions other than for the hexaquocomplex in the supercage it allows to follow the trend of the localization of Ni ions in the various catalysts (ref. 6).

The field dependence of the measured susceptibility following the reduction of the catalysts gave evidence of ferromagnetic nickel particles. The size of the reduced Ni particles can be roughly estimated to be 20 - 30 Å according to the X-ray data and in order to exhibit ferromagnetism (ref. 7). Particles of this size can be contained within the zeolite framework. Following the catalytic reaction cycle all catalysts exhibit increased ferromagnetic susceptibilities,which could be due to additional reduction and/or sintering of small not yet ferromagnetic particles.

TABLE 3

Magnetic moments of Ni^{2+} ions in hydrated and dehydrated catalysts and tentative distribution of Ni^{2+} ions with O_h and T_d symmetry

Sample	Effective number of Bohr magnetons of Ni^{2+} ions in		Ni^{2+} ions in dehydrated samples with symmetry			
	hydrated sample	dehydrated sample	O_h		T_d	
			%	no /u.c.	%	no/u.c.
NiCaX 3.1	3.4	3.9	49	4.2	51	4.4
NiCaX 1.1	3.3	3.9	49	5.0	51	5.1
NiCaX 1.2	3.4	3.6	69	13.9	31	6.4
NiCaX 5	3.3	3.7	65	6.2	35	3.4
NiX (57)	3.3	3.6	71	17.6	29	7.0
NiX (17)	3.3	3.8	58	4.4	42	3.2

(iii) <u>DTA curves</u>. Ion exchanged and hydrated samples with high nickel contents (NiCaX 1.2, NiX (57)) show sharp dehydration peaks in the DTA spectrum (Fig. 2) at $165^{\circ}C$. A less pronounced maximum of dehydration at $165^{\circ}C$ can also be identified for some samples with lower nickel content (NiX(17), NiCaX 1.1, NiCaX 5). Hydrated samples containing only or mainly calcium (CaX 1(26 % exchange), CaX 3 (62% exchange), NiCaX 3.1) show maxima of dehydration between $120^{\circ}C$ and $145^{\circ}C$. After the activation rehydration cycle only the samples with the highest nickel content (NiX (57), NiCaX 1.2) have still dehydration maxima at $165^{\circ}C$.

The higher dehydration temperature and the sharper dehydration peak of Ni^{2+} compared to Ca^{2+} is in agreement with the wellknown complex chemical behaviour of these ions. It can be assumed, however, that the nickel dehydration process is affected by high calcium contents and by diffusion processes from "open places" to "hidden places" during dehydration and activation. Ni ions in "hidden positions" cannot be fully rehydrated.

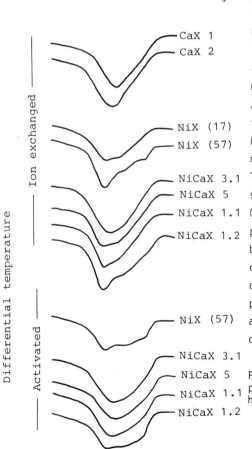

Fig. 2 DTA curves showing temperature programmed dehydration of modified zeolites; heating rate 5°/min.

4. DISCUSSION

High initial catalytic activity (NiCaX 1.2, NiCaX 5, NiX (17), NiX (57)) might be due to a relatively large portion of Ni ions in positions easily accessible for reduction. According to Table 3 these positions should be the SI sites in the hexagonal prisms, in support of the hypothesis of Egerton and Vickerman (ref. 5). For samples exchanged in the sequence calcium followed by nickel, the Ca ions seem to hinder the Ni ions from reaching SI sites. This effect is balanced by either a high Ni content (NiCaX 1.2) or a simultaneous exchange (NiCaX 5).

The samples with the lowest initial catalytic activity (NiCaX 3.1, NiCaX 1.1) exhibit the highest relative increase of activity, raising the temperature from 250^0 to 300^0C. This effect might be due to especially pronounced additional reduction of Ni ions in other than SI sites. These samples also have a strong tendency to sinter, according to the activity decline at 350^0C and the X-ray analysis data.

The highest final activity and stability at 350^0C is observed for samples NiCaX 5, NiX (17) and NiX (57), i.e. for catalysts, which also had the highest initial activity. This means that the initial reduction condition determines the quality of the catalyst.

In the light of this conclusion the outstanding properties of NiCaX 5 can be understood by assuming, that most of the nickel ions had been in positions easily accessible for reduction resulting in a catalyst with high metal dispersion and little tendency to sinter. The assumption is supported by the results of the characterization methods applied. The number of Bohr magnetons indicates a high occupation of SI sites after dehydration and prior to reduction, the DTA curve has lost the nickel dehydration maximum at 165^0C for the activated sample, sintering stability is indicated by X-ray analysis, yielding no Ni signal and the ferromagnetic susceptibility shows the smallest relative increase during the reaction cycle of all samples.

Acknowledgment: The authors wish to thank G. Ernst for carrying out the X-ray analysis of the samples.

REFERENCES

1 M.A. Vannice, J.Cat. 44 (1976) 152
2 S. Bhatia, J.F. Mathews and N.N. Bakhshi,
   Acta Phys. et Chem., Szeged 24 (1978) 83
3 M. Briend Faure, M.F. Guilleux, J. Jeanjean, D. Delafosse
   G. Djega Mariadassou and M. Bureau-Tardy
   Acta Phys. et Chem., Szeged 24 (1978) 99
4 F. Schmidt, H. Kacirek, W. Gunßer, Ch. Minchev, V. Kanazirev,
   L. Kosova and V. Penchev,
   Proc. 4th Intern. Symp. Het. Catalysis, Varna 2 (1979) 331
5 T.A. Egerton and J.C. Vickerman,
   J. Chem. Soc. Faraday Trans. I 69 (1973) 39
6 A.Andreev, D. Shopov, E.C. Haß, N.Jaeger, R. Nowak, P. Plath
   4th Intern. Symp. Het. Catalysis, Varna 1979
   Commun. Dept. Chem. Bulgar. Acad. Scie. 1980 in print
7 W. Romanowski, Z.anorg.allg. Chem. 351 (1967) 180

B. Imelik *et al.* (Editors), *Catalysis by Zeolites*
© 1980 Elsevier Scientific Publishing Company, Amsterdam — Printed in The Netherlands

THE ROLE OF ELECTROLYTIC PROPERTIES OF ZEOLITES IN CATALYSIS

JULE A. RABO
Union Carbide Corp., Tarrytown, NY (U.S.A.)

ABSTRACT

The intracrystalline pore-cavity system in zeolites, often called the zeolitic surface, is surrounded by the zeolite crystal lattice and it is consequently strongly influenced by the zeolite crystal field. This crystal field which pervades the intracrystalline pore-cavity system renders zeolites solid electrolites. Depending on the ionic character of the crystal which is mainly controlled by the alumina content, zeolites show properties of weak to very strong electrolites.

The adsorption properties of zeolites demonstrate strong electrostatic interaction between zeolites and occluded polar molecules. The redox chemistry of zeolites testifies that this interaction between zeolites and adsorbed species often goes beyond physical inter action, even to the extent of ionization of adsorbed atoms or molecules. Several of these ionization reactions readily proceed in zeolites at low temperatures, aided by the zeolite electrolyte. Significantly, most of these redox reactions are endothermic by one to several eV/mol when they occur in the gas phase outside the zeolite crystal. Several examples show that the electrolitic character of zeolites is displayed by both alkalication zeolites as well as by H-zeolites.

Mechanistic studies of n-hexane cracking over alkali cation-zeolites show the absence of ionization of hexane or its cracked fragments. The absence of skeletal isomerization and the product composition shows that the cracking of hydrocarbons proceeds by a radical type mechanism. Thus, in spite of abundant ionization phenomena in the chemistry of alkali zeolites, they do not ionize hydrocarbons. It is worth noting that the ionization energy of hexanes and their radical fragments is $\geqq$ 170 kcal/mole, significantly higher than the free energy change of several redox reactions observed in zeolites ($NO + NO_2 \quad NO^+ + NO_2^-$, $Na + NaY \quad Na_4^{3+}Y$, etc.). Even in the absence of ionization phenomena the cracking of hexanes over alkali zeolites shows major differences in product distribution from the non-catalysed, thermal cracking process. These differences can be explained, and the different product composition can be fully accounted for, on the asis that zeolites concentrate hydrocarbon reactants within the zeolite crystal, resulting in substantially higher reactant concentrations within the zeolite crystal relative to the surrounding gas phase. The high reactant concentration within zeolite crystals results in a strong enhancement in the rate of bimolecular reaction steps (5 to 10 fold) over uni olecular reaction steps. Specifically, in the cracking of hydrocarbons the alkali zeolite "catalyst" emphasizes the bimolecular H-transfer reactions over the unimolecular C-C bond split.

In the cracking of hydrocarbons over the strong acid H-Y zeolite, an unusually strong Bronsted acid type catalytic activity is observed. Here, very high cracking activity and skeletal isomerization are apparent. In addition, by comparison with amorphous silica-alumina, the product distribution obtained with the H-type zeolites indicates a great enhancement of H redistribution, presumably through H⁻ shifts. Thus, in the cracking of hydrocarbons over H zeolites, similar to the cracking over alkali cation zeolites, the bimolecular reaction steps (H redistribution) are favored over the unimolecular reaction steps (fragmentation).

It is concluded that the common cause of these catalytic phenomena over both alkali and H-type zeolites is the high concentration of the hydrocarbon reactants in the zeolite pore-and-cavity system. An enhanced reactant concentration on catalyst surfaces is, of course, characteristic of any heterogeneous catalyst. However, the reaction mechanistic evidence available for zeolites and amorphous silica-alumina gel catalysts indicates that the reactant concentration effect is far greater with zeolites relative to other, amorphous acid catalysts.

REFERENCES

1. J.A. Rabo in "Zeolite chemistry and catalysts" (edited by J.A. Rabo) ACS monograph, ACS Washington 1976
2 J.A. Rabo, R.D. Bezman and M.L. Poutsma, "Symposium on zeolites Hungary", Acta Physica Et Chemica 24, 1978, 39

B. Imelik *et al.* (Editors), *Catalysis by Zeolites*
© 1980 Elsevier Scientific Publishing Company, Amsterdam — Printed in The Netherlands

WHY ARE HY ZEOLITES SO MUCH MORE ACTIVE THAN AMORPHOUS SILICA-ALUMINA

J. FRAISSARD

Laboratoire Chimie des Surfaces

Université P. et M. Curie — Paris

I.   INTRODUCTION

There have been several reviews on the acidity of solid surfaces (1-4) showing
thus the importance of the subject. Heterogeneous acid catalysis plays an all important role in areas as important as the cracking (5) and isomerisation (6) of hydrocarbons, the alkylation of paraffins (7) and aromatics (8) by olefins and the polymerisation of olefins (9). Carbonium ions are thought to be involved in the mechanisms of these acid-catalysed reactions (10-13).

Studying the acidity of a surface consists in measuring the acid strength of the
various sites and determining how many there are. Several classical methods are now
applied to this sort of study. From the literature we shall show that they prove
that the amorphous silica-aluminas and HY zeolite are of similar acidity. Next we
shall recapitulate attempts to study Brönsted acidity by NMR and we shall propose
a new acidity scale based on a classification of the chemical shifts of hydrogens
with acidic properties. This criterion does not seem to be able to distinguish the
acidities of silica-aluminas and HY zeolite either at least not sufficently to understand why the catalytic activities of crystalline alumino-silicates are one to
four orders of magnitude greater than those of amorphous alumino-silicates.

NMR study of the adsorption of xenon on HY and on silico-aluminas (14) makes
it possible to propose the hypothesis that the high catalytic activity of solids
with cavities can be explained in part by a structural effect which imposes upon
the molecule a much greater number of molecule-active centre collisions than in the
case of an amorphous solid.

II.   INDICATOR MEASUREMENTS

The Brönsted acid strength of a compound is its aptitude to transfer a proton.
In a homogeneous medium this aptitude is expressed by the pK. For several reasons :
leveling effect of the water, modifications of the properties of the surface or
the structure,the titrations of an aqueous suspension of a powdered solid against
a base is a very bad method. Consequently the acidities is studied in non-aqueous

media. This aptitude to transfer a proton can be expressed quantitatively by the Hammett acidity function (15,16) :

$$Ho = -\log \frac{a_{H^+} \cdot f_B}{f_{HB^+}} \qquad (1)$$

where a_{H^+} is the acidity of the surface proton, f_B and f_{BH^+} are the activity coefficients of the acid and base forms of the adsorbed Hammett indicator B. This indicator reacts with the Brönsted acid HA as follows :

$$B + HA = BH^+ + A^- \qquad (2)$$

By using the colours of the different indicators adsorbed it is possible to bracket the Ho value of a solid surface in the same way as for the pH of an aqueous solution.

To define the acidity of a compound it is necessary to determine not only the acid strength of the sites but also their number. Johnson (17), Benessi (18), Hirschler (19) and Drushel and Sommers (20) have studied in detail the titration of a catalyst suspension against a chosen amine solution in an inert solvant using indicators of the above type to determine the end-point. The methods differ as to the nature of the amines and the indicators and even the titration procedures.

The results obtained by various authors concerning the acid strength of silica-alumina and HY zeolite are fairly consistent.

II.1 ACIDE STRENGTH

II.1.1 Silica-alumina

According to Parera et al. the Ho values lie mainly between -8.2 and $+3.3$ (21); but there are also stronger sites (Ho <-8.2). This is confirmed by Take et al. (22) who find sites with Ho between -10.5 and -12.8, and even a few stronger sites (Ho <-12.8).

However, U.V spectroscopie studies on p–Nitrotoluene and 2,4– dinitrotoluene lead Ikemoto et al. (23) to conclude that the strong acid sites have a Ho value betwen -8.2 and -10.8. Similar Ho value distributions are found in the results of other authors (18–20).

II.1.2 HY Zeolite

The Ho values for HY zeolite treated at about 450°C are very close to those found for silica-aluminas. According to Otouma et al. (24) most of the sites have a Ho value between -6 and -8.2. About 20 % are uniformly distributed in the -6 to $+3.3$ range. The remaining 20 % are more strongly acid (Ho $\leqslant-8.2$). Ikemoto (23) states that these strongly acid sites are in the -8.2 to -10.8 range ; his results are very similar to those for silica-alumina. Morita (25), on the other hand, found no sites with an acidity greater than Ho ~-8.2. Moreover, according to L.G. Karakchiev et al. (26,27) the electronic spectra of different adsorbed compounds such as diphenil carbinol, benzene, DNT, etc., indicate that the proton centres in amorphous silica-

aluminas (10 % Al_2O_3) and HY zeolite are almost identical strength, the stron-
gest corresponding to Ho close to -8 to -9.

II.2 NUMBER OF ACID CENTRES

The problem of counting the acid centres of different strength seems to be
more difficult. Usually each author finds that there is a good correlation between the catalytic activity of the solid for a certain reaction and the variation (determined by his method) of the number of centres with a given acid
strength. Thus one finds a good correlation between the total number of acid sites (Ho +3.3) in silica-aluminas and their activity in the methylation of me-
thylaniline by methanol(21). There is an analogous correlation for the dispro-
portion of toluene on HNaY zeolites (25). Also in the cracking of cumene, the
activity of silica-aluminas depends on the number of strong acid centres
(Ho $\leqslant$ -8.2) (21) ; that of HNaY zeolites depends linearly on the concentration
of mediums and strong acid sites (Ho $\leqslant$ -3.0) (25,28 a).

Generaly the number of acid sites measured by those methods is greater for HY
zeolites (2-3 meq/g) than for amorphous silica-alumina (14 % Al_2O_3 : 0.5 meq/g.
21,23,24). However, it is sometimes difficult to compare different author results.
Whereas the acid strength distribution are usually similar, the total acidities
measured can vary enormously. Thus for HNaY, the results in references 23-25 differ by 200 % although the samples are very much alike in their composition (between 87.5 and 94.0 % H ; silica/alumina ratio between 4.7 and 5.0) These differences can sometimes be explained. In fact depending on the nature of the amines
and the indicators they can be more or less adsorbed on sites other than Brönsted
acids. The acidity detected must therefore often be higher than the Brönsted acidity alone. Some workers have also considered the problem of the accessibility of
the acid sites (relative sizes of the cavities, the organic bases and the indicator molécules).(28b, 29, 30).

III. BASE TITRATION AND SPECTROSCOPIC STUDY

Another method for determining the acidity of a surface consists in adsorbing
a base and assuming that the amount chemisorbed corresponds to the number of active sites in the surface. In this way Mills et al. (31) found a good correlation
between the amount of chemisorbed quinoline and the cracking activity of catalysts
with a large range of SiO_2/Al_2O_3 composition and catalytic activity. The numbers
of Brönsted sites determined by this method are moreover very close to those obtained by the indicator method (0.5 meq/g). But in general these measurements
give the total acidity (Brönsted + Lewis). Furthermore, adsorption can occur on
inactive parts of the surface. For this reason such chemisorption studies can
hardly be used as precise references for catalytic activity if they are not done
spectroscopically (32). Brönsted adsorption centres (protonated form of the amine)

and Lewis centres (coordinated form) can be determined selectively by I.R. spectroscopy in particular. The use of stronger or weaker bases and their desorption as the temperature is raised makes it possible to choose the acidities. Although measurements of I.R. band intensities are not very accurate the results are the same order of magnitude as those given by indicators. (0.1 meq/g for silica-alumina 13 % (34,35) ; 2 meq/g for HY zeolite (36).

It should be pointed out that the positions of the I.R. band of the OH has sometimes been successfully related to the Brönsted acid strength of the corresponding site (37). We do not however believe such relation ships to be generally valid, the ν(OH) frequency of a solid depending on numerous factors (location for example). Despite this criticism we note that the I.R. bands of OH in silica-aluminas and HY zeolite are quite close (about 3650 and 3550 cm^{-1}).

IV. NMR STUDY

The Brönsted and strength distribution obtained by classical methods are therefore very similar for amorphous silica-aluminas and for decationised HY zeolites. Quantitatively, there are on average 4–8 times as many sites in HY as in silica-aluminas. Let us now examine what information NMR can provide concerning the acidity of solids.

The Brönsted acidity of a compound being its aptitude to transfer a proton, it seemed normal to relate the acidic property to the proton mobility. Mestdagh(42) et al. attribute the temperature dependence of the second moment and the relaxation times of protons in HY to the movement of the latter by jumps between the oxygen atoms of the lattice. The jump frequency

$$\nu = 3.3.10^{10} \exp(-10^4/RT) sec^{-1}$$

is about 10^7 sec^{-1} at 450°C. According to Fripiat (43), for a proton lifetime of the order of 10^{-7} sec, the residence time of a molecule on the surface site occupied by a proton should be at least of the order of 10^{-6} sec in order to have an appreciable chance of capturing this proton.

Freude et al (42) determined the life-time of the protons on the oxygens in the zeolite lattice and on the adsorbed molecules. These authors reach approximately the same conclusions as the previous ones. They consider that the total acidity of a solid depends on the number of OH groups and on τ_A^{-1} where τ_A is the mean life-time of a proton on the lattice in presence of the adsorbate.

However we do not think that proton mobility is a reliable criterion for comparing the intrinsic acidities of solids. It can be affected by numerous factors. For example, the mobility of a proton depends on the number of neighbouring sites able to accept it. As a result of the spatial distribution of the oxygens in faujasite this number must be much greater for HY than for a plane surface.

The width of NMR signals depends on dipolar interactions and spin motion.

Because of their low mobility the protons of acid SOH groups distributed on the surface S of a catalyst are generally associated with a broad signal (44). Bonardet et Fraissard (45) have however shown that it is possible to increase proton mobility by adsorption of a base AH likely to accept the protons in the equilibrium :

$$SOH + AH = SO^- + AH_2^+ \qquad (3)$$

If proton exchange between sites SO^- and AH is sufficiently rapid, the characteristic signal of the system is much narrower than in the absence of AH. Pearson (46) tried to exploit this possibility in order to measure the number of Brönsted acid sites at an Al_2O_3 surface by adsorption of deuterated pyridine. However, as Knözinger (47) has rightly remarked " the experimental observations of Pearson may be discussed in terms of an increased mobility of surface protons which is induced by the adsorbed pyridine, rather than in terms of protonic acidity which accounts for carbonium ion reactions on the surface of aluminas". We would add that, even accepted Pearson optimistic assumption, this assay must in our opinion correspond to the number of pyH^+ formed by equation (3) and not to the total number of Brönsted sites. In order to determine the values of the latter the equilibrium constant has to be measured. This can be done by studying the chemical shift (45).

If rapid exchange occurs between the proton of the surface SOH and those of the adsorbed molecule AH($^{15}NH_3$ for example) the acid proton must affect the chemical shift of the adsorbed phase. The ^1H spectrum should contains only one line at frequency νe due to the coalescence of the lines at frequencies ν_{AH}, $\nu_{AH_2^+}$ and ν_{OH}. This the observed chemical shift $d_{obs.}$ is (45)

$$d_{obs.}(^1H) = p(OH) \, d(OH) + p(AH) \, d(AH) + p(AH_2^+) \, d(AH_2^+) \qquad (2)$$

For the same reason

$$d_{obs.}(A) = p^1(AH) \, d(AH) + p^1(AH_2^+) \, d(AH_2^+) \qquad (3)$$

where p_i and p_i^1 are the concentration of H and A atoms respectively in the group i.

Knowing the chemical shift of A and H in the species AH and (AH_2^+), equation 3 can be used to calculate the relative concentration p^1_{AH} and $p^1_{AH_2^+}$ and the dissociation coefficient of SO – H in the presence of AH. Gay (51 has studied type (3) equilibria qualitatively by ^{13}C-NMR of bases adsorbed on silica, alumina and silica-alumina.

The "intrinsic" Brönsted acid strengh of a compound depends on the polarization of X–H linkage (X = O,N...), therefore on the electronic environnement of the proton which will be measured by the NMR chemical shift $d(H)$. Consequently, $d(H)$ is characteristic of the Brönsted acid strengh and must be considered as a new acidity scale, regardless the state of the coumpounds.

Until very recently the chemical shift of SOH group proton could not be measured directly because of dipolar interactions. It could however be determined by means of equations (3) and (2) after AH adsorption. Although there are as yet no results on protons it could in theory be measured with the help of new line-narrowing techniques.

Whatever the case, we note that average Brönsted acidity strength of zeolite HY (7–8 ppm in comparaison with gazeous TMS) is higher than that of silica-alumina ($\sim$ 5 ppm) ; but in our opinion this difference is not very large.

V.   A POSSIBLE CAUSE FOR THE GREAT ACTIVITY OF ZEOLITE

One can therefore ask ourself why HY zeolites are often 3–4 order of magnitude more active than amorphous silica-aluminas (49). In our view this very high activity which does not seem to be attributable to a particular chemical property can be explained in part by the structure of the solid which allows a large number of collisions between the actives sites and the reagents.

Let us compare the number of collisions, assumed elastic, at ambient temperature, between a Xe atom and either the surface of a classical catalyst (assumed planar) or the inner surface of a Y supercage.

We consider only one Xe atom per supercage. If we neglect the contact time of xenon with the wall of the cavity, the number of Xe-wall collisions is $\bar{v}/\bar{l}$ where $\bar{v}$ is the mean velocity of Xe and $\bar{l}$ the mean free path in the cavity. The size of the cavity and of Xe being respectively 13 Å and 4,4 Å, the maximum allowed path for Xe atom is 8.6 Å. At 300 K, $\bar{v}$ (Xe) = $2.2.10^4$ cm.s^{-1}. So the Xe-wall collision number is higher than $2.6.10^{11}$ times/s/cavity that is :

$$N_{(y)} = 4.8.10^{24} \text{ times/s/cm}^2$$

The pressure corresponding to an average of one Xe atom per Y zeolite supercage, at room temperature, is about 300 torrs. At the same pressure the number of collision with an open surface is

$$Np = 4.2.10^{22} \text{ times/s/cm}^2$$

Consequently, for the same pressure, the collision number is about 100 times greater for the Y zeolite compared to the open surface.

In the case of a chemical reaction catalysed by certain active centres of the solids, it would be better to calculate the number of collisions per active site rather than per unit area. One should also take into account in each case the life-time of the reagent in the site considered. However, the previous comparisons are pointless if one does not consider sites of the same chemical characteristics, and with identical concentrations on the planar surface and on the inner surface of the cavity. Under these conditions the above ratio Ny/Np is conserved.

In order to check that the calculation of the number of collisions of Xenon situated in the supercage was justified we have performed an NMR study of Xe adsorbed on various Y zeolites (50). When the supercage contains only one Xe, the chemical shift $d(Xe) = 58 \pm 4$ ppm towards high frequency. This shift is practically independent of the nature of the H and Na cation, and also of the Si/Al ratio. It is due to collisions between the Xe atoms and the cage walls. For pure gaseous xenon this value corresponds to a pressure equivalent to a density of about a hundred amagats .

VI CONCLUSION

The various methods for studying the acidity of solids show that the Brönsted acid strength distributions for amorphous silica–aluminas and HY zeolites are closely similar. The number of Brönsted acid sites on the other hand is greater for HY. But this difference does not seem to be able to explain the relative "super-activity" of HY.

NMR study of adsorbed xenon reveals that for a given gas pressure the molecule adsorbed in a zeolite is subjected to an "apparent pressure" about 100 times greater (at least) than the pressure upon a molecule in contact with a planar surface. This structural effect could explain why, for active sites of the same chemical nature and of the same concentration, zeolites have much greater catalytic activity than surface of conventional catalysts.

REFERENCES

1 M.S. Goldstein, in "Experimental Methods in Catalytic Research" (R.B. Anderson Ed.),Academic Press, New York 1968, pp 361.
2 K. Tanabe, Solid Acids and Bases, Academic Press, New York 1970.
3 F. Forni, Catal. Rev. 8(1973), 69.
4 H.A. Benessi and B.H.C. Winquist, Advances in Catalysis, 27 (1978).
5 H.H. Voge, Catalysis, 6 (1958) 407.
6 F.E. Condon, Catalysis 6 (1958) 43.
7 R.M. Kennedy, Catalysis 6 (1958) 1.
8 A.G. Olah,"Friedel–Crafts Chemistry", New York 1973, pp. 43, Wiley.
9 A.G. Oblad, G.A. Mills and H. Heinemann, Catalysis 6 (1958) 341.
10 F.C. Whitmore, J. Am. Chem. Soc. 54(1932) 3274.
           Ind. Eng. Chem. 26 (1934) 94.
11 C.L. Thomas, Ind. Eng. Chem. 41 (1949) 2564.
           J. Am. Chem. Soc. 66 (1944) 1586.
12 B.S. Greensfelder, H.H. Voge and G.M. Good, Ind. Eng. Chem. 39 (1947)1032.
           ibid. 41 (1949) 2573.
13 L.B. Ryland, M.W. Tamele and J.N. Wilson, Catalysis (P.H. Emmett Ed.), Vol. VII, Reinhold, New York 1960.
14 T. Ito and J. Fraissard, Fifth International Conference on Zeolites, Italia, Napoli, June 1980.
15 L.P. Hammett and A.J. Deyrup, J. Am. Chem. Soc. 54 (1932) 2721.
16 L.P. Hammett, Physical Organic Chemistry, Mc. Graw-Hill, New York 1940, pp. 251.
17 O. Johnson, J. Phys. Chem. 59 (1955)827.
18 H.A. Benessi, J. Phys. Chem. 61 (1957) 970.
19 A.E. Hirschler, J. Catal. 2 (1963) 428.
20 H.V. Drushel and A.L. Sommers, Anal. Chem. 38 (1966) 1723.
21 J.M. Parera, S.A. Hillar, J.C. Vincenzini and N.S. Figoli, J. Catal. 21 (1971) 70.

22 J. Take, T. Tsuruya, T. Sato and Y. Yoneda, Bull. Chem. Soc. Japan 45 (11) (1972) 3409.

23 M. Ikemoto, K. Tsutsumi and H. Takahashi, Bull. Chem. Soc. Japan 45 (1972) 1330.

24 H. Otouma, Y. Arai and H. Ukihashi, Bull. Chem. Soc. Japan 42 (1969) 2449.

25 Y. Morita, T. Kimura, F. Kato and M. Tamagrawa, Bull. Jpn. Pet. Inst. 14 (1972) 192.

26 L.G. Karakchiev, N.S. Kotsarenko, E.A. Paukshtis and V.G. Shinkarento, Kinetika i Kataliz 16, N°5, September–October (1975) pp 1305–1312.

27 N.S. Kotsarenko, L.G. Karakchiev and A. Dzis'Ko, Kinetika i Kataliz 9 (1968) 158.

28a M.D. Navalikhina and N.A. Kuzin, Kinetika i Kataliz 16, N°1, January–February (1975) pp 202–210.

28b W.F. Klading, J. Phys. Chem. 83, N°6 (1979) pp 765.

29 D. Barthomeuf, J. Phys. Chem. 83, N°6 (1979) pp 767.

30 D. Barthomeuf, A.C.S. Symp. Ser 453, N°40 (1977)

31 G.A. Mills, E.R. Boedeker and A.G. Oblad, J. Am. Chem. Soc. 72 (1950) 1554.

32 E.P. Parry, J. Catalysis 2 (1962) 371.

33 L.H. Little, Infrared Spectra of Adsorbed Species, Academic Press. N.Y 1966.

34 J. W. Ward and R.C. Hansford, J. Catal. 13 (1969) 154–160.

35 D. Ballivet, D. Barthomeuf and P. Pichat, J. Chem. Soc., Faraday Trans, I, 68 (1972) 1712.

36 P.A. Jacobs and C.P. Heylen, J. Catal. 34 (1974) 267.

37 P.G. Rouxhett and R.E. Sempels, J. Chem. Soc., Faraday Trans. 70 (1974) 2021.

38 Z.A. Markova, Kinetika i Kataliz, 2 (1961) 435.

39 P. Pichat, J. Kermarec, J. Fraissard and M.V. Mathieu, Bull. Soc. Chim. France 11 (1966) pp 3652.

40 J.B. Uytterhoeven , L.G. Christner and W.K. Hall, J. Phys. Chem. 69 (1965) 2117.

41 M.M. Mestdagh, W.E. Stone and J.J. Fripiat, J. Phys. Chem. 76, N°8 (1972).

42 D. Freude, W. Oehme, H. Schmiedel and B. Staudte, J. Catal. 49 (1977) 123–134.

43 J.J.Fripiat, Catal. Rev. 5,(2) (1971) 269.

44 J. Fraissard, I. Solomon, R. Caillat, J. Elston and B. Imelik, J. Chim. Phys. (1963) pp 676.
    J. Kermarec, J. Fraissard and B. Imelik, J. Chim. Phys. (1967) pp 911 ; (1968) pp 920.

45 J.L. Bonardet and J. Fraissard, Ind. Chim. Belg. 38 (1973) pp 370–374 – Japan J. Applied Physics. Suppl. 2, Pt 2, (1974 pp 319.

46 R.M. Pearson, J. Catal. 46 (1977) 279–288.

47 H. Knözinger, J. Catal. 53 (1978) 171–172.

48 M. Mehring,"High Resolution NMR Spectroscopy in Solids", Springer Verlag (1976).

49 J.N. Miale, N.Y. Chen and P.B. Weisz, J. Catal. 6 (1966) 278–287.

50 A.K. Jameson, C.J. Jameson and H.S. Gutowsky, J. Chem. Phys. 53 (1970) 2310.

51 I.D. Gay and S. Liang, J. Catal. 44 (1976) 306–313.

AUTHOR INDEX